Advances in Soil Science

Advances in Soil Science

B.A. Stewart, Editor

Advances in Soil Science

Volume 18

Edited by B.A. Stewart

With Contributions by
M.S. Aulakh, M.S. Cresser, J.W. Doran,
A.C. Edwards, Li Shengxiu, A. Matar,
A.R. Mosier, J. Ryan, J. Torrent, Xiao Ling

With 47 Illustrations

Springer-Verlag
New York Berlin Heidelberg London Paris
Tokyo Hong Kong Barcelona Budapest

B.A. Stewart
USDA Conservation & Production Research Laboratory
Bushland, TX 79012, USA

ISBN-13: 978-1-4612-7693-7 e-ISBN-13: 978-1-4612-2844-8
DOI: 10.1007/978-1-4612-2844-8

Printed on acid-free paper.

Softcover reprint of the hardcover 1st edition 1992

Production managed by Henry Krell; manufacturing supervised by Jacqui Ashri.
Typeset by Best-set Typesetter Ltd., Hong Kong.

9 8 7 6 5 4 3 2 1

Preface

Advances in Soil Science was conceived in 1982 to provide a forum for leading international scientists to analyze and summarize the available scientific information on a subject, assessing its importance and identifying additional research needs. This goal seems even more appropriate today. Much is known about our soil resources. The principles learned and the technology developed need to be used to increase food production, particularly in developing countries, and sustain the productivity of the resource base. *Advances in Soil Science* fills a gap between the scientific journals and the comprehensive reference books. Scientists can delve in depth on a particular subject relating to soil science. Contributors are asked in particular to develop and identify principles that have practical applications to both developing and developed agricultures.

Advances in Soil Science was established to be international in scope and cover all subjects relating to soil science. This volume certainly fulfills that objective. The first chapter is concerned with denitrification. The significance of denitrification in nitrogen use efficiency in agricultural, grassland, and forest ecosystems, and in global nitrogen cycling is now well recognized. Therefore, an understanding of denitrification is essential for the efficient use of nitrogen fertilizers and for protecting the environment. The second chapter looks at the effects of subzero temperatures on the physical, biological, and chemical processes. This is an important area of soil science that has not been studied in great detail. The third chapter discusses soil and fertilizer phosphorus and crop responses in the dryland Mediterranean zone, a major food-deficient region of the world. The lack of phophorus in many of the soils in that region prevents the efficient use of the limited water resources. It is crucial in dryland regions that whatever water is available be used as efficiently as possible. When phosphorus is applied to deficient soils, it accelerates crop development and hastens maturity. The fourth contribution is a very thorough and interesting review of the dryland regions of the People's Republic of China. With the growing interest in sustaining the soil resource base, that review offers unusual insights of the battles waged by the Chinese people over centuries to produce food and fiber while trying to control erosion.

The acceptance of *Advances in Soil Science* by both authors and readers has been very gratifying and confirms our perception that a need did exist for a medium to publish soil science reviews. I want to thank the authors for their excellent contributions and cooperation. I also want to thank members of the Editorial Board for their help in selecting such competent authors and the Springer-Verlag staff for their kind assistance and counsel. Lastly, and most importantly, I want to thank the readers for their acceptance and use of *Advances in Soil Science*.

B.A. Stewart

Contents

Contributors

M.S. Aulakh, Department of Soils, Punjab Agricultural University, Ludhiana 141004, India.

M.S. Cresser, Department of Plant and Soil Science, University of Aberdeen, Aberdeen AB9 2UE, United Kingdom.

J.W. Doran, U.S. Department of Agriculture, Agricultural Research Service, Department of Agronomy, University of Nebraska, Lincoln, Nebraska 68583, USA

A.C. Edwards, Macaulay Land Use Research Institute, Craigiebuckler, Aberdeen AB9 2QJ, United Kingdom.

Li Shengxiu, Northwestern Agricultural University, Yangling, Shaanxi, People's Republic of China.

A. Matar, International Center for Agricultural Research in Dry Areas, Aleppo, Syria.

A.R. Mosier, U.S. Department of Agriculture, Agricultural Research Service, Ft. Collins, Colorado 80522, USA

J. Ryan, Mid-American International Agricultural Consortium, Settat, Morocco.

J. Torrent, University of Cordoba, Cordoba, Spain.

Xiao Ling, Northwestern Agricultural University, Yangling, Shaanxi, People's Republic of China.

Soil Denitrification—Significance, Measurement, and Effects of Management

M.S. Aulakh, J.W. Doran, and A.R. Mosier

Advances in Soil Science, Volume 18

I. Introduction

Denitrification is a form of anaerobic bacterial respiration during which nitrogen (N) oxides, principally nitrate and nitrite, are reduced sequentially through NO and N_2O to N_2. The commonly accepted reductive pathway for denitrification is as follows:

(+5)	(+3)	(+2)	(+1)	(0)
NO_3	$\rightarrow NO_2^-$	$\rightarrow NO$	$\rightarrow N_2O$	$\rightarrow N_2$
Nitrate	Nitrite	Nitric oxide	Nitrous oxide	Dinitrogen

The majority of denitrification is biologically catalyzed and closely linked to bacterial respiratory metabolism. Some N_2O is produced by nonrespiratory N-oxide reduction, but the environmental significance of this source has yet to be demonstrated (Tiedje, 1988). In chemodenitrification generation of N gas is catalyzed by abiotic agents, but this process may only be of importance in acidic or frozen soils (Christianson and Cho, 1983). Chemodenitrification and other processes such as ammonium production during dissimilatory nitrate reduction (Tiedje, 1988), and N_2O production during nitrification have been reviewed by Knowles (1981) and Aulakh (1986a). Microbial ecology of denitrification, growth kinetics of organisms involved, and factors regulating denitrification in the environment have been documented by Payne (1981), Firestone (1982), Tiedje (1988), and Tiedje et al. (1989).

Delwiche and Bryan (1976), Payne (1981), Knowles (1981, 1982), and Beauchamp et al. (1989) summarized factors affecting denitrification and especially the individual effect of each in monoculture or mixed-culture studies and in soils under controlled laboratory conditions. Few of these reports, however, have provided an in-depth review of denitrification regulation under field conditions where the interacting role of regulatory factors may vary considerably from the simple sum of effects of individual factors.

Several direct and indirect approaches have been used to measure denitrification in laboratory studies (Knowles, 1981). Methods to measure denitrification in natural habitats have been limited (Tiedje, 1988; Tiedje et al., 1989). In recent years, techniques for measuring denitrification in the field have been developed but their relative accuracy needs further evaluation. Some of these techniques have been employed to measure the influence of different management practices on denitrification.

The aims of this paper are (i) to discuss briefly the significance of denitrification to the N cycle, including fertilizer use efficiency and ground water pollution; (ii) to assemble available information on factors controlling denitrification under field conditions; (iii) to identify the influence of agricultural management systems; (iv) to assemble recent developments in field methodology; and (v) to identify the gaps in knowledge for future research directions.

Table 1. Global flows in the available N budget. (Adapted from Tiedje (1988) and FAO, 1986)

Inputs/outputs	N transfer, million metric tons ($Tg\,N\,year^{-1}$)		
	Land	Sea	Total
Inputs			
Biological N_2 fixation	139–180	1–130	140–310
Industrial fixation	70	—	70
Combustion and atmospheric fallout	19–25	5	24–30
Total	—		234–410
Outputs			
Denitrification	105–185	25–250	130–435
Aquatic sediments	—	10	10
Total			140–445
Net balance			−94–+35

II. Significance to N Cycle

Although enveloped in an atmosphere of 78% N_2, production of food and fiber on earth is limited more by availability of N than by that of any other element. Atmospheric N inputs to terrestrial and aquatic ecosystems through gaseous N_2 fixation and atmospheric fallout roughly balance return of gaseous N to the atmosphere via denitrification (Table 1). In meeting the food and shelter needs of an ever-increasing world population, the amount of industrially fixed fertilizer N (which is applied to about 11% of the earth's land surface) has increased dramatically. Between 1950 and 1985 fertilizer N use increased from 3.5 to almost 70 Tg (million metric tons) and is anticipated to increase to 100 Tg by the year 2000 (Hauck, 1988). The increased input of industrially fixed N is thought to be balanced in the global N budget by increased losses of soluble N to surface and groundwaters and by increased denitrification of this aquatic N (Tiedje, 1988).

In addition to its central role in the global N balance, denitrification is important as a mediator of environmental quality through removal of nitrate N from surface waters and addition of N oxides to the atmosphere. Denitrification also results in less efficient use of fertilizer N. Understanding the relative importance of each of these roles will largely determine agricultural management practices for reduced or enhanced denitrification.

A. Fertilizer Use Efficiency

Efficiency of fertilizer N use by most crops ranges from 20% to 60% and commonly averages 50%. Denitrification is one of the major mechanisms for N loss from the soil. Nitrogen lost through denitrification can range from negligible to as high as 100 kg N ha^{-1} (Aulakh, 1986a, 1986b). In a classic review of long-term lysimeter trials across the USA, Allison (1955) showed that, depending on cropping, from 12% to 20% of the available N added yearly was unaccounted for and presumed lost via denitrification. In a later review of ^{15}N field studies Hauck (1981) estimated an average fertilizer N deficit of 30% which he attributed largely to denitrification.

Estimates indicate that denitrification recycles from 52% to 100% of the total global nitrogen inputs (Table 1). Because of the lack of direct measurements and extreme temporal and spatial variability, estimates of global denitrification losses range from 83 Tg yr^{-1} (Stevenson, 1982) to 390 Tg yr^{+1} (Hauck and Tanji, 1982).

B. Atmospheric Pollution

Nitrous oxide evolved during denitrification diffuses into the upper atmosphere. The stratosphere contains a few parts per million of ozone which is the only atmospheric gas to protect the biosphere from harmful ultraviolet radiation. Nitrous oxide contributes to the destruction of the protective ozone layer in the stratosphere (Crutzen, 1981). According to one estimate (Crutzen and Enhalt, 1977), a doubling of the atmospheric N_2O would cause a 10% decrease in the ozone layer which would increase the ultraviolet radiation reaching the earth surface by about 20%. This could result in increased incidence of skin cancer and other health hazards as well as warming of the global climate (Crutzen, 1976; Wang et al., 1976; McElroy et al., 1977; Lijinsky, 1977). Recent estimates of the effects of ozone depletion suggest a 4% to 6% increase in cases of skin cancer with each 1% drop in ozone (Shea, 1988).

Atmospheric N_2O also absorbs thermal radiation. An increase of 0.2% to 0.3% in N_2O concentration in the atmosphere contributes about 5% to the supposed greenhouse warming (Enquete-Komission, 1989). The tentative global estimate of N_2O emissions is 9.7 to 12.0 Tg N yr^{-1} for natural terrestrial ecosystems and 2.3 to 3.7 Tg N yr^{-1} for cultivated lands (Bouwman, 1990a). Eichner (1990) summarized N_2O-emission data from 104 field experiments reported between 1979 and 1987 and estimated that the global release of N_2O from fertilized soils to the atmosphere ranged from 0.2 to 2.1 Tg N_2O-N yr^{-1}.

Crutzen (1981) elaborated the photochemical role of nitrogen oxides, chemical reactions involved in stratospheric and tropospheric chemistry, and changes in the vertical distribution of ozone due to variations in N_2O. Crutzen (1981) estimated that the maximum increase in atmospheric N_2O (0.4% per year) would not result in a doubling of the atmospheric N_2O content until the end of the next century at the earliest. Current measurements of atmospheric N_2O indicate an annual concentration increase of 0.25% (Bouwman, 1990b). A number of laboratory and field studies (Aulakh et al., 1983a, 1983b, 1984a; Aulakh and Rennie, 1985a, 1986, 1987; Sahrawat and Keeney, 1986; Arah and Smith, 1990) indicate that the mole fraction of N_2O produced during terrestrial denitrification can vary from 0 to 1 depending on soil conditions. There is also a pronounced negative relationship with soil moisture content; the higher the water content the lower the relative proportion of N_2O in denitrification products. Therefore, the use of a single ratio in the estimation of N_2O liberated from agricultural soils could lead to erroneous estimates. Secondly, the extreme temporal and spatial variability of the denitrification process, lack of methodology, and very few direct field measurements make global budget estimates for N_2O uncertain at best (Bouwman, 1990a).

C. Water Quality

Humankind's use of N reserves (and fertilizer N) has produced both desirable and undesirable effects on the environment. Widespread cultivation practices and fertilizer N applications have greatly increased food and fiber supplies for an expanding world population. However, N associated with eroded sediments from agricultural lands, leaching associated with excess irrigation and rainwater, as well as N discharges from urban and industrial areas in sewage effluents and other wastes are major causes of pollution of surface and groundwaters. The domestic water supply for 90% of the rural population and 50% of the total USA population comes from groundwater (Power and Schepers, 1989). Keeney et al. (1987) summarized US Geological Survey data from analyses of well samples across the United States and identified regions where ground water NO_3^--N levels exceeded drinking water standards (Table 2). Increasing levels of NO_3^--N in groundwater resulting from agricultural production is cause for concern since NO_3^--N levels in drinking water exceeding $10\,mg\,N\,L^{-1}$ are considered unsafe for human consumption. Inside the body NO_3^- can be reduced to NO_2^- which reduces hemoglobin to methemoglobin and interrupts oxygen transfer to cells of infants (Paul and Clark, 1989).

Where high NO_3^--N levels in soil or water pose a pollution hazard, denitrification may be a desirable process to reduce NO_3^--N. Since the increased N load (from leached N from lands and urbanization) into lakes, bays, and coastal environments is usually accompanied by a decrease in oxygen, the conditions for denitrification are also favorable (Ronner, 1983). An interesting example of biological control was provided by Ellis et al. (1975) who found that anaerobic conditions produced by livestock tramping and waste additions in cattle feedlots resulted in denitrification of NO_3^--N produced at feedlot surfaces and prevented nitrate pollution of groundwater. However, when feedlots were abandoned, aeration of soil profiles increased and NO_3^- accumulated in

Table 2. States with groundwater nitrate contamination problems. (Adapted from Keeney et al., 1987)

State	NO_3-N in well water 3–10 mg L^{-1} (% of wells tested)	>10 mg L^{-1}	Average N Fertilizer rate[a] (kg ha^{-1})
	New England		
Rhode Island	9	36	5.6
	Mid-Atlantic		
New York	29	11	6.5
Pennsylvania	24	6	5.8
Delaware	26	9	32.5
Maryland	22	7	20.4
	East North Central		
Illinois	13	5	65.0
	West North Central		
Minnesota	11	9	23.3
Iowa	13	5	65.0
South Dakota	8	7	5.8
Nebraska	23	9	31.0
Kansas	34	20	25.0
	West South Central		
Texas	14	9	10.9
	Mountain		
Colorado	17	6	6.4
Arizona	24	14	2.7
	Pacific		
California	22	10	13.9

[a] Fertilizer-N consumption for 1984 divided by the total land area of each state.

ground water to levels as high as 66 mg N L^{-1}. Considerable efforts have been made to improve the designs and conditions for the efficient and economical removal of NO_3^- from municipal waste water through denitrification (Brar et al., 1978; Christensen and Harremoës, 1978; Knowles, 1982).

III. Factors Controlling Denitrification

The major factors controlling soil denitrification are soil pH, texture, organic C supply and crop residue management, temperature, soil nitrate and fertilizer N, soil aeration and water status, and certain agrichemicals.

A. Soil pH

The optimum pH for denitrification varies with species and age of organisms and NO_3^- concentration (Delwiche and Bryan, 1976), but most denitrifying organisms have optimum growth pH between 6 and 8. Although the process is favored at slightly alkaline pH, it proceeds to pH as low as 3.5 and can account for significant N losses in naturally acid soils (Gilliam and Gambrell, 1978; Knowles, 1981, 1982; Waring and Gilliam, 1983; Parkin et al., 1985a; Weier and Gilliam, 1986). Long-term acidic soil conditions appear to select for denitrifier populations adapted to low-pH environments (Parkin et al., 1985a). Klemedtsson et al. (1977) and Müller et al. (1980) found a direct positive relationship between the rate of denitrification and soil pH in acid soils from Sweden and 22 locations in Finland, respectively. Few studies have demonstrated whether denitrifier activity is limited by the direct effects of hydrogen ion activity or indirectly through nutrient deficiencies and/or toxicities induced by low pH. Decreasing pH reduces the availability of molybdenum which in turn may reduce the synthesis of NO_3^- reductase, a molybdo-protein enzyme. Bremner and Shaw (1958) however, did not obtain an increase in rate of denitrification with additional molybdenum in a soil at pH 3.6. Koskinen and Keeney (1982) reported that pH does not directly control denitrification but exerts an indirect effect by controlling C availability to the denitrifying organisms. This mechanism of denitrification control in acid soils did not account for the results observed by Parkin et al. (1985a) in an agricultural soil that had a 20-year history of low pH (4.0). They speculated that indirect effects of low pH, such as C availability, may limit the size of the denitrifying population in the acid soil. Alternatively, Firestone (1982) speculated that with decreasing pH, NO_2^- formed by NO_3^- reduction would become toxic and solubilization of aluminum or manganese might cause toxicity effects. However, the actual mechanism for decreased denitrification under acidic conditions is unknown and needs further investigation.

Decreasing pH increases the mole fraction of N_2O in the products of denitrification (Nömmik, 1956; Blackmer and Bremner, 1978; Firestone et al., 1980) even where the total rate of gas production ($N_2O + N_2$) does not decrease (Tiedje et al., 1981). For instance, Koskinen and Keeney (1982) found that N_2O accounted for 83% of the gaseous N products at pH 4.6 and 5.4 when 90% of the NO_3^- was denitrified. As pH increased, N_2O was still produced, but was more readily reduced to N_2. They suggested that the presence of increasing amounts of NO_2^- at the lower pH levels may have been partly responsible for increased mole fraction of N_2O. Later, Weier and Gilliam (1986) suggested a possible relationship between NO_2^--N accumulation and mole fraction of N_2O in denitrification products.

B. Soil Texture

Soil texture influences the denitrification process in several ways. Exposed soil surfaces provide attachment sites for microbial cells. Negatively charged soil colloids may concentrate nutrients in the soil solution (Focht and Verstraete, 1977). Greater rates of denitrification have been reported in finer-textured soils (Lund et al., 1974; Chaterpaul et al., 1980). The greater rate may not be a direct effect of the clay on microorganisms since van der Staay and Focht (1977) were unable to show any differences in denitrification rates between clays of different particle sizes after removal of indigenous organic matter and subsequent addition of substrate and inoculum.

The effect of soil texture on denitrification likely results from physical variations in soil structure, pore size, aggregation, and water infiltration rates that affect aeration, water holding/absorption capacity, and micro-environment, and may be due to other natural differences in capacity of soils to supply substrate (NO_3^- and organic C) and other intrinsic soil factors. Field studies of Chaterpaul et al. (1980) revealed increased rates of denitrification as fineness of soil texture increased (sandy loam < loam < clay loam). In these studies, soil moisture content was directly related and soil aeration inversely related to degree of fineness of soil texture. Groffman and Tiedje (1989a; 1989b) used soil texture and drainage class (general indicator of soil wetness) as predictors of denitrification activity in forest landscapes.

In most soil denitrification research, soil water content is usually expressed on a gravimetric or volumetric basis or as a proportion of soil water holding capacity without consideration of differences in density or porosity for soils of varying textures. Therefore, it is difficult to obtain conclusive results on the effects of soil texture because the same soil water content in texturally different soils can result in vastly different conditions of soil aeration and associated denitrification activity. A water content which results in aerobic conditions in a clay loam may result in anaerobic conditions in a sand. To overcome such problems and to normalize the effects of varying soil texture, a parameter such as water-filled pore space (WFPS) or alternatively air-filled pore space (100-WFPS), which incorporates both soil water content and bulk density, may be a better index of aeration-dependent biological processes such as denitrification than soil water content (Doran et al., 1990). Soil WFPS is synonymous with relative degree of saturation and is equal to soil volumetric water content divided by total soil porosity.

Aulakh et al. (1991c) recently found that at similar water contents, rates of denitrification differed greatly in soils of varying texture, but when varying water-holding capacity and bulk density were accounted for using WFPS, all soils behaved similarly (Fig. 1). Use of WFPS as an index of aeration status enabled the identification of differences in denitrification

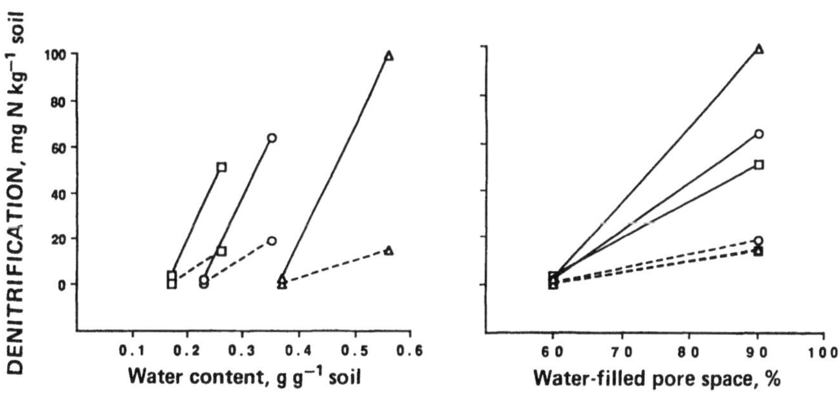

Figure 1. Relationship between total soil gaseous N losses via denitrification over 30 days and two measures of soil water status for three soils of varying texture with (——) and without (----) incorporated hairy vetch (*Vicia villosa*) crop residue. Adapted from Aulakh et al. (1991c). □, Tolewal SL; ○, Gurdaspur SiL; △, Houston black SiC.

rates in soils of varying texture and results of their study suggested the influence of soil texture on denitrification at equivalent states of soil aeration was largely due to the complex interaction between the soils' inherent capacity to supply NO_3^- and metabolize organic residues (C supply).

C. Organic C Supply and Crop Residues

Denitrifying organisms use organic C compounds as electron donors for energy and for synthesis of cellular constituents. Therefore, denitrification is strongly dependent on the availability of organic compounds such as native soil organic matter, crop residues, root exudates, and green and farmyard manures. In addition to the amount of C, its position in the soil profile relative to NO_3^- will largely influence the rate of denitrification.

The important relationship between soil organic C and denitrification was established three decades ago (Bremner and Shaw, 1958; McGarity, 1961). Subsequently, procedures were developed to determine the specific portion of soil organic C which influences or determines the rate of denitrification in soils. Initially water-soluble C (Burford and Bremner, 1975; Reddy et al., 1980a), available or glucose equivalent C (Stanford et al., 1975b), and mineralizable C assessed under aerobic conditions (Burford and Bremner, 1975) showed good correlation with denitrification potential (DNP) where NO_3^- and water are not limiting in soil. However, in later research Beauchamp et al., (1980) found poor correlations between denitrification and the forementioned C fractions. In most of the previous studies, air-dried or oven-dried samples of soil had been used.

Drying of soil can significantly affect its microbial, enzymatic, physical, and chemical properties (Storrier, 1966; Skujins, 1976). Myers and McGarity (1971) showed that drying of three subsurface samples of a solodized solonetz soil at 60°C for 14 days caused a two- to threefold increase in their capacity for denitrification. Patten et al. (1980) and Germon et al. (1981) observed several-fold increases in denitrification due to air-drying of soils. Therefore, the observed relationships between different forms or fractions of organic C and DNP may depend on the soil water status before incubation. In a recent study with 32 field-moist and air-dried soils from a range of arable and grasslands, Bijay-Singh et al. (1988) demonstrated that while air-dried soils showed the greatest correlations between DNP and water-soluble C and aerobically mineralized C, the field-moist soils showed closest correlation of DNP with C mineralized under anaerobic conditions. The authors suggested that, in the field, denitrification may often be limited by the amount of C susceptible to mineralization under anaerobic conditions.

Similarly, biomass C has been reported to give high correlations with DNP especially in soils that were not air-dried prior to experimentation (Germon et al., 1981; Myrold and Tiedje, 1985). It is speculated that either (i) biomass C becomes readily available to denitrifiers following soil saturation or (ii) the availability of C to denitrifiers is proportional to the active biomass mineralizing the organic matter or (iii) part of the biomass C of facultative anaerobic denitrifiers is released by osmotic lysis of microorganisms unable to adjust to rapid increases in soil water content (Kieft et al., 1987). In fact, anaerobiocally mineralizable C, as suggested by Bijay-Singh et al. (1988), most probably measures the same fraction of C that is tied up in biomass C under aerobic conditions.

Because the majority of denitrifying bacteria are heterotrophs, the nature and availability of C sources are significant factors influencing denitrification. In several pure- and mixed-culture studies and laboratory experiments using soil, readily metabolized compounds such as glucose, formate, sucrose, mannitol, citrate, succinate, etc., have been used as sources of C and found highly effective in promoting denitrification (Beauchamp et al., 1989). In growth-chamber and field studies, plant residues, and green and farmyard manures have been reported to increase rates of denitrification (Aulakh et al., 1983a, 1984b; de Catanzaro and Beauchamp, 1985; Aulakh and Rennie, 1987; Aulakh, 1988; Beauchamp et al., 1989; Buresh et al., 1989; Webster and Goulding, 1989; Aulakh et al., 1991a, 1992). However, if soil N is limiting, denitrification losses may be reduced due to immobilization of N caused by decomposition of crop residues having a wide C:N ratio (Craswell, 1978; Fredrickson et al., 1981). Therefore, when crop residues are incorporated in conjunction with fertilizer N, denitrification N losses should increase. This was shown in a field study (Aulakh et al., 1984b) where 3000 kg ha^{-1} wheat (*Triticum aestivum* L.) residues incorporated with 100 kg N ha^{-1} (as ammonium sulfate) resulted in twofold greater denitrification losses during growth of wheat than where no N was applied. In their study, surface-placed crop residues in no-till soil helped conserve rain and snowmelt water and increased denitrification losses due to increasing the proportion of soil pore-space filled with water.

Decomposition of organic residues and hence supply of C to denitrifiers depends upon the composition of specific residue added to the soil. Recently Ford et al. (1989) found the C:N ratio a better predictor of crop residue decomposition than other residue characteristics such as % N, % C, or % lignin. Aulakh et al. (1991a) examined the influence of incorporated and surface placed wheat, corn (*Zea mays*, L.), soybean [*Glycine max* (L.) Merr.], and hairy vetch (*Vicia villosa*) residues (that varied in C:N ratio from 8 to 82) on denitrification and found that cumulative denitrification loss with time and the initial intensity of denitrification were, in general, inversely related to initial C:N ratio of residue. In this study denitrification losses for 35 days were maximum (102 mg N kg^{-1}) with hairy vetch (C:N = 8) and lowest (70 mg N kg^{-1}) with wheat (C:N = 82). They further observed that incorporation of crop residue into soil resulted in several-fold higher initial denitrification rates compared with surface placement. However, after 4 to 5 weeks, the cumulative denitrification losses between placement methods did not differ significantly.

The contributions of plant roots to denitrification have also been studied both in controlled laboratory studies and under field conditions. In C-limited soil systems, denitrification in the rhizosphere may be enhanced by the supply of root exudates and sloughed root surface cells (Garcia, 1975; Bailey, 1976; Volz et al., 1976; Smith and Tiedje, 1979) and through respiratory O_2 consumption provided soil conditions (especially water and NO_3^-) are congenial for denitrification (Stefanson, 1972; Bailey, 1976; Mosier et al., 1988). However, studies of Haider et al. (1985, 1987) indicated that corn or wheat plants did not stimulate denitrification under soil water and NO_3^- conditions usually encountered during plant growth. They showed that rhizo-deposited C was not immediately available for mineralization or denitrification but entered an organic matter fraction that turned over more slowly. Under field conditions, growing plants compete with denitrifiers for NO_3^- and water and can reduce the rates of denitrification (Lance et al., 1978; Aulakh et al., 1982, 1983b). In a 2-year field study in two chernozemic soils of Canada, denitrification losses were two to seven times lower where the same soil was seeded to wheat as compared with fallow (Fig. 2) since WFPS (soil water) and NO_3^- remained significantly lower in the cropped compared with fallowed soil. In contrast, in flooded rice paddies, availability of O_2 near plant roots may be higher than in the surrounding soil due to transport of O_2 to the root zone via arenchyma tissues in the plant stem and root (van Raalte, 1941). Thus, ammonium is oxidized to nitrate and

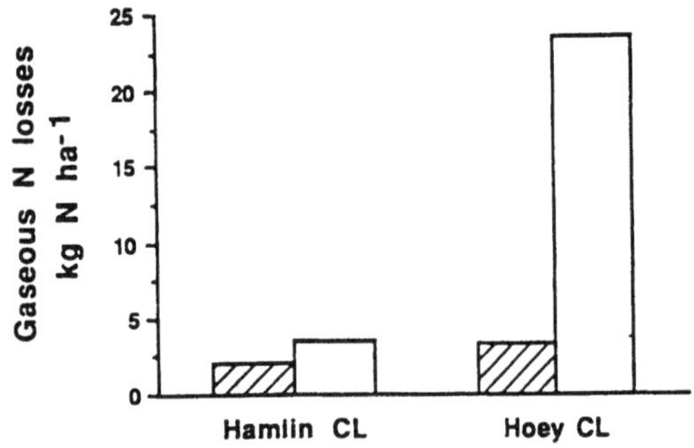

Figure 2. Gaseous N loss from April through November under cropped and fallow conditions for two conventionally tilled soils. Hoey CL soil received normal rainfall during the study period whereas the Hamlin CL soil received below normal rainfall. Adapted from Aulakh et al. (1983a). ▨, wheat; □, fallow.

the nitrate is either taken up by the plant root or denitrified. Reddy and Patrick (1986a, 1986b) demonstrated that a significant amount of ammonium, injected into the rooting zone of rice, was lost from the system. They indirectly determined that denitrification was the major loss mechanism. Later, Mosier et al. (1990b), however, found that when the N gas trapped in the soil was accounted for, denitrification N gas production in soil planted to rice was less ($5.6\,mg\,N\,plot^{-1}$) than in unplanted soil ($7.2\,mg\,N\,plot^{-1}$).

The organic C content decreases with depth in most mineral soils. Although leaching of NO_3^- into lower horizons is a common phenomenon in most agricultural soils, the availability of organic C is usually one of the main factors limiting denitrification activity in subsoils (Aulakh et al., 1983a, 1983b, 1984a; Parkin and Meisinger, 1989; Weier and Doran, 1987). Denitrification potential in peat soils can remain high to a depth of 100 cm but in mineral soils denitrification is often limited to the surface 10- to 20-cm layer (Colbourn et al., 1984); even when water and NO_3^--N are added DNP continues to decrease with increasing soil depth (Khan and Moore, 1968; Bailey and Beauchamp, 1973; Brar et al., 1978; Aulakh et al., 1983a, 1983b, 1984a). However, in certain solonetzic and other soils, soluble C may sometimes move down to the deeper horizons (100 to 150 cm) and may support appreciable denitrification activity (Myers and McGarity, 1971; Rolston et al., 1976; Gilliam et al., 1978). Meek et al. (1969), when relating soil NO_3^--N decrease to redox potential, suggested that soluble organic matter carried in water percolating to the water table might provide microorganisms with sufficient energy to produce conditions suitable for denitrification. Volz et al. (1976) hypothesized that organic C produced in the root zone and percolating with the NO_3^--solution may increase NO_3^- loss both in and below the root zone.

D. Temperature

As with most biological processes, the denitrification rate increases with increasing temperature until an optimum is reached above which it decreases. In glucose-amended soils of varying organic matter, physical, and chemical characteristics, Stanford et al. (1975a) found denitrification rate minimal at 0° to 5° C, increasing ten fold between 5° and 10° C. Between 15° and 35° C the temperature coefficient of denitrification (Q_{10}) was approximately 2. Between 35° and 45° C denitrification rates varied little. Focht and Chang (1975) concluded that the Q_{10} for denitrification in soil was in the range of 5 to 16 for temperatures below 12° to 15° C. Others, however, found denitrification rate increases with temperature to a maximum between 65° and 75° C (Bremner and Shaw, 1958; Keeney et al., 1979); above the maximum, denitrification rate decreases rapidly to zero with further temperature increase, probably due to thermal denaturation of proteins (Stanier et al., 1976). It is not well established whether the high optimum temperature for denitrification results from activity of thermophilic bacteria (Focht and Verstraete, 1977), from biological reactions (Knowles, 1981), or from their combined action (Keeney et al., 1979).

Bailey and Beauchamp (1973) observed a rapid decrease of denitrification rate below 5° C. However, the minimum temperature for denitrification may vary widely, from 0° to 10° C (Cho et al., 1975; Bailey, 1976; Smid and Beauchamp, 1976) because denitrifiers have the capability of adapting to their thermal environment (Gamble et al., 1977). Since decreasing temperature could increase O_2 solubility in soil water and decrease O_2 consumption (Craswell, 1978) a consistent exponential relationship for denitrification is sometimes difficult to demonstrate.

In field studies with different tillage and cropping systems, Aulakh et al. (1983a, 1983b, 1984a; Aulakh and Rennie, 1984) observed little influence of temperature on denitrification rate during the crop-growing season when temperatures ranged from 10° to 30° C. However, in early spring and late fall when temperatures were 5° C or below, virtually no denitrification was detectable even in wet soils with high NO_3^- content.

In some cases, decreasing soil temperature can dramatically change the proportion of N oxide gases produced during denitrification. At temperatures of 15° to 30° C, the major product of denitrification in wet or water-logged soils is N_2. Under similar conditions at temperatures of 4° to 8° C, N_2O and NO often predominate (Bailey, 1976; Melen and Nömmik, 1983).

E. Soil Mineral N and Fertilizer N

Under laboratory conditions, denitrification rates are substrate-dependent (first order reaction) at nitrate concentrations below 100 mg NO_3^--N kg^{-1}

soil. Given the very low apparent Km for NO_3^- and NO_2^- reducing enzyme systems (Yoshinari et al., 1977) it is not surprising that soil denitrification is often unaffected by NO_3^- concentration (zero order) in the range between 40 and 600 mg NO_3^--N kg^{-1} soil (Bowman and Focht, 1974; Starr and Parlange, 1975; Kohl et al., 1976; Knowles, 1981). Reddy et al. (1980b) and Cho (1982) suggested that the rate of NO_3^- disappearance can be expressed as a first order reaction in conditions where the rate of electron acceptor supply (NO_3^-, NO_2^-) is smaller than the maximum rate of electron production (biological activity). Since the rate of denitrification is also dependent on soil water (anaerobiosis) and available C concentrations, kinetic constants for NO_3^- and NO_2^- have little quantitative meaning on their own. This is the reason that under laboratory conditions where soils are incubated with high water and C supply, rate of denitrification depends upon NO_3^- supply (Aulakh and Rennie, 1987, Aulakh et al., 1991a, 1991c). However, under natural conditions where several factors such as water content, C availability, and NO_3^- are interrelated, NO_3^- concentration often does not show a significant correlation with denitrification losses (Aulakh et al., 1983a, 1983b, 1984a). Further wetting and drying cycles prevalent in the field increase soil C levels available for denitrification. For example, Paul and Myers (1971) noted highest losses of the indigenous organic N from soil (through mineralization-nitrification-denitrification) where drying was more severe. Thus, the interpretation of denitrification kinetics which is difficult in natural soils because of inherent heterogeneity in NO_3^- concentrations, C availability, aeration, and microbial distributions is further complicated by frequency and severity of wetting and drying cycles.

F. Soil Aeration and Water Status

Soil water can directly and indirectly influence denitrification through: (i) provision of suitable conditions for microbial growth and activity; (ii) restricting supply of O_2 to microsites by filling soil pores; (iii) release of available C and N substrates through wetting and drying cycles; and (iv) providing a diffusion medium through which substrates and products are moved to and away from soil microorganisms.

Previous research demonstrates the importance of soil air/water balance upon aerobic and anaerobic microbial activities and formation and loss of soil nitrate nitrogen. Aerobic microbial activities increase with soil water content until a point is reached where further water restricts the diffusion and availability of oxygen. The rate of O_2 diffusion through water is about 10^4 times less than through air. Maximum rates of predominantly aerobic processes such as respiration, nitrification, and mineralization occur at the highest soil water content at which O_2 remains nonlimiting (Bhaumik and Clark, 1948; Miller and Johnson, 1964; Parker and Larson, 1962; Stanford and Epstein, 1974). When soil water content approaches or exceeds field capacity, the percentages of soil pore space filled with air or water are better indicators of aerobic and anaerobic microbial activity than either water content, matric tension, or water potential (Miller and Johnson, 1964; Sommers et al., 1981).

The summary of results from many studies, involving a wide range of soil types and tillage management practices, indicate that aerobic microbial activities are greatest at a soil water content equivalent to 60% of a soil's water-holding capacity (Doran et al., 1990; Linn and Doran, 1984a). The results of several laboratory studies suggested that denitrification is negligible at water content below 60% of maximum water-holding capacity, regardless of NO_3^- supply (Aulakh, 1986a, 1989; Aulakh et al., 1984c, Aulakh and Rennie, 1987; Aulakh et al., 1991a, 1991c).

Doran and his associates (Linn and Doran, 1984a, 1984b; Doran et al., 1988, 1990; Aulakh et al., 1991a, 1991c) have shown that WFPS depicts a better relationship with the rate of denitrification (Fig. 3). Use of WFPS overcomes several problems associated with varying water satu-

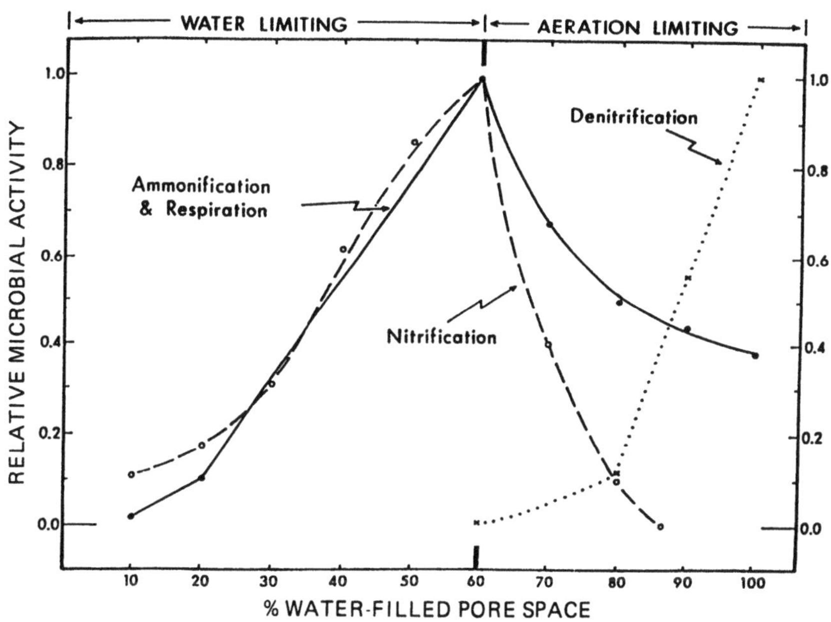

Figure 3. Relationship between soil water-filled pore space (relative saturation) and the aerobic microbial processes of respiration and nitrification and anaerobic denitrification in carbon-amended soil. Divergence of relationship between ammonification/respiration and nitrification above 60% to 70% water-filled pore space results from production of CO_2 and ammonification by anaerobic microbial fermentations. Adapted from Linn and Doran (1984a).

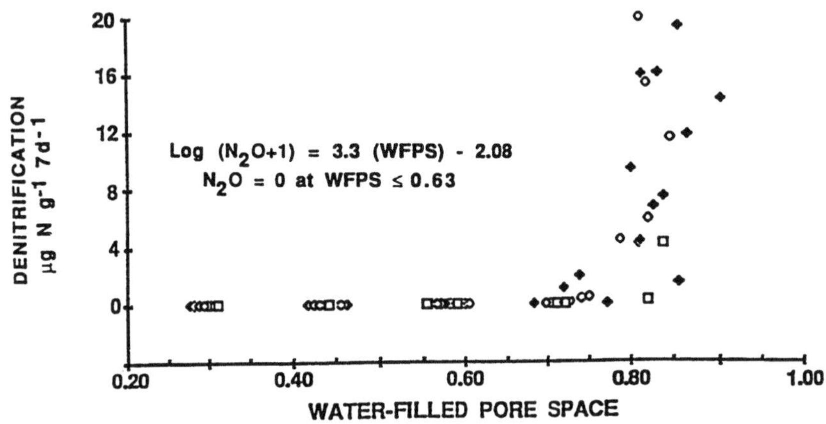

Figure 4. Relationship between denitrification and water-filled pore space for repacked cores of 18 benchmark reference soils of varying textures and organic matter contents. Adapted from Doran et al. (1990). ◆, medium-fine; ◇, coarse; □, other.

ration levels and bulk density for soils differing in texture (as discussed earlier in Section III.B) and tillage regimes. Moreover, WFPS is a more practical index of soil aeration than measurement of oxygen diffusion rates because it requires only a knowledge of gravimetric soil water content and soil bulk density (assuming the particle density of mineral soils = $2.65\,Mg\,m^{-3}$). Technology for measurement of both these parameters should be readily available to anyone conducting soil management research.

In field studies with several soils, Aulakh and Rennie (1985a) revealed that the rate of denitrification often remains negligible but increases dramatically when soil volumetric water content exceeds a certain critical level between 40% and 50% which was equivalent to from 65% to 75% WFPS. Doran et al. (1990) found that this critical threshold for denitrification was between 70% and 80% WFPS for 18 soils of varying physical and chemical characteristics (Fig. 4).

Several studies (Aulakh et al., 1982, 1983a, 1983b, 1984a, 1984b; Mosier et al., 1986a, 1986b; Smith and Tiedje, 1979; Sexstone et al., 1985a) show that a short-term increase in soil water due to rainfall (or irrigation) results in dramatic increases in gaseous N fluxes within hours, suggesting that levels of denitrifying organisms and denitrifying enzymes present in natural agricultural soils are sufficient to respond quickly to changes in aeration.

The simultaneous measurement of soil water, temperature, and mineral N along with denitrification losses from natural soils under various management systems makes it possible to estimate their individual and

Table 3. Effect of wheat residue incorporation on denitrification losses ($mg\,N\,kg^{-1}$) in a 96-day incubation. (Adapted from Aulakh and Rennie, 1987)

	Water-filled pore space (%)		
Crop residues ($mg\,kg^{-1}$)	60	90	120
0	2.5a*	53.7a	61.6a
2500	5.1b	78.0c	75.4b
5000	7.9c	75.5c	81.0b
10,000	10.1d	67.5b	83.9c

* Values within each column followed by the same letter do not differ significantly ($P \geq 0.05$).

combined effects on denitrification. Results of field studies by Aulakh et al. (1983a, 1983b, 1984a) and Aulakh and Rennie (1985a) have shown that variations in denitrification losses were significantly correlated with volumetric soil water content (or WFPS) followed by temperature and then soil nitrate-N. In evaluating the combined effects of these factors, combining water and temperature accounted for more variation in denitrification losses as evidenced by a higher correlation coefficient. Adding nitrate-N or ammonium-N did not further significantly increase the R value. These results indicated that either NO_3^- was not limiting denitrification in these field soils or that other factors such as WFPS and temperature were more important controls of denitrification. However, in some soils substantial denitrification may occur at low water contents due to O_2 depletion in metabolically active soil microsites (Currie, 1961; Greenwood, 1975; Sexstone et al., 1985b; Parkin, 1987) which are associated with enhanced supply of C (Aulakh and Rennie, 1987; Aulakh et al., 1991c). Decomposition of incorporated plant residues in soil increases denitrification through increased microbial activity and uptake of O_2 at a rate exceeding its replacement via diffusion (Table 3). Under field conditions, management and incorporation of crop residues and soil disturbance can create localized microhabitats of greater bioactivity and results in several-fold greater rates of denitrification even at WFPS of 60% or below (Table 4).

G. Agrichemicals

A variety of chemicals used in agriculture such as pesticides, herbicides, and nitrification inhibitors, are known to affect microbial processes. Few pesticides, however, have been studied for their effect on denitrification (Knowles, 1981; Goring and Laskowski, 1982). Nitrapyrin (2-chloro-6 [trichloromethyl]-pyridine), a nitrification inhibitor, did not affect soil denitrification at the rate of $10\,mg\,kg^{-1}$ (Sacks and Barker, 1952; Bundy and Bremner, 1973; Henninger and Bollag, 1976) but when applied at the

Table 4. Denitrification losses (using AI soil-core method) and soil respiration in corn field 1 week after planting under ridge-tillage cover crop and conventional disk management without a cover crop. (Unpublished data of J.W. Doran)

Management system	Position in field[c]	Denitrification ($gN\,ha^{-1}\,day^{-1}$)	Soil respiration ($kg\,C\,ha^{-1}\,day^{-1}$)	Water filled pore space (%)
Ridge tillage[a]	Row	4	3	33
and cover crop	Interrow	256	140	44
	Total or average	260	143	40
Disk[b]	Row	21	5	35
	Interrow	67	14	40
	Total or average	88	19	38

[a] Soybean and winter-cover crop (Hairy vetch and rye) residues were moved to interrow area at time of planting on May 8, 1989.

[b] Crop residues from previous soybean crop incorporated in soil by fall disking in 1988 and again in spring 1989.

[c] Relative area under row and interrow represented 33% and 67% of the total surface area, respectively.

rate of $16\,mg\,kg^{-1}$ to a paddy soil (Mitsui et al., 1964), it caused 30% inhibition of the process. Inhibition of nitrification with nitrapyrin results in NH_4^+ accumulation from soil mineralization and ammonical fertilizers. Thus, the chances of NO_3^- formation and loss through denitrification are reduced (Aulakh and Rennie, 1984). Similarly other nitrification inhibitors such as AM (2-amino-4-chloro-6-methyl-pyremidine), ST (2-sulfanilamide-thiazole), ASU (1-amidino-2-thiourea), and DCS (N-2,5-dichloro phenylsuccsinamic acid) also have been found very effective in reducing N_2O emission in soils (Minami et al., 1990). However, potassium azide, another very effective nitrification inhibitor increases N_2O emissions when added to soil (Freney et al., 1979; Aulakh and Rennie, 1985b). The increased N_2O emissions are presumably due to azide-induced stimulation of denitrifying enzyme synthesis (Burke et al., 1980; Calder et al., 1980). Elemental sulfur which is used for controlling several pathogens and as a source of S nutrition for plants, stimulates the activity of *Thiobacillus denitrificans* (Mann et al., 1972) and may increase denitrification losses from soils.

IV. Methodology for Field Measurements

Knowles (1981) summarized direct and indirect methods available to measure denitrification from soil in the laboratory. Only a few of these methods are applicable for field use. The most common indirect method used in the field is the N balance approach using ^{15}N-labeled fertilizer (Allison, 1955; Hauck, 1981; Aulakh 1986a). The lack of direct methods to quantify denitrification products (N_2O and N_2) has been a long-standing problem in field soil N research. Nitrous oxide production can be precisely measured using gas chromatography (GC) and easily identified over relatively low atmospheric background concentrations (310 ppb). The amounts of N_2 produced during denitrification, however, are not large enough to be easily distinguished from the normal ambient atmospheric concentration of 78%. As a result of interest in quantifying the contribution of N_2O to the atmosphere, several approaches have been taken for in situ measurement of N_2O evolutions in the field (Aulakh, 1986a):

1. Measurement of N_2O concentration gradients in the soil profile
2. Micrometeorological technique estimating surface gas fluxes
3. Soil cover or chamber methods

Among these methods, the soil cover method, which is both inexpensive and practical, has been more extensively used for measuring in situ N_2O fluxes and also for N_2 fluxes using ^{15}N or acetylene-inhibition techniques. There are two general approaches to containment and measurement of evolved soil gases using covers or chambers:

1. "Open soil chambers" as described by Ryden et al. (1979) where outside gases flow continuously through chambers, and evolved soil N gases are continuously removed.
2. "Closed soil chambers" which have closed-loop air circulation or no forced air circulation, either of which allows accumulation of evolved N gases (Matthias et al., 1980; Hutchinson and Mosier, 1981; Colbourn et al., 1984).

Both approaches in using soil covers overcome problems associated with soil diffusion and micrometeorological methods. There is no need to know the diffusion coefficient of N_2O in the soil and its changes with time as concentrations inside the chambers are high enough for reliable estimation of relatively small gas fluxes. Potential problems associated with different soil cover methods and rationale for choice of a particular approach are described in several review articles (Denmead, 1979; Jury et al., 1982; Mosier and Heinemeyer, 1985; Mosier, 1989). Consequently we will only briefly discuss them below.

Open chamber methods are sensitive to pressure deficits inside the chamber caused by the induced air flow which may result in artificially high fluxes and overestimation of gas loss. Use of the closed chamber technique permits measurement of undisturbed N gas fluxes and has been used to estimate field denitrification in several recent studies (Colbourn et al., 1984; Hallmark and Terry, 1985; Terry et al., 1986; McConnaughey and Duxbury, 1986; Mosier et al., 1986a, 1986b, 1990a; Aulakh et al., 1991b). Appropriate venting of closed chambers allows pressure equilibrium with the external atmosphere and minimizes losses of gaseous N by diffusion. However, it is necessary to prevent the development of abnormally high N_2O concentrations in the chamber and/or soil environment which may reduce the rate of diffusion of N_2O towards the soil surface from deeper in soil (Denmead, 1979). Thus, the rate of N gas emission may be monitored by periodically covering chambers and withdrawing air samples at short intervals of 15 min to 60 min depending upon its rate of production.

The finding that acetylene (C_2H_2) catalytically inhibits the reduction of N_2O to N_2 during denitrification (Balderston et al., 1976; Yoshinari and Knowles, 1977) and verification that NO_3^--N is stoichimetrically converted to N_2O by soil denitrifiers in the presence of C_2H_2 (Yoshinari et al., 1977) resulted in development of the acetylene-inhibition technique for measuring denitrification. Another direct field method of denitrification involves the use of highly enriched ^{15}N-labeled fertilizers and subsequent measurement of $^{15}N_2O$ and $^{15}N_2$ gases (Hauck et al., 1958; Hauck and Bouldin, 1961; Siegel et al., 1982; Strong et al., 1987).

A. ^{15}N Balance Approach

In conducting an N balance the amount of applied ^{15}N-labeled fertilizer N is measured in various soil N pools and in the plant (if a crop is raised) and the amount of unaccounted-for N is assumed to have been lost from the soil-plant system. However, before assuming that this loss resulted from denitrification, it is necessary to verify that no N has been lost either by leaching or by ammonia volatilization. With the advent of precise techniques for ^{15}N measurements in soil, water, and plant samples and with necessary precautions (Jenkinson et al., 1985) the ^{15}N balance method can provide reliable information on the transformation and efficient use of fertilizer N. However, it cannot be used to assess short-term response of denitrification to management practices such as rainfall events, irrigation, tillage, and incorporation of crop residues. This results from the lack of precision due to involvement of several facets such as NO_3^-, NH_4^+, organic N, fixed NH_4^+-N in soil, and N recovery by the plant. Hauck (1978) discussed in detail the dominant sources of error while calculating fertilizer ^{15}N recovery in the soil-plant system. Fertilizer N in tracer studies is assumed to include fertilizer N that has pooled with labile organic N in soil and is subsequently made available for plant use. The immobilization-mineralization sequence can cause error in the interpretation of tracer data (Hauck, 1978; Jenkinson et al., 1985). Similarly, relative error in calculating the percentage recovery of applied N by plants increases as the ^{15}N concentration of labeled fertilizer decreases.

B. ^{15}N Chamber Method

This method involves applying a ^{15}N-enriched fertilizer source (>20 atom % ^{15}N) followed by sampling the N gases evolved through use of a chamber (or enclosure) to isolate the atmosphere above the soil for a specified period. The rate of change of ^{15}N atoms (N-gases) in the chamber atmosphere over time can be determined using isotope ratio mass spectrometry (MS). The theory underlying the direct measurement of $^{15}N_2O$ and $^{15}N_2$ evolution was developed three decades ago (Hauck et al., 1958; Hauck and Bouldin, 1961; Hauck, 1966). Only recently has the concept been actually applied to field studies (Siegel et al., 1982; Mosier et al., 1986a, 1986b; Mulvaney and Vanden Heuvel, 1988; Mosier et al., 1989). Isotope dilution equations based on a normal distribution of isotopes cannot be used to accurately calculate the amount of ^{15}N-labeled N_2 evolving into the atmosphere that contains relatively large amounts of N_2 of natural isotopic composition. This is because the ^{14}N and ^{15}N atoms are not randomly distributed in a gaseous mixture formed when N_2 derived from isotopically enriched NO_3^--N evolves into an atmosphere already containing N_2. Therefore, the method of calculating the N_2 and N_2O derived from isotopically enriched NO_3^- assumes that two sets of molecular species ($^{14}N_2$, $^{14}N^{15}N$, $^{15}N_2$) retain their separate isotopic compositions. Small amounts of evolved N_2 can be determined because owing to the low abundance of $^{15}N_2$ in a natural atmosphere, its concentration can be readily changed by small amounts of N_2 of mass 30 derived from substrate highly enriched in ^{15}N. By analyzing the chamber contents on an isotope ratio mass spectrometer for the increases in m/e $^{30}N_2$ and $^{29}N_2$, the emissions of N_2 or N_2O are quantified. Rolston et al. (1982) used measurement of m/e 29/28 and 30/(28 + 29) for calculating the atom % ^{15}N of the N_2 in the atmosphere above soil surface which measures denitrification N losses from applied fertilizer N only. The method proposed by Siegel et al. (1982), however, can be used to estimate total N gas ($^{14}N_2$ + $^{15}N_2$) emissions. Also, Siegel's method takes into account the changes in ^{15}N enrichment of soil NO_3^- with time; e.g., dilution of NO_3^- pool with ^{14}N-NO_3^- derived from mineralization of soil organic matter or its depletion due to plant uptake, denitrification, or immobilization of NO_3^-.

For the details on equations used to calculate the total N gas flux from the soil using a double collector or triple collector isotope ratio mass spectrometer, assumptions and associated problems, and for description of a gas sampling inlet system, the reader is referred to work by Siegel et al. (1982), Mulvaney (1984, 1988), Mulvaney and Boast (1986), Smith (1987), and Mosier and Schimel (1992). We briefly discuss assumptions and associated problems below.

1. Uniformity of ^{15}N-NO_3^- Distribution

An underlying assumption of the Seigel method is that applied ^{15}N mixes with the soil mineral N pools to produce a uniformly labeled ^{15}N-NO_3^- pool that is uniformly subject to denitrification. The initial ^{15}N-NO_3^- content is required to calculate the total amount of N_2 evolved from the site but it is not necessary to disturb the soil in the plot during the course of the experiment. If gaseous denitrification products are derived from NO_3^- pools with markedly different ^{15}N composition, the contribution from unlabeled soil NO_3^--N can lead to a serious underestimation of denitrification (Focht, 1985; Duxbury, 1986; Mulvaney, 1988). For example, with band-placed fertilizer N and in no-till management, uniform mixing of fertilizer and soil N pools may not be achieved. Mulvaney (1988), however, demonstrated, that although the nonuniformity of ^{15}N-NO_3^- was theoretically a problem, in practice the technique provides good N loss estimates from nonuniform ^{15}N-NO_3^- pools. Mosier and Schimel (1991) also gave examples of accurate N-gas production estimates when the soil NO_3^- pool was not uniformly labeled.

2. Nitrous oxide and N_2 from Different Pools

If the ^{15}N label is the same for N_2O and N_2 the N_2O can be converted to N_2 in the MS inlet system, and only one analysis is required (Mosier et al., 1986a, 1986b). But problems may arise if the ^{15}N labels of the N_2O and N_2 evolved from the soil are not the same; then a separate MS analysis may be made for N_2O and N_2 to determine the ^{15}N content of the separate gases (Focht, 1978; Mulvaney and Kurtz, 1982; Poth and Focht, 1985; Mulvaney and Boast, 1986). However, in the field N_2O is only a small part of the total denitrification N gases when soil conditions are anoxic and N losses are important.

C. Aceteylene-Inhibition Chamber Method

There are four advantages in using the acetylene-inhibition (AI) method for assessment of denitrification under field conditions:

1. The method is relatively simple and straightforward.
2. Equipment for collection and analysis (gas chromatography) of samples for N_2O is relatively inexpensive and accessible to most researchers.
3. The method can be used in nonfertilized ecosystems.
4. Instrumentation is available to permit detection of N_2O fluxes as low as $1\,g\,N\,ha^{-1}\,day^{-1}$.

Appropriate and accurate use of the AI method requires consideration of several factors. First, since C_2H_2 also inhibits NH_4^+ oxidation (Walter et al., 1979; Mosier, 1980; Aulakh et al., 1984c), the concurrent inhibition of nitrification where soil NO_3^- supply is limiting will result in underestimation of denitrification rates. In such cases this technique should be used for short-term (1 to 3 days) denitrification measurements only. Second, when soils are incubated for a prolonged period (over 7 days), C_2H_2 in the soil atmosphere may disappear or be metabolized (Yeomans and Beauchamp, 1982; Terry and Duxbury, 1985). However, these complications of C_2H_2 decomposition/metabolism in soils and nitrification inhibition by C_2H_2 in field denitrification studies can be overcome by intermittent exposure of soil to C_2H_2 or by rotating field plots (Terry et al., 1986).

In the AI chamber method, C_2H_2 is injected into the soil at selected depths through use of diffusion probes which are used for either a continuous flow of C_2H_2 (Ryden et al., 1979; Colbourn et al., 1984; Hallmark and Terry, 1985; McConnaughey and Duxbury, 1986; Terry et al., 1986; Aulakh et al., 1991b) or one-time injection before measurement of denitrification (Mosier et al., 1986b). Research indicates that a C_2H_2 concentration of >0.1% in the soil atmosphere is required to insure complete inhibition of N_2O reduction to N_2 (Yoshinari et al., 1977; Ryden et al., 1979; Germon, 1980; Duxbury, 1986). The absolute amount of C_2H_2 required, however, varies from soil to soil and should be quantified before denitrification studies are initiated for a particular soil.

D. Calcium Carbide Acetylene-Inhibition Chamber Method

When calcium carbide (CaC_2) is placed in moist soil, it is rapidly hydrolyzed to calcium hydroxide and acetylene:

$$CaC_2 + 2H_2O \rightarrow Ca(OH)_2 + C_2H_2$$

Therefore, use of CaC_2 granules lightly coated with paraffin prolongs the delivery of C_2H_2 to soil and provides an excellent alternative to the labor-intensive maintenance of C_2H_2 levels in the soil through use of diffusion probes for measuring in-field denitrification. Aulakh et al. (1991b) demonstrated that insertion of lightly coated CaC_2 granules in soil around denitrification measurement chambers produced C_2H_2 levels which effectively blocked N_2O reduction for a 48-h period in the field. This method can be easily adopted for longer term denitrification studies with a few precautions. For instance, CaC_2 granules are inserted after each 48-h period to maintain C_2H_2 production. Because CaC_2 granules are inserted outside the chambers, production of $Ca(OH)_2$ and resultant increase in soil pH may not affect denitrification and other biochemical processes within the experimental plot (confined inside the chamber). The problem of C_2H_2 decomposition/metabolism in soils due to prolonged incubation (over 7 days), can he overcome by intermittent exposure of soil to C_2H_2 or by rotating field plots (Terry et al., 1986).

E. Acetylene-Inhibition Soil-Core Method

Another approach to measuring field denitrification N gas flux is the AI soil-core method. In this method intact soil cores are taken by pushing a small cylinder or probe of an appropriate size into the soil and incubation in an enclosure to which C_2H_2 is added. Two types of "soil-core methods" have been used. In the first, intact soil cores are contained in rubber-stoppered plastic tubes (4.7 cm diameter × 20 cm long) through which air containing C_2H_2 is circulated (Parkin et al., 1984, 1985b) or mixed with soil air by repeated pumping with a syringe (Robertson et al., 1987). The linear rates of N_2O production obtained within 1 to 2 h of incubation are used to estimate the denitrification rates. In the second, intact soil cores of appropriate size (6 cm diameter × 7.5 to 20 cm long) are removed using metal cylinders and immediately placed in air-tight jars (often fruit-preserving jars of 1- or 2-L capacity are used), the lids of which are fitted with rubber septums for gas sampling (Aulakh et al., 1982; Ryden et al., 1987). Five to ten percent of the jar's headspace air is replaced with C_2H_2. Narrow slits in the side of the metal (often aluminum) cylinder walls encourage rapid diffusion of C_2H_2 into the soil core. Each core-containing jar is incubated in the field for 24 h in the soil at the same depth from which the core was sampled. Air samples are periodically removed from the headspace and analyzed for N_2O to estimate denitrification N gas production rates.

The AI soil-core method has the following advantages:

1. Less complicated procedures in the field and laboratory.
2. Permits integration of rates of denitrification over a 24-h period, thus accounting for daily variability in denitrification rate due to fluctuations in soil temperature (Blackmer et al., 1982; Mosier, 1989; Aulakh et al., 1991b).
3. Denitrification can be studied in systems when farm operations would normally restrict the installation of AI chambers.
4. Soil cores can be used for additional analysis for mineral N, water content, and bulk density. Since NO_3^- is not increased during incubation via nitrification (due to inhibition of NH_4^+ oxidation by C_2H_2) and is depleted only through denitrification, NO_3^- present at the end of incubation plus the N denitrified indicates the amount of NO_3^- at the start of incubation.
5. The AI soil-core method performs better in poorly drained soils where C_2H_2 diffusion into soil out of soil is restricted because relatively rapid diffusion of C_2H_2 across the 3-cm radius of the soil cores can be achieved (Ryden et al., 1987).

For accurate estimates of the field denitrification rates with the AI soil-core method, an assessment of the appropriate sampling depth is necessary. Also, intact soil cores must be removed taking care to avoid disturbing and compacting the samples.

F. Relative Performance of Different Methods

Comparison of the different methods for measurement of field denitrification are summarized in Table 5. Ryden et al. (1979) compared the AI chamber and ^{15}N chamber methods, using open chambers, and found

Table 5. Comparison of field denitrification losses by several methods of measurement

Soil/site(s) Cropping system[a]	N Application rate (kg ha^{-1}) and source	Duration and unit of measurement	Methods used ^{15}N Balance	 ^{15}N Chamber	 AI Chamber	 CC-AI Chamber	 AI Soil-Core	Reference(s)
Silinas CT Irrigated fallow	6.3 KNO_3	170 h mg N	—	355a*	392a	—	—	Ryden et al., 1979
Hoey, Hamlin CT wheat	50 $(NH_4)_2SO_4$	TCS[b] kg N ha^{-1}	1.2–2.3a	—	—	—	1.2–1.8a	Aulakh et al., 1983b
Elstow CT wheat	100 $(NH_4)_2SO_4$	TCS kg N ha^{-1}	4.6–6.6a	—	—	—	4.3–7.2a	Aulakh et al., 1984b
Elstow ZT wheat	100 $(NH_4)_2SO_4$	TCS kg N ha^{-1}	11.2–20.9a	—	—	—	11.9–22.5a	Aulakh et al., 1984b
Spinks CT Fallow	14.2 NH_4NO_3	30 days mg N kg^{-1}	1.8a	—	—	—	1.1a	Parkin et al., 1985b
Capac CT Fallow	4.4 KNO_3	30 days mg N kg^{-1}	6.7a	—	—	—	2.2a	Parkin et al., 1985b
Nunn CT Irrigated barley	200 $(NH_4)_2SO_4$	TCS kg N ha^{-1}	4.0a	1.5a		—	—	Mosier et al., 1986a
Nunn CT Fallow	200 $(NH_4)_2SO_4$	16 days kg N ha^{-1}	—	9.0a	12.0a	—	—	Mosier et al., 1986b
Calauan Wetland rice	80 Urea	20 days % of applied N	52a	0.1b	—	—	—	DeDatta et al., 1987; Buresh and DeDatta, 1990

Table 5. *Continued*

Soil/site(s) Cropping system[a]	N Application rate (kg ha^{-1}) and source	Duration and unit of measurement	Methods used					Reference(s)
			^{15}N Balance	^{15}N Chamber	AI Chamber	CC-AI Chamber	AI Soil-Core	
8 (Munoz) Wetland rice	44–58 Urea	17 days % of applied N	26–46a	0.2–1.1b	—	—	—	Buresh and Austin, 1988
2 (Los Banos) Wetland rice	29 Urea	20 days % of applied N	40–41a	0.5b	—	—	—	John et al., 1989
1 (Los Banos) Wetland rice	27 KNO_3	20 days % of applied N	53a	20b	—	—	—	John et al., 1989
5 soils Wetland rice	87 Urea	10–15 days % of applied N	10–52a	<0.1–2.2b	—	—	—	Buresh and DeDatta, 1990
Crete-Butler ZT Fallow	50 KNO_3	4 days g N ha^{-1}	—	190b	173b	240b	1015a**	Aulakh et al., 1991b
Crete-Butler CT Fallow	100 KNO_3	2 days kg N ha^{-1}	—	—	—	3.5a	3.6a	Aulakh et al., 1991b

[a] Cropping system: CT, conventional-till; ZT, zero-till; wetland, flooded or irrigated or lowland.

[b] TCS, total cropping season.

* Values between columns followed by same letter do not differ significantly ($P \geq 0.05$).

** Soil-cores incubated for 4 days.

good agreement between them. Later Mosier et al. (1986b), using closed chambers, also obtained comparable results between AI chamber and ^{15}N chamber methods. Aulakh et al. (1983b, 1984b) and Parkin et al. (1985b) obtained a good agreement between cumulative denitrification losses and the amount of unaccounted N by the ^{15}N-balance approach. In the study of Parkin et al. (1985b) average denitrification losses with ^{15}N-balance in a Capac CL soil were threefold higher than those measured with AI soil-core method; measurements using either method, however, were still within the 95% confidence limits due to large spatial variability (a major problem in denitrification studies as discussed later). Ryden et al. (1987) obtained a close agreement between the AI soil-core and the AI-chamber methods.

In several studies with wetland rice soils (DeDatta et al., 1987; Buresh and Austin, 1988; John et al., 1989; Buresh and DeDatta, 1990), field measurements of denitrification losses using the ^{15}N chamber method were much lower than estimated using the ^{15}N-balance technique (Table 5). However, since the fertilizer N source in these studies was urea, losses of N could have occurred via ammonia volatilization (Fillery et al., 1986; DeDatta et al., 1989; Freney et al., 1990; Buresh and DeDatta, 1990). Other possible reasons for differences between ^{15}N chamber and ^{15}N-balance techniques include transmission of soil N gases to the atmosphere through rice plants (Bhadrachalm et al., 1991; Mosier et al., 1990b) and soil entrapment of N gases (Holt et al., 1988; Lindau et al., 1988; Mosier et al., 1989; Aulakh and Doran, 1990) which are discussed in detail in the following sections.

Recently Aulakh et al. (1991b) compared four methods for measuring denitrification and found that use of chambers with either ^{15}N or acetylene inhibition was reliable if proper sampling intervals were chosen to account for daily fluctuations in the N gas flux rate. Denitrification estimates using the AI soil-core method gave results comparable with those measured using chamber methods provided cores were incubated no longer than 1 day.

G. Enigma of Denitrification Measurement

Estimating annual or long-term denitrification losses from agricultural soils using the previously discussed methods is complicated by diffusion restrictions and entrapment of N gases in wet soil, the transmission of N gases through rice plants, and diurnal, temporal, and spatial variability. Each of these problems will be discussed briefly.

1. Entrapment of N Gases in Soil

Reliable estimation of denitrification N losses from soils requires a quantitative relationship between N gas production rate and their flux rate at the soil surface as an estimate of all of the N gas produced. In relatively dry or moist soils, surface measurements of N gas flux using chamber or soil-core methods, could be quantitatively related to the actual N gas production rate because diffusion of N gases from the soil surface is rapid (Jury et al., 1982; Aulakh and Doran, 1990). In wet soils (for instance, after rainfall or irrigation, and flooded rice soils) N gases produced during denitrification may not diffuse to soil surface because gases diffuse about 10,000 times more slowly through water than air. In such soils, a week or more may be required to collect all $N_2 + N_2O$ produced within the soil, especially in chamber-confined soils (Jury et al., 1982). However, the gas diffusion limitation is likely to be less severe in soil cores because of smaller dimensions.

In wet soils the denitrification N gases remain entrapped in the soil pore water. Entrapment increases with increasing rate of denitrification and WFPS (Holt et al., 1988; Lindau et al., 1988; Mosier et al., 1990, 1990b; Aulakh and Doran, 1990; Samson et al., 1990; Reddy et al., 1990). When soil cores are used, the entrapped N gases in soil are quickly released by physical disturbance of the cores at the end of incubation and then quantitatively measured. For example, Holt et al. (1988) observed that shaking soil incubation vessels for 1 h resulted in a 6- to 12-fold increased recovery of gaseous ^{15}N denitrification products. Similarly, Aulakh and Doran (1990), using the AI soil-core method, observed that for soils incubated at 120% WFPS, about 65% of the total N_2O produced was entrapped in the soil. Mosier et al. (1990, 1990b) reported a several-fold increase in entrapment of N gases in an unplanted soil which was flooded for 30 days prior to the addition of urea (Table 6). These studies indicate that N gas flux rates measured with chamber methods may lead to underestimation of total denitrification losses due to entrapped N gases in wet or flooded soils.

2. Transmission of N Gases Through Plants

The transmission of gas through internal voids (aerenchyma) in roots and stems occurs in several plants including rice and wheat (Armstrong, 1979; Prade and Trolldenier, 1990). Transport of oxygen, carbon dioxide, or

Table 6. Relative amount of denitrification N gases ($^{15}N_2 + {}^{15}N_2O$) measured in chamber headspace air and trapped in flooded soil without plants. (Adapted from Mosier et al., 1990b)

Days after fertilization	Denitrification N gas: Chamber headspace (mg ^{15}N plot^{-1})	Denitrification N gas: Trapped in soil (mg ^{15}N plot^{-1})
7	110	190
14	140	930
21	40	1100
28	23	940

methane between the atmosphere and soil via aerenchyma in rice plants has been observed (van Raalte, 1941; Cicerone and Shetter, 1981; Seiler et al., 1984; Sebacher et al., 1985). In a recent denitrification study, Mosier et al. (1990b) obtained a greater recovery of N gases ($^{15}N_2$ + $^{15}N_2O$) when chambers were placed over rather than between rice plants (Fig. 5) and demonstrated that N gases produced in the soil can be transported through rice plants to the atmosphere. In their study, young rice plants facilitated the efflux of N_2O and N_2 from the soil-water system. The N gas flux rate was not only greater but was also faster in the planted system. Similarly Bhadrachalm et al. (1992) found that inclusion of rice plants inside measurement chambers increased N gas flux by about 30%. Methane evolves almost exclusively through the rice plant (Bouwman, 1990b).

To reliably estimate denitrification rates from rice-field soils, growing plants must be included in chambers or soil cores. This is not feasible, however, when plants are at an advanced growth stage. Even where chambers can be placed over young plants, the environment surrounding the plant, especially air temperature and relative humidity, may be altered and affect the transmission of soil gases through the plant to the atmosphere. Nitrogen losses at advanced crop growth should be small since little mineral N is left in the soil. Information on emission of N gases by plants, other than rice, is lacking, however.

3. Temporal Variability

Daily variability in denitrification rates generally results from day-night changes in soil temperature (Blackmer et al., 1982; Mosier, 1989; Aulakh et al., 1991b). However, in some cases there may be no relationship

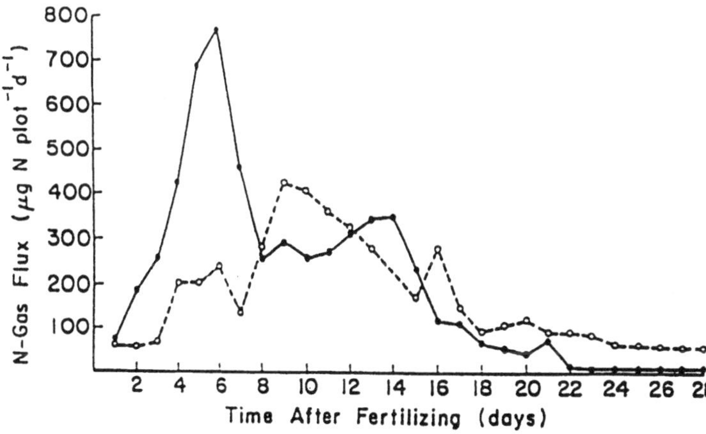

Figure 5. Total N gas flux (N_2 + N_2O) from urea-fertilized, flooded soils in the glasshouse which were planted to rice (●——●) and unplanted (O- - -O). Adapted from Mosier et al. (1990b).

between soil temperature fluctuations and denitrification (Mosier, 1989). Recently, Aulakh et al. (1991b) observed that frequent measurement of N gas flux over 24 h with the chamber method was needed to obtain denitrification rates comparable with those measured with the soil-core method where cores were enclosed for 24 h. Daily changes in N gas flux rate are accounted for in the AI soil-core method, but when chamber methods are used, the sampling schedule must be representative of flux rates that occur within a 24-h period. For example, when water is added to soil, gas efflux is increased for several hours to days and then returns to a lower baseline level. Simply measuring the maximum and minimum flux rates will not suffice in estimating daily denitrification losses unless the flux rates oscillate uniformly over the 24-h period.

Due to the daily temporal variability of denitrification rates, the effect of various management practices or of environmental changes may not be accurately characterized by weekly or biweekly measurements. Therefore, annual denitrification loss estimates may require daily measurements.

4. Spatial Variability

Soil denitrification, like the physical, chemical, and microbiological processes which regulate it, exhibits a high degree of spatial variability. Coefficients of variability for emissions of N_2O and N_2 from agricultural fields of a single soil type have ranged from 6% to 1800% (Matthias et al., 1980; Mosier et al., 1981; Robertson and Tiedje, 1984; Folorunso and Rolston, 1984; Aulakh and Rennie, 1985a; Robertson et al., 1988; Myrold, 1988). Thus, spatial variability appears a major problem in estimating gas flux from a given field. Attempts have been made to find soil parameters that correlate well with soil denitrification. These parameters include WFPS, water content or air-filled porosity, available C, respiration rate, soil aggregation, NO_3^-, NH_4^+, temperature, soil bulk density, and air permeability. However, in general, results have been inconsistent across sites or fields and only a few statistically significant relationships have been found (Burton and Beauchamp, 1985; Grundmann and Rolston, 1987; Sexstone et al., 1988; Robertson et al., 1988; Myrold, 1988; Groffman and Tiedje, 1988, 1989a, 1989b; Doran et al., 1990). For instance, Myrold (1988) used multiple regression analysis to show that 43% of the variability in denitrification rates was associated with variations in soil water content, respiration rate, NO_3^-, and soil temperature. In studies of forest soils (Groffman and Tiedje, 1989a, 1989b), less than 50% of daily variability in denitrification rates at the field scale was explained by soil texture and drainage class. But the same authors found that soil texture and drainage class explains over 80% of the variability in annual denitrification losses on the landscape scale (a relatively large area covering several types of soils and catenas representing a regional ecosystem).

Denitrification rates have been observed as log-normal distributions (Folorunso and Rolston, 1984; Parkin et al., 1985b). A variable is considered to be log-normally distributed if the logarithm of the variable is normally distributed. To accurately estimate denitrification rates and to account for the stochastic nature of the process, using replicated experimental data with high coefficients of variation, several statistical approaches or mathematical models have been developed (Parkin et al., 1988; 1990; Parkin and Robinson, 1989).

V. Impact of Management Systems

The rate of denitrification varies greatly between soil ecosystems due to the many environmental factors controlling this microbially mediated process. In the past decade, denitrification losses from different agricultural soil systems have been reported in several studies. The denitrification losses reported from arable crop soils, flooded rice soils, grassland and prairies, and forest soils are summarized in the following sections.

A. Arable Crops and Soil Systems

Until recently, a major cropping pattern of agricultural land in the Great Plains of North America and other areas of the world with dryland agriculture involved a year of summer fallow followed by one or more years of cropping. During fallow, undesirable plant growth (which uses water needed for crop plants) is controlled on conventional-tillage (CT) fallow fields by frequent tillage. During fallow, water is stored as a hedge against drought, mineralization of soil organic N is accelerated by tillage, and soil-available N reserves are replenished. However, the summer-fallow practice is not only an inefficient way of water storage (only 10% to 25% of the precipitation is stored) but also accelerates wind and water erosion, salinization, and long-term losses of soil organic matter (fertility). Gaseous N losses through denitrification from CT fallow fields are two to seven times higher than from CT cropped fields (Fig. 2), due largely to greater soil WFPS (Aulakh et al., 1982, 1983b, 1984a). For irrigated crops in northeast Colorado, Mosier et al. (1986a) reported that total cropping season denitrification losses, measured using the ^{15}N chamber method, were 4.0 and 1.5 kg N ha^{-1} from corn and barley (*Hordeum vulgare*), respectively. Myrold (1988), using the AI soil-core method, reported denitrification losses of 1.7 and 0.7 kg N ha^{-1} yr^{-1} in the winter wheat and ryegrass (*Lolium multiflorum* Lam.), respectively, for the climatic conditions of the Willamette Valley in Oregon. Rolston et al. (1982), using the ^{15}N chamber method, reported a range of denitrification losses over a 40-day period of 1.7 kg N ha^{-1} in an infrequently irrigated soil with no crop residue, to 14.9 kg N ha^{-1} in a frequently irrigated soil receiving 10 Mg ha^{-1} chopped barley residue. Large denitrification losses ranging from 19.6 to 41.8 kg N ha^{-1} yr^{-1} were reported from seven heavily fertilized and frequently irrigated soils cropped to different vegetables (Ryden and Lund, 1980). Dowdell and Webster (1984), using the ^{15}N balance approach, reported 10% to 20% of the N applied to spring barley in England was lost through denitrification. Losses from farmyard manure treated plots were six times greater (29 kg N ha^{-1}) than from untreated plots and 4.5 kg N ha^{-1} in fallow fields (Webster and Goulding, 1989).

Reduced or no-till (NT) farming systems which maintain crop residues on the soil surface are becoming more prevalent because of increased soil productivity and environmental quality resulting from reduced wind and water erosion and conservation of soil moisture. Thus CT fallow is being replaced by NT fallow where weeds are controlled with herbicides in place of tillage. No-till farming results in soils which are often cooler, wetter, and more compact than CT soils (Gauer et al., 1982; Mielke et al., 1986; Aulakh and Rennie, 1986). These differences in the soil physical environment of NT soils are associated with less aerobic microbial activity (Doran, 1980a; Linn and Doran, 1984a), lower net mineralization of N (Carter and Rennie, 1982; Doran, 1987), and greater gaseous N losses through denitrification as compared with CT soils (Rice and Smith, 1982; Aulakh et al., 1982, 1984a, 1984b; Linn and Doran, 1984a, 1984b; Fox and Bandel, 1986).

Results from several field studies (Table 7) revealed that, although gaseous N losses vary greatly, they are generally least for CT cropped soils and greatest for NT fallow fields. In general, rates of gaseous N loss were associated with rainfall events and were several-fold greater from CT and NT fallow fields than from cropped fields. In one study (Aulakh and Rennie, 1985a), where comparisons were made from three slope positions of a catena topo-sequence, N losses were greatest for lower slope positions and least for the upper slope positions.

Significantly greater gaseous N losses via denitrification from NT than from CT soil systems are a result of the following factors:

1. Greater soil bulk density resulting in reduced diffusion of air in NT surface soil (Lindstrom et al., 1981; Mielke et al., 1986).
2. Larger and more anaerobic soil aggregates in NT soils (Grevers and de Jong, 1982).
3. Greater soil water content and WFPS due to greater conservation of rain and snowmelt and reduced evaporation (Rice and Smith, 1982; Linn and Doran, 1984b; Aulakh and Rennie, 1986).
4. Greater concentrations of organic matter and crop residues near the surface of NT soils that increase C available to drive denitrification (Aulakh et al., 1991b).
5. All of the above encourage a more favorable environment for denitrifiers which is confirmed by several-fold greater populations in

Table 7. Range of gaseous N losses[a] reported from soils under conventional-till (CT) and no-till (NT) management systems

No. of soils/sites (Country)	Crop	Duration of study	Losses ($kg\,N\,ha^{-1}$) CT	NT	Reference(s)
1 (Denchworth, England)	Wheat/rape	2 years	1–6	5–9	Burford et al., 1981
1 (Lawford, England)	Wheat/rape	2 years	0.5–1	1.5–2	Burford et al., 1981
1 (Elstow, Canada)	Wheat	June to Sept.	3.1	6.3	Aulakh et al., 1982
1 (Elstow, Canada)	Wheat	2 years	7–14	23–32	Aulakh et al., 1984a
1 (Elstow, Canada)	Fallow	1 year	14.3	33.6	Aulakh et al., 1984b
1 (Athens, Ga., USA)	Rye/sorghum	1 year	3.7[b]	5.1	Groffman, 1985
1 (Athens, Ga., USA)	Rye/sorghum	1 year	22.4[c]	12.3	Groffman, 1985
1 (Dijon, France)	Wheat	March to May	0.2	2.5	Germon, 1985
3 (Yorkton, Canada)	Fallow	May to Sept.	11–86	25–98	Aulakh and Rennie, 1985a
3 (Oxbow, Canada)	Fallow	May to Sept.	8–68	28–84	Aulakh and Rennie, 1985a
3 (Weyburn, Canada)	Fallow	May to Sept.	9–60	18–68	Aulakh and Rennie, 1985a

[a] Measured by acetylene-inhibition soil-core technique except in a study of Burford et al. (1981) where only N_2O emission recorded by chamber method.
[b] Losses from intact cores 0–5 cm soil layer.
[c] Losses from intact cores at 5–13 cm soil depth.

the surface of NT as compared with CT soils (Doran, 1980a, 1980b; Aulakh et al., 1984a); potential denitrification rates are also 1.4- to 2-fold greater in NT as compared with CT fields (Aulakh and Rennie, 1986).

Alternative agricultural management systems have recently been developed that use winter legume crops and reduced tillage to control soil erosion, NO_3^- leaching and reduce N fertilizer inputs (Ott and Hargrove, 1989). When winter cover crops are killed with chemicals or tillage (ridge tillage or cultivation) prior to crop planting in the following spring, they provide substantial amounts of mineral N and organic C which enhance the microbial activity and under certain soil conditions, denitrification.

Doran (1990) found that denitrification in a ridge tillage field 1 week after planting and incorporation of hairy vetch and rye (*Secale cereale*) cover crops averaged 260 g N ha^{-1} day^{-1} which was threefold greater than yields from an adjacent area managed by conventional tillage without cover (Table 4). These results illustrate the importance of available C in denitrification as indicated by the elevated rates of respiration and denitrification in the interior area with ridge tillage where crop residues were concentrated after planting. It is important to note that these soils were dry at time of sampling and below the threshold (60% to 65% WFPS) above which denitrification normally occurs in soil.

B. Flooded Rice Soil Systems

Seventy-five percent of the world's rice-growing area of 143 million ha occurs in lowlands or wetlands where rice is grown in flooded fields during part or all of the cropping season (Watanabe et al., 1987). Management by flooding and puddling of the soil favors growth of rice by maintaining soil pH near neutrality, increasing the availability of nutrients (especially P, Mn, and Fe), stabilization of available N as NH_4^+, stimulating N_2-fixation by algae, depressing soil-borne diseases and weeds, and by preventing water percolation. However, flooded soils present a favorable environment for denitrification especially when NO_3^- and organic C supplies are adequate. Fortunately, in such flooded soils denitrification is determined by NO_3^- formation which is limited by lack of nitrification under anoxic conditions and mineralized N exists mainly as NH_4^+-N (Savant and DeDatta, 1982; Aulakh, 1988).

Since over 90% of the total fertilizer N used for global rice production is urea and other ammoniacal forms (Vlek and Byrnes, 1986; DeDatta and Buresh, 1989), the rate of nitrification is a primary determinant of denitrification losses in flooded rice soils (John et al., 1989). Generally nitrification may not be completely inhibited in wet or flooded rice soils because NH_4^+ could be oxidized to NO_3^- in the thin O_2-containing surface soil layer and in the overlying water phase of flooded soils (Engler

Table 8. Nitrification of applied $^{15}NH_4^+$ in an arable soil at 60% water-filled pore space (WFPS) and flooded soil (120% WFPS) over a 4-day period. (Modified from Aulakh, 1989)

Soil system	NH_4^+-N in soil mg N kg^{-1}	NO_3^--N[a] mg N kg^{-1}	NO_3^--N[a] % of applied
Arable	21.0	17.1	34.2
Flooded	35.2	5.7	11.6

[a] Includes amount of NO_3^--N denitrified.

and Patrick, 1974). Aulakh (1989), for example, found that over 11% of NH_4^+-N added to a flooded soil was nitrified within 4 days (Table 8). Under field conditions the O_2 transported from the atmosphere to the root zone by growing rice plants (van Raalte, 1941) helps to enhance nitrification. Watanabe et al. (1981), for example, reported a high rate of nitrification (1.2 kg N ha^{-1} day^{-1}) after application of urea to a flooded soil.

Rice is also grown on a fairly large proportion of coarse-textured soils which have high percolation rates. Since it is difficult to maintain continuous flooding in these soils, production of nitrates through nitrification during "dry spells" and their subsequent reduction to N gases through denitrification when soils are reflooded is a common phenomenon (Reddy and Patrick, 1975). Several ^{15}N balance studies conducted for intermittently flooded soils provide indirect evidence that gaseous N losses via nitrification-denitrification may range from 42% to 51% of applied N (Krishnappa and Shinde, 1980; Rekhi et al., 1982; Katyal et al., 1985).

Previously, lack of suitable methodology for direct field measurement of denitrification in flooded soils has resulted in most researchers using the ^{15}N balance approach to estimate N loss (Table 9). Estimated N losses from soils planted to rice from around the world range from none to as high as 58% of the applied fertilizer N. Recently many researchers have reported direct measurement of denitrification using the ^{15}N chamber method (Buresh and Austin, 1988; John et al., 1989; Mosier et al., 1989, 1990a, 1990b; Banerjee et al., 1990; Buresh and DeDatta, 1990; Mohanty and Mosier, 1990; Freney et al., 1990; Bhadrachalm et al., 1991). In several of these studies denitrification losses obtained by N gas flux measurements represented only a fraction (1% to 10%) of the total gaseous N losses as measured using the ^{15}N-balance approach (Table 5). In a field study John et al. (1989) accounted for 20% of the applied KNO_3 (27 kg N ha^{-1}) as denitrification N gas loss using the ^{15}N chamber method as compared with 53% unaccounted for N using a ^{15}N balance approach during the same 19-day period. Possible causes for the low recovery of labeled N gases are:

Table 9. Summary of gaseous N losses reported from flooded rice soils, using the ^{15}N-balance approach

No. of soils/sites (country)	N application rate ($kg\,ha^{-1}$) and source	Method and time of N application	Gaseous N loss (% of applied N)	Reference(s)
4 (Philippines)	58–87 Urea	Basal	3–40	Fillery et al., 1986; Buresh and Austin, 1988; Buresh and DeDatta, 1990
1 (China)	90 Urea	Basal	39	Cai et al., 1986
1 (Thailand)	87 Urea	Basal	22	Buresh and DeDatta, 1990
2 (Indonesia)	87 Urea	Basal	10–31	Buresh and DeDatta, 1990
1 (India)	180 Urea	Basal	34–50	Rekhi et al., 1982
1 (India)	180 Urea	3 Splits	17–28	Rekhi et al., 1982
1 (India)	58–116 Urea, $(NH_4)_2SO_4$	Basal	46–50	Katyal et al., 1985
1 (India)	100 Urea	Basal	58	Mohanty and Mosier, 1990
4 (Philippines)	53–80 Urea	Basal, Bd DT, 10 days	3–70	Freney et al., 1990
1 (USA)	133 NH_4Cl	Basal	22	Meyer et al., 1989
9 (Philippines)	29–80 Urea	Bd DT 10–21 days	0–45	Fillery and DeDatta, 1986; Fillery et al., 1986; DeDatta and Buresh, 1989; John et al., 1989; Buresh and DeDatta, 1990
1 (Philippines)	87 Urea	Bd	31–34	Buresh et al., 1991
1 (Philippines)	87 Urea	Incorporated	10–22	Buresh et al., 1991
1 (Malaysia)	44 Urea	Bd DT 10–15 days	46	Buresh and Austin, 1988

Bd DT, broadcast fertilizer, days after planting.

1. Entrapment of N gases in soil pore water and flood water
2. Transmission of N gases through rice plants
3. Nitrogen loss through NH_3 volatilization of applied ammoniacal fertilizer N reportedly ranges from 27% to 56% of applied N (Fillery et al., 1986; Fillery and DeDatta, 1986; DeDatta et al., 1989; Reddy et al., 1990)

In a few studies (Mosier et al., 1989, 1990b) where all the possible mechanisms of N loss were considered, the recovery of applied fertilizer N was 94% ± 0.2%. These studies indicate that estimates of denitrification using the ^{15}N balance approach tend to be greater than those with the ^{15}N chamber method; either due to the analytical error in determining the various components and/or N loss by unknown mechanisms. Several studies indicate that NH_3 volatilization could be an important mechanism for gaseous N loss from flooded rice soils and in preparing a ^{15}N balance, it must be measured. Recent studies by Freney et al. (1990) and Reddy et al. (1990) demonstrated that denitrification and NH_3 volatilization processes in flooded rice soils are interdependent. In the study of Freney et al. (1990) reducing NH_3 losses by incorporating urea into flooded soil resulted in increased denitrification losses, whereas denitrification losses were low when NH_3 volatilization was high from urea broadcast into the flood water. The efficiency of fertilizer N use could only be improved when both N-loss processes were controlled simultaneously. Few recent studies (Banerjee and Mosier, 1989; Banerjee et al., 1990; Bronson and Mosier, 1990) have demonstrated the possibility of substantial reduction in denitrification losses through use of nitrification inhibitors (Table 10). Among two nitrification inhibitors used, encapsulated calcium carbide appears to be the more effective in reducing denitrification losses and increasing rice yield.

C. Grasslands

Many aspects of N transport and transformations including denitrification in native grasslands, short or tall grass prairie ecosystems, and grass pastures remain poorly investigated.

Table 10. Effect of nitrification inhibitors on denitrification losses and rice grain yield in a flooded soil in India. (Adapted from Banerjee et al., 1990)

Treatment	Rice yield ($kg\,ha^{-1}$)	Denitrification ($g\,N\,ha^{-1}\,h^{-1}$) on days after fertilization				
		2	4	6	8	10
Urea alone	4660	27.3	81.3	52.0	87.0	62.0
Urea + DCD	5360	10.4	15.9	16.7	10.8	12.5
Urea + ECC	6130	7.8	3.5	5.0	2.0	2.0

DCD, dicyandiamide; ECC, encapsulated calcium carbide.

Table 11. Denitrification potential rate (DPR) and field denitrification rate (mean ± SD) in a permanent grassland and adjacent arable soil in United Kingdom. (Adapted from Bijay-Singh et al., 1989)

	DPR	Denitrification rate (g N ha^{-1} day^{-1})				NO_3^--N
	mg N $kg^{-1}h^{-1}$	31 July	1 August	6 August	7 August	on 7 August (mg N kg^{-1})
Grassland	1.46 ± 0.05	23 ± 4	32 ± 6	29 ± 9	37 ± 16	4.2 ± 0.4
Arable	0.31 ± 0.01	136 ± 17	165 ± 31	237 ± 37	230 ± 29	7.4 ± 0.4

Preliminary studies (Denmead et al., 1979; Webster and Dowdell, 1982; Mosier et al., 1981, 1983) in grassland indicate that N_2O is the primary gaseous N product of nitrification-denitrification. Although very low N_2O emissions (2 g N ha^{-1} day^{-1}) have been observed, these losses account for about 10% of the typically low total N input to natural grassland ecosystems.

In general, natural grassland soils have greater porosity and enhanced infiltration and permeability which result in rapid movement and dissipation of soil water after rainfall (Garwood et al., 1977; Mosier and Parton, 1985; Bijay-Singh et al., 1989). Therefore, the rate of denitrification from grassland soils is likely to remain high only when rainfall occurs for extended periods. In a few studies (Colbourn et al., 1984; Bijay-Singh et al., 1989; Groffman, 1990) laboratory measurements of potential denitrification rates from natural grassland soils were several-fold higher than those from similar arable soils. These results are presumably due to the greater potential availability of organic C in the stable undisturbed grassland ecosystems. However, in the field, actual denitrification rates from grasslands are often very low (Table 11).

Denitrification losses from permanent and temporary grass pastures, which are often fertilized at N rates ranging from 50 to 450 kg N ha^{-1} may be greater than cropland receiving less fertilizer N (Church, 1979; Forbes et al., 1980; Mosier and Parton, 1985). Ryden (1983), using the AI chamber method, measured annual gaseous N denitrification losses from a grassland in the United Kingdom of 1.6, 11.1, and 29.1 kg N ha^{-1} from fields receiving 0, 250, and 500 kg N ha^{-1}, respectively. He observed that more than 60% of the annual losses occurred during a period of 8 weeks when fertilizer was applied to relatively wet soil following rainfall (145 mm). Colbourn et al. (1984) studied denitrification losses (using the AI chamber method) in a clay soil under permanent grass. They measured a total gaseous N loss of 2.1 kg N ha^{-1} over 21 days following irrigation (24 mm) and the application of 70 kg N ha^{-1} (ammonium nitrate). Mosier and Parton (1985) recorded short N_2 fluxes in native shortgrass prairie in eastern Colorado only after heavy rainfall events, even where soil was fertilized with 450 kg N ha^{-1} as urea. Parton et al.

Table 12. Denitrification losses in soils differing in texture and drainage class in temperate forests of Michigan. (Adapted from Groffman and Tiedje, 1989a)

Soil	Denitrification losses ($kg\,N\,ha^{-1}\,yr^{-1}$)
Loam	
Well-drained	10
Somewhat poorly drained	11
Poorly drained	24
Clay loam	
Well-drained	18
Somewhat poorly drained	17
Poorly drained	40
Sand	
Well-drained	0.6
Somewhat poorly drained	0.8
Poorly drained	0.5

(1988) computed a mean flux of $104\,g\,N\,ha^{-1}\,yr^{-1}$ from the shortgrass prairie ecosystem.

In summary, laboratory measurements of denitrification potential for grassland soils may range from 20 to $40\,kg\,N\,ha^{-1}\,day^{-1}$ but actual denitrification in the field is usually low ($<0.1\,kg\,N\,ha^{-1}\,day^{-1}$). Grassland ecosystems are often NO_3^- limited and soils dry rapidly after each rainfall or irrigation. It appears that although grasslands act as sources of N_2O, on many occasions, they may also act as a sink for N_2O under very wet conditions (Ryden, 1981).

D. Forest Soils

Few studies have focused on denitrification losses in natural forest ecosystems. In a recent study involving soils of varying texture and drainage status in a temperate forest landscape in Michigan, Groffman and Tiedje (1989a) estimated annual denitrification losses ranging from $<1\,kg\,N\,ha^{-1}\,yr^{-1}$ in a well-drained sand to $40\,kg\,N\,ha^{-1}\,yr^{-1}$ in a poorly-drained clay loam (Table 12).

While irrigation, fertilization, and tillage are important determinants of denitrification in agricultural systems, other factors assume greater importance in forest lands. Important factors regulating denitrification in forest soils include soil texture and organic matter which both influence the water-holding capacity and supply of NO_3^- and available C. In the studies of Davidson and Swank (1987) and Groffman and Tiedje (1989a,

1989b) lack of soil NO_3^- and water content were the primary factors limiting denitrification in summer. Available C was also occasionally a limiting factor. In 12 soils ranging from young sand dunes to mature old-growth hardwood forest stands, Robertson and Tiedje (1984) found that denitrification rates were generally controlled by soil water. Hussey et al. (1985) further observed that optimal soil moisture and high organic matter near streams in the riparian zone of a mountain watershed, resulted in elevated denitrification rates. In a study with riparian, meadow and forest sites in eastern Sierra Nevada, Hixon et al. (1990) found that during summer increasing denitrification rates were directly correlated with soil organic matter (carbon) levels but during October denitrification rates were strongly influenced by soil temperature. In addition to soil organic matter content, C availability and its effect on denitrification in forest soils was also found to be regulated by freezing and thawing and wetting and drying (Melillo et al., 1983; Edwards and Killham, 1986; Groffman and Tiedje, 1988).

VI. Conclusions and Future Research Needs

During the past two decades much information on denitrification appeared in the literature which enhanced our understanding of specific microorganisms that mediate the process and the many factors that regulate it. The role played by several gaseous N products (NO, N_2O, N_2) in the denitrification process is now better understood. The NO_x gases produced in denitrification affect atmospheric chemistry while N_2O is important both as a greenhouse gas and in contributing to stratospheric ozone depletion. It is now well recognized that the proportion of gaseous denitrification products represented by N_2O [$N_2O/(N_2O + N_2)$] is extremely variable (none to 100%) and a single ratio cannot be used to estimate global denitrification losses based on N_2O emissions from different management systems. The significance of denitrification in N use efficiency in agricultural, grassland, and forest ecosystems, and in global N cycling is now well recognized.

Denitrification is regulated by a number of physical and chemical factors such as soil pH, temperature, organic C supply, nitrate concentration, aeration, and water status. The effect of each individual parameter on denitrification under controlled conditions is well established. Hence the control and regulation of these factors has been effectively used in commercial and municipal waste water treatment processes which use denitrification to remove NO_3^- and protect water quality. The occurrence of denitrification in various soil environments depends on the interaction of many limiting factors. As a result, high rates of denitrification from soil are frequently observed as "short pulses" which occur when soil conditions become congenial/optimal. Conversely, lack of understanding of these complex interactions limits our ability to effectively manage denitrification in different management systems.

Quantification of denitrification losses in agricultural, grassland, and forest soils has been limited mainly to point in time determinations. Results of various field studies suggest that denitrification N losses could account for significant proportions of the total N inputs in different management systems. However, in general, a reliable quantification of total denitrification N losses remains a goal yet to be achieved.

Field methodology and instrumentation have developed substantially in recent years. Several new methods for estimating field denitrification have been devised. For example, the approach of using ^{15}N to estimate denitrification as proposed by Hauck et al. (1958) was revitalized for field application (Siegel et al., 1982) and several improvements in assumptions and calculations have been made thereafter (Mulvaney and Boast, 1986). Similarly, the acetylene-inhibition technique to block N_2O reduction to N_2 has been applied in the field using chambers and intact soil cores. A fast and efficient method of using lightly coated CaC_2 as a source of C_2H_2 in the field has been proposed. We now know that NO_3^- lost from the soil through denitrification but previously unaccounted for in measurement of N gas emissions can be accounted for as N gas trapped in soil pore water and floodwater, and/or transmitted to the atmosphere through plants such as rice. These newer advances in methodology and understanding should permit us to study and interpret the complex effects of management and identify approaches in agricultural and nonagricultural ecosystems to minimize the loss of N and aid in the efficient use of N. It appears that the best way to reduce denitrification losses from soils is by increasing N-use efficiency in crop production. Denitrification N emissions in rice soils fertilized with urea and ammoniacal N forms are large but may be reduced through use of nitrification inhibitors.

Inadequacies in understanding spatial and temporal variations in denitrification still exist. However, several mathematical models have been developed to predict denitrification rates using log-normal distributed data and to account for the stochastic nature of the process (Arah and Smith, 1989).

Given our current state of knowledge we see the following research needs:

1. Accurate prediction of the proportion of N inputs (atmospheric deposition, N fixation, N fertilization) to agricultural and natural ecosystems lost via denitrification; such information is more lacking from tropical and subtropical ecosystems than temperate regions.
2. Better understanding of the relationship and relative importance of denitrification compared with other mechanisms for N loss such as N leaching and ammonia volatilization.
3. Estimate global N_2O emissions and the relative predominance of N_2O in N gas emissions from ecosystems where denitrification is dominant.

4. Evaluate the effect of changing land use (e.g., conversion of tropical forests to pastures) on denitrification and the relative predominance of N_2O in gaseous emissions.
5. Scale up point source measurements to predict N gas flux for field, landscape, regional, and ultimately global levels.
6. Development of quantitative predictive models for denitrification losses from different management systems; modeling the currently available data would identify the kind of information needs for more reliable prediction of total denitrification losses and N_2O emissions at the global scale.

Acknowledgments. The authors wish to thank Drs. T.B. Parkin, W.W. Wilhelm, and K.L. Weier for helpful comments and suggestions on an earlier version of this manuscript.

References

Allison, F.E. 1955. The enigma of soil nitrogen balance sheets, pp. 213–250. In: Norman, A.G. (ed.) *Advances in Agronomy*, vol. 3. Academic Press, N.Y.

Arah, J.R.M., and K.A. Smith. 1989. Steady-state denitrification in aggregated soils: a mathematical model. *J. Soil Sci.* 40: 139–149.

Arah, J.R.M., and K.A. Smith. 1990. Factors influencing the fraction of the gaseous products of soil denitrification evolved to the atmosphere as nitrous oxide, pp. 475–480. In: Bouwman, A.F. (ed.) *Soils and the Greenhouse Effect.* Wiley, Chichester.

Armstrong, W. 1979. Aeration in higher plants. *Adv. Bot. Res.* 7: 225–332.

Aulakh, M.S. 1986a. Gaseous losses of nitrogen from soils through biological processes: a review, pp. 99–115. In: Mishra, M.M., Kappor, K.K. (eds.) *Soil Biology*. Haryana Agric. Univ. Hissar, India.

Aulakh, M.S. 1986b. Nitrogen losses from "available N pool" of soil through denitrification, ammonium fixation, immobilization, and leaching, pp. 146–147. Proc. Symp. Soil management in relation to environment and crop growth under intensive agriculture. Seventy-third session of Indian Science Congress, Delhi, India.

Aulakh, M.S. 1988. Influence of plant residues incorporation on microbiological transformations of N in soils, and crop production, pp. 691–702. In: Chang, S., Chan, K., Woo, N.Y.S. (eds.) *Recent Advances in Biotechnology and Applied Biology*. Chinese University, Hong Kong.

Aulakh, M.S. 1989. Transformations of ammonium nitrogen in upland and flooded soils amended with crop residues. *J. Indian Soc. Soil Sci.* 37: 248–255.

Aulakh, M.S., and J.W. Doran. 1990. Effectiveness of acetylene inhibition of N_2O reduction for measuring denitrification in soils of varying wetness. *Comm. Soil Sci. Plant Anal.* 21: 2223–2243.

Aulakh, M.S., J.W. Doran, and A.R. Mosier. 1991b. In-field evaluation of four methods for measuring denitrification. *Soil Sci. Soc. Am. J.* 55: 1332–1338.

Aulakh, M.S., J.W. Doran, D.T. Walters, A.R. Mosier, and D.D. Francis. 1991a. Crop residue type and placement effects on denitrification and mineralization. Soil *Sci. Soc. Am. J.* 55: 1020–1025.

Aulakh, M.S., J.W. Doran, D.T. Walters, and J.F. Power. 1991c. Legume residues and soil water effects on denitrification in soils of different textures. *Soil Biol. Biochem.* 23: 1161–1167.

Aulakh, M.S., and D.A. Rennie. 1984. Transformation of fall-applied, nitrogen-15-labeled fertilizers. *Soil Sci. Soc. Am. J.* 48: 1184–1189.

Aulakh, M.S., and D.A. Rennie. 1985a. Gaseous nitrogen losses from conventional and chemical summer fallow. *Can. J. Soil. Sci.* 65: 195–204.

Aulakh, M.S., and D.A. Rennie. 1985b. Azide effects upon N_2O emission and transformations of N in soils. *Can. J. Soil Sci.* 65: 205–211.

Aulakh, M.S., and D.A. Rennie. 1986. Nitrogen transformations with special reference to gaseous N losses from zero-tilled soils of Saskatchewan, Canada. *Soil Tillage Res.* 7: 157–171.

Aulakh, M.S., and D.A. Rennie. 1987. Effect of wheat straw incorporation on denitrification of N under anaerobic and aerobic conditions. *Can. J. Soil Sci.* 67: 825–834.

Aulakh, M.S., D.A. Rennie, and E.A. Paul. 1982. Gaseous nitrogen losses from cropped and summer-fallowed soils. *Can. J. Soil Sci.* 62: 187–195.

Aulakh, M.S., D.A. Rennie, and E.A. Paul. 1983a. The effect of various clover management practices on gaseous N losses and mineral N accumulation. *Can. J. Soil Sci.* 63: 593–605.

Aulakh, M.S., D.A. Rennie, and E.A. Paul. 1983b. Field studies on gaseous nitrogen losses from soils under continuous wheat versus a wheat-fallow rotation. *Plant Soil* 75: 15–27.

Aulakh, M.S., D.A. Rennie, and E.A. Paul. 1984a. Gaseous nitrogen losses from soils under zero-till as compared with conventional-till management systems. *J. Environ. Qual.* 13: 130–136.

Aulakh, M.S., D.A. Rennie, and E.A. Paul. 1984b. The influence of plant residues on denitrification rates in conventional and zero tilled soils. *Soil Sci. Soc. Am. J.* 48: 790–794.

Aulakh, M.S., D.A. Rennie, and E.A. Paul. 1984c. Acetylene and N-serve effects upon N_2O emissions from NH_4^+ and NO_3^- treated soils under aerobic and anerobic conditions. *Soil Biol. Biochem.* 16: 351–356.

Bailey, L.D. 1976. Effects of temperature and root on denitrification in a soil. *Can. J. Soil Sci.* 56: 79–87.

Bailey, L.D., and E.G. Beauchamp. 1973. Effects of moisture, added NO_3^- and macerated roots on NO_3^- transformation and redox potential in surface and subsurface soils. *Can. J. Soil Sci.* 53: 219–230.

Balderston, W.L., B. Sherr, and W.J. Payne. 1976. Blockage by acetylene of nitrous oxide reduction in *Pseudomonas perfectomarinus. Appl. Environ. Microbiol.* 31: 504–508.

Banerjee, N.K., and A.R. Mosier. 1989. Coated calcium carbide as a nitrification inhibitor in upland and flooded soils. *J. Indian Soc. Soil Sci.* 37: 306–313.

Banerjee, N.K., A.R. Mosier, K.S. Uppal, and N.N. Goswami. 1990. Use of encapsulated calcium carbide to reduce denitrification losses from urea-fertilized flooded rice. Proc. Int. Denitrification Workshop, Giessen, FRG, March 1989. Mitt. Dtsch Boden Ges. 60: 245–248.

Beauchamp, E.G., C. Gale, and J.C. Yeomans. 1980. Organic matter availability for denitrification in soils of different textures and drainage classes. *Commun. Soil Sci. Plant Anal.* 11: 1221–1233.

Beauchamp, E.G., J.T. Trevors, and J.W. Paul. 1989. Carbon sources for bacterial denitrification. *Adv. Soil Sci.* 10: 113–142.

Bhadrachalm, A., S.P. Chakravorti, N.K. Banerjee, S.K. Mohanty, and A.R. Mosier. 1992. Denitrification in intermittently flooded rice fields and N-gas transport through rice plants. In: Svensson, B., Ojima, D. (eds.) *Trace Gas Exchange in a Global Perspective*. Ecol. Bull. (Stockholm). Sweden. (in press).

Bhaumik, H.D., and F.E. Clark. 1948. Soil moisture tension and biological activity. *Soil Sci. Soc. Am. Proc.* 12: 234–238.

Bijay-Singh, J.C. Ryden, and D.C. Whitehead. 1988. Some relationships between denitrification potential and fractions of organic carbon in air-dried and field-moist soils. *Soil Biol. Biochem.* 20: 737–741.

Bijay-Singh, J.C. Ryden, and D.C. Whitehead. 1989. Denitrification potential and actual rates of denitrification in soils under long-term grassland and arable cropping. *Soil Biol. Biochem.* 21: 897–901.

Blackmer, A.M., and J.M. Bremner. 1978. Inhibitory effect of nitrate on reduction of N_2O to N_2 by soil microorganisms. *Soil Biol. Biochem.* 10: 187–191.

Blackmer, A.M., S.G. Robbins, and J.M. Bremner. 1982. Diurnal variability in rate of emission of nitrous oxide from soils. *Soil Sci. Soc. Am. J.* 46: 937–942.

Bouwman, A.F. 1990a. Analysis of global nitrous oxide emissions from terrestrial natural and agro-ecosystems. Trans. 14th Congress of Soil Science 2: 261–266.

Bouwman, A.F. 1990b. Soils and the Greenhouse Effect. Wiley, Chichester.

Bowman, R.A., and D.D. Focht. 1974. The influence of glucose and nitrate concentrations upon denitrification rates in a sandy soil. *Soil Biol. Biochem.* 6: 297–301.

Brar, S.S., R.H. Miller, and T.J. Logan. 1978. Some factors affecting denitrification in soils irrigated with waste water. *J. Water Pollut. Control Fed.* 50: 709–717.

Bremner, J.M., and K. Shaw. 1958. Denitrification in Soil. II. Factors affecting denitrification. *J. Agric. Sci.* 51: 40–52.

Bronson, K.F., and A.R. Mosier. 1990. Control of denitrification and methane emissions in flooded rice with encapsulated calcium carbide. *Agron. Abstr.* Am. Soc. Agron., Madison, Wis., p. 244.

Bundy, L.G., and J.M. Bremner. 1973. Inhibition of nitrification in soils. *Proc. Soil Sci. Soc. Am.* 37: 396–398.

Buresh, R.J., and E.R. Austin. 1988. Direct measurement of dinitrogen and nitrous oxide flux in flooded rice fields. *Soil Sci. Soc. Am. J.* 52: 681–688.

Buresh, R.J., and S.K. DeDatta. 1990. Denitrification losses from puddled rice soils in the tropics. *Biol. Fertil. Soils.* 9: 1–13.

Buresh, R.J., S.K. DeDatta, M.I. Samson, S. Phongphan, P. Suitwongse, A.M. Fogi, and R. Tejasarwana, 1991. Dinitrogen and nitrous oxide flux from urea basally applied to puddled rice soils. *Soil Sci. Soc. Am. J.* 55: 268–273.

Buresh, R.J., T. Woodhead, K.D. Shepherd, E. Flordelis, and R.C. Cabangon. 1989. Nitrate accumulation and loss in a mungbean/lowland rice cropping system. *Soil Sci. Soc. Am. J.* 53: 477–482.

Burford, J.R., and J.M. Bremner. 1975. Relationships between the denitrification capacities of soils and total, water-solube and readily decomposable soil organic matter. *Soil Biol. Biochem.* 7: 389–394.

Burford, J.R., R.J. Dowdell, and R. Cress. 1981. Emission of nitrous oxide to the atmosphere from direct-drilled and ploughed clay soils. *J. Sci. Food Agric.* 32: 219–223.

Burke, K.A., K. Calder, and J. Lascelles. 1980. Effects of molybdenum and tungsten on induction of nitrate reductase and formate dehydrogenase in wild type and mutant *Parococcus denitrificans*. *Arch. Microbiol.* 126: 155–159.

Burton, D.L., and E.G. Beauchamp. 1985. Denitrification rate relationships with soil parameters in the field. *Commun. Soil Sci. Plant Anal.* 16: 539–549.

Cai, Gui-Xin, Zhu Zhao-Liang, A.C.F. Trevitt, J.R. Freney, and J.R. Simpson. 1986. Nitrogen loss from ammonium bicarbonate and urea fertilizers applied to flooded rice. *Fert. Res.* 10: 203–215.

Calder, K., K.A. Burke, and J. Lascelles. 1980. Induction of nitrate reductase and membrane cytochromes in wild type and chlorate-resistant *Paracoccus denitrificans*. *Arch. Microbiol.* 126: 149–153.

Carter, M.R., and D.A. Rennie. 1982. Changes in soil quality under zero tillage farming systems: distribution of microbial biomass and mineralizable C and N potentials. *Can. J. Soil Sci.* 62: 587–597.

Chaterpaul, L., E.A. Paul, and W. Calaco. 1980. Denitrification in Saskatchewan soils under field conditions. Abstracts of the 80th Annual Meeting, American Society of Microbiology, Miami Beach, Florida, May 11–16, 1980.

Cho, C.M. 1982. Oxygen consumption and denitrification kinetics in soil. *Soil Sci. Soc. Am. J.* 46: 756–762.

Cho, C.M., L. Sakdinan, and C. Chang. 1975. Denitrification intensity and capacity of three irrigated Alberta soils. *Soil Sci. Soc. Am. J.* 43: 945–950.

Christensen, M.K., and P. Harremoës, 1978. Nitrification and denitrification in wastewater treatment, pp. 391–414. In: Mitchell, R. (ed.) *Water Pollution Microbiology*, vol. 2. Wiley, New York.

Christianson, C.B., and C.M. Cho. 1983. Chemical denitrification of nitrite in frozen soils. *Soil Sci. Soc. Am. J.* 47: 38–42.

Church, B.M. 1979. Survey of fertilizer practice. In: *Fertilizer Use on Farm Crops in England and Wales*. MAFF Report SS/SAF/29, HMSO, London, U.K.

Cicerone, R.J., and J.D. Shetter. 1981. Sources of atmospheric methane: measurements in rice paddies and a discussion. *J. Geophys. Res.* 86: 7203–7209.

Colbourn, P., M.M. Iqbal, and I.W. Harper. 1984. Estimation of the total gaseous nitrogen losses from clay soils under laboratory and field conditions. *J. Soil Sci.* 35: 11–22.

Craswell, E.T. 1978. Some factors influencing denitrification and nitrogen immobilization in a clay soil. *Soil Biol. Biochem.* 10: 241–245.

Crutzen, P.J. 1976. Upper limits on atmospheric ozone reductions following increased applications of fixed nitrogen to the soil. *Geophys. Res. Lett.* 3: 169–172.

Crutzen, P.J. 1981. Atmospheric chemical processes of the oxides of nitrogen including nitrous oxide, pp. 17–44. In: Delwiche, C.C. (ed.) *Denitrification, Nitrification and Atmospheric Nitrous Oxide*. Wiley, New York.

Crutzen, P.J., and D.H. Enhalt. 1977. Effects of nitrogen fertilizers and combustion on the stratospheric ozone layer. *Ambio* 6: 112–117.

Currie, J.A. 1961. Gaseous diffusion in porous media. II. Wet granular materials. *Brit. J. Appl. Phys.* 12: 275–281.

Davidson, E.A., and W.T. Swank. 1987. Factors limiting denitrification in soils from mature and disturbed south-eastern hardwood forests. *Forest Sci.* 33: 135–144.

de Catanzaro, J.B., and E.G. Beauchamp. 1985. The effect of some carbon substrates on denitrification rates and carbon utilization in soil. *Biol. Fert. Soils.* 1: 183–187.

DeDatta, S.K., and R.J. Buresh. 1989. Integrated nitrogen management in irrigated rice. *Adv. Soil Sci.* 10: 143–169.

DeDatta, S.K., R.J. Buresh, W.N. Obcemea, and J.G. Real. 1987. Direct field measurement of ammonia volatilization and denitrification in lowland rice soils, p. 201. *Agron. Abstr.* Am. Soc. Agron. Madison, Wis.

DeDatta, S.K., A.C.F. Trevitt, J.R. Freney, W.N. Obcemea, J.G. Real, and J.R. Simpson. 1989. Measuring nitrogen losses from lowland rice using bulk aerodynamic and nitrogen-15 balance methods. *Soil Sci. Soc. Am. J.* 53: 1275–1281.

Delwiche, C.C., and B.A. Bryan. 1976. Denitrification. *Annu. Rev. Microbiol.* 30: 241–262.

Denmead, O.T. 1979. Chamber systems for measuring nitrous oxide emissions from soils in the field. *Soil Sci. Soc. Am. J.* 43: 89–95.

Denmead, O.T., J.R. Freney, and J.R. Simpson. 1979. Studies of nitrous oxide emission from a grass sward. *Soil Sci. Am. J.* 43: 726–728.

Doran, J.W. 1980a. Microbial changes associated with residue management with reduced tillage. *Soil Sci. Soc. Am. J.* 44: 518–526.

Doran, J.W. 1980b. Soil microbial and biochemical changes associated with reduced tillage. *Soil Sci. Soc. Am. J.* 44: 765–771.

Doran, J.W. 1987. Microbial biomass and mineralizable nitrogen distributions in no-tillage and plowed soils. *Biol. Fert. Soils* 5: 68–75.

Doran, J.W. 1990. Microbial activity and N transformations in sustainable systems, pp. 109–116. In: *Proc. Conf. "Extending Sustainable Systems."* Minnesota Extension Service and Farm Management, St. Cloud, Minn.

Doran, J.W., L.N. Mielke, and J.F. Power. 1990. Microbial activity as regulated by soil water-filled pore space. Trans. 14th Internat. *Congress. Soil Science* 3: 94–99.

Doran, J.W., L.N. Mielke, and S. Stamatiadis. 1988. Microbial activity and N cycling as regulated by soil water-filled pore space. In: Proceedings 11th Conference Internat. *Soil Tillage Res. Organisation* 1: 49–54.

Dowdell, R.J., and C.P. Webster, 1984. A lysimeter study of the fate of fertilizer nitrogen in spring barley crops grown on a shallow soil overlying chalk: denitrification losses and the nitrogen balance. *J. Soil Sci.* 35: 183–190.

Duxbury, J.M. 1986. Advantages of the acetylene method of measuring denitrification. In: Hauck, R.D. Weaver, R.W. (eds.) Field Measurement of Dinitrogen Fixation and Denitrification. *Soil Sci. Soc. Am. Pub.* 18: 73–91.

Edwards, A.C., and R. Killham. 1986. The effect of freeze/thaw on gaseous nitrogen loss from upland soils. *Soil Use Manage.* 2: 86–91.

Eichner, M. 1990. Nitrous oxide emissions from fertilized soils: Summary of available data. *J. Environ. Qual.* 19: 272–280.

Ellis, J.R., L.N. Mielke, and G.E. Schuman. 1975. The nitrogen status beneath beef cattle feedlots in eastern Nebraska. *Soil Sci. Soc. Am. Proc.* 39: 107–111.

Engler, R.M., and W.H. Patrick. 1974. Nitrate removal from flooded water overlying flooded soils and sediments. *J. Environ. Qual.* 3: 409–413.

Enquete-Kommission. 1989. *Vorsorge zum Schutz der Erdatmosphäre. Schutz der Erdatmosphäre: eine internationale Herausforderung*. Zwischenbericht de Enquete-Kommission des II. Deutschen bundestages Vorsorge zum Schutz der Erdatmosphäre. Dt. Bundestag, Referat Offentlichkeitsarbeit, Bonn.

FAO. 1986. *Production Yearbook 1985*. Food and Agriculture Organization, Rome, Italy.

Fillery, I.R.P., and S.K. DeDatta. 1986. Ammonia volatilization from nitrogen sources applied to rice fields. I. Methodology, ammonia fluxes, and nirogen-15 loss. *Soil Sci. Soc. Am. J.* 50: 80–86.

Fillery, I.R.P., J.R. Simpson, and S.K. DeDatta. 1986. Contribution of ammonia volatilization to total nitrogen loss of the applications of urea to wetland rice plants. *Fert. Res.* 8: 193–202.

Firestone, M.K. 1982. Biological denitrification, pp. 289–326. In: Stevenson, F.J. (ed.) Nitrogen in Agricultural Soils. *Agronomy monograph No. 22*. Amer. Soc. Agron. Madison, Wis.

Firestone, M.R., R.B. Firestone, and J.M. Tiedje. 1980. Nitrous oxide from soil denitrification. Factors controlling its biological production. *Science* 208: 749–751.

Focht, D.D. 1978. Methods for analysis of denitrification in soils, pp. 433–490. In: Nielsen, D.R., MacDonald, J.G. (eds.) *Nitrogen in the Environment*. Academic Press, New York.

Focht, D.D. 1985. Differences in nitrogen-15 enrichments of evolved nitrous oxide and dinitrogen and the question of a uniform nitrate-15N pool. *Soil Sci. Soc. Am. J.* 49: 786–787.

Focht, D.D., and A.C. Chang. 1975. Nitrification and denitrification processes related to waste-water treatment. *Adv. Appl. Microbiol.* 19: 153–186.

Focht, D.D., and M. Verstraete. 1977. Biochemical ecology of nitrification and denitrification, pp. 135–214. In: Alexander, M. (ed.) *Advances in Microbial Ecology*, vol. 1. Plenum Press, New York.

Folorunso, O.A., and D.E. Rolston. 1984. Spacial variability of field measured denitrification gas fluxes. *Soil Sci. Soc. Am. J.* 48: 1214–1219.

Forbes, T.J., C. Dibb, J.O. Green, A. Hopkins, and S. Peel. 1980. *Factors Affecting the Productivity of Permanent Grassland*. A national farm study. Grassland Research Institute, Hurley, Berks., U.K.

Ford, P.B., W.L. Hargrove, and C.L. Neely. 1989. Crop residue decomposition under conventional and no-tillage. *Agron. Abstr.* Am. Soc. Agron., Madison, Wis., p. 327.

Fox, R.H., and V.A. Bandel. 1986. Nitrogen utilization with no-tillage, pp. 117–148. In: Sprague, M.A., Triplett, G.B. (eds.) *No-tillage and Surface-tillage Agriculture*. The Tillage Revolution. Wiley, New York, N.Y.

Fredrickson, J.K., F.E. Koehler, and H.H. Cheng. 1981. Transformations of ^{15}N-labeled nitrogen in fertilizer and wheat straw in no-till soil. *Agron. Abstr.* Am. Soc. Agron., Madison, Wis., p. 176.

Freney, J.R., O.T. Denmead, and J.R. Simpson. 1979. Nitrous oxide emission from soils at low moisture contents. *Soil Biol. Biochem.* 11: 167–173.

Freney, J.R., A.C.F. Trevitt, S.K. DeDatta, W.N. Obcemea, and J.G. Real. 1990. Interdependence of ammonia volatilization and denitrification as nitrogen loss processes in flooded rice fields in the Philippines. *Biol. Fert. Soils* 9: 31–36.

Gamble, T.N., M.R. Betlach, and J.T. Tiedje. 1977. Numerically dominant denitrifying bacteria from world soils. *Appl. Environ. Microbiol.* 33: 926–939.

Garcia, J.L. 1975. Effect rhizosphere du riz sur la denitrification. *Soil. Biol. Biochem.* 7: 139–141.
Garwood, E.A., K.C. Tyson, and C.R. Clement. 1977. A comparison of yield and soil conditions during 20 years of grazed grass and arable cropping. *Technical Report No. 21*. Grassland Research Institute, Hurley, Berks., U.K.
Gauer, E., C.F. Shaykewich, and E.H. Stobbe. 1982. Soil temperature and soil water under zero tillage in Manitoba. *Can. J. Soil Sci.* 62: 311–325.
Germon, J.C. 1980. Etude quantitative de la denitrification biologique dons le sol à l'aide de l'acetylene. *Ann. Microbiol.* Paris 131B: 69–80.
Germon, J.C. 1985. Denitrification in cropped soils. *Fertillzers Agric.* 89: 3–13.
Germon, J.C., M. Pocket, R. Chaussod, and G. Catroux. 1981. Relations entre l'activité dénitrifiante potentielle et les differentes formes de carboné du sol, pp. 338–341. In: Dutil, P., Jacquin, F. (eds.) *Colleque humus-azote.* Int. Soil Sci. Soc. and Inst. Nat. Recher, Agron., Reims, France.
Gilliam, J.W., S. Dasberg, L.J. Lund, and D.D. Focht. 1978. Denitrification in four California soils: effect of soil profile characteristics. *Soil Sci. Soc. Am. J.* 42: 61–66.
Gilliam, J.W., and R.P. Gambrell. 1978. Temperature and pH as limiting factors in loss of nitrate from saturated Atlantic coastal plain soils. *J. Environ. Qual.* 7: 526–532.
Goring, C.A.I., and D.A. Laskowski. 1982. The effects of pesticides on nitrogen transformations in soil, pp. 689–720. In: Stevenson, F.J. (ed.), Nitrogen in Agricultural Soils, no. 22 *Agronomy Series*. American Society of Agronomy, Madison, Wis.
Greenwood, D.J. 1975. Measurement of soil aeration in soil physical conditions and crop production, pp. 261–272. *Tech Bull. No. 29*, Minist. Agric. Fish. Food., H.M.S.O., London, England.
Grevers, M.C.J., and E. de Jong. 1982. The effects of traditional versus innovative farming systems on soil texture, pp. 49–57. Proc. Soils and Crops Workshop, Extension Div. University Saskatchewan, Saskatoon, Canada. Ag Dex 510, Publ. No. 464.
Groffman, P.M. 1985. Nitrification and denitrification in conventional and no-tillage soils. *Soil Sci. Soc. Am. J.* 49: 329–334.
Groffman, P.M. 1990. Microbial nitrogen gas fluxes in native and cultivated tallgrass prairie soils. *Agron. Absts.* Am. Soc. Agron. Madison, Wis., p. 249.
Groffman, P.M., and J.M. Tiedje. 1988. Denitrification hysteresis during wetting and drying cycles in soil. *Soil Sci. Soc. Am. J.* 52: 1626–1629.
Groffman, P.M., and J.M. Tiedje. 1989a. Denitrification in north temperate forest soils: spatial and temporal patterns at the landscape and seasonal scales. *Soil Biol. Biochem.* 21: 613–620.
Groffman, P.M., and J.M. Tiedje. 1989b. Denitrification in north temperate forest soils: relationships between denitrification and environmental factors at the landscape scale. *Soil Biol. Biochem.* 21: 621–626.
Grundmann, G.L., and D.E. Rolston. 1987. A water function approximation to degree of anaerobiosis associated with denitrification. *Soil Sci.* 144: 437–441.
Haider, K., A. Mosier, and O. Heinemeyer. 1985. Phytotron experiments to evaluate the effect of growing plants on denitrification. *Soil Sci. Soc. Am. J.* 49: 636–641.

Haider, K., A. Mosier, and O. Heinemeyer. 1987. The effect of growing plants on denitrification at high soil nitrate concentrations. *Soil Sci. Soc. Am. J.* 51: 97–102.

Hallmark, S.L., and R.E. Terry. 1985. Field measurement of denitrification in irrigated soils. *Soil Sci.* 140: 35–44.

Hauck, R.D. 1966. N-isotope distribution in nitrogen gas evolved from soil during denitrification: possible application to nitrogen transformation studies. pp. 447–456. In: FAO/IAEC (ed.) *The Use of Isotopes in Soil Organic Matter Studies* (Proc. Int. Conf. Brunswick-Völkenrode, 1963) Pergamon Press, Oxford.

Hauck, R.D. 1978. Critique of field trials with isotopically labeled nitrogen fertilizer, pp. 63–77. In: Nielsen, D.R., MacDonald, J.G., (eds.) *Nitrogen in the Environment*, vol. 1. Nitrogen Behaviour in Field Soil. Academic Press, New York.

Hauck, R.D. 1981. Nitrogen fertilizer effects on nitrogen cycle processes. In: Clark, F.E., Rosswall, T. (eds.) Terrestrial N Cycle. *Ecol. Bull. Stockholm* 33: 551–562.

Hauck, R.D. 1988. A human ecosphere perspective of agricultural nitrogen cycling, pp. 3–19. In: Wilson, J.R. (ed.) *Advances in Nitrogen Cycling in Agricultural Ecosystems.* C.A.B. International, Oxon, U.K.

Hauck, R.D., and D.R. Bouldin. 1961. Distribution of isotope nitrogen in nitrogen gas during denitrification. *Nature* 191: 871–872.

Hauck, R.D., S.W. Melsted, and P.E. Yankwich. 1958. Use of N-isotope distribution in nitrogen gas in the study of denitrification. *Soil Sci.* 86: 287–291.

Hauck, R.D., and K.K. Tanji. 1982. Nitrogen transfers and mass balances, pp. 891–925. In: Stevenson, F.J. (ed.) *Nitrogen in Agricultural Soils.* Agronomy monograph no. 22, American Society of Agronomy, Madison, Wis.

Henninger, N.C., and J.M. Bollag. 1976. Effect of chemicals used as nitrification inhibitors on the denitrification process. *Can. J. Microbiol.* 22: 668–672.

Hixon, S.E., R.F. Walker, and C.M. Skau. 1990. Soil denitrification rates in four subalpine plant communities of the Sierra Nevada. *J. Environ. Qual.* 19: 617–620.

Holt, L.S., C.B. Christianson, E.R. Austin, and J.C. Katyal. 1988. A laboratory technique for releasing and measuring denitrification products trapped in soil. *Soil Sci. Soc. Am. J.* 52: 1510–1511.

Hussey, M.R., Q.D. Skinner, J.C. Adams, and A.J. Harvey. 1985. Denitrification and bacterial numbers in riparian soils of a Wyoming mountain watershed. *J. Range Manage.* 38: 492–496.

Hutchinson, G.L., and A.R. Mosier. 1981. Improved soil cover method for field measurement of nitrous oxide flux. *Soil Sci. Soc. Am. J.* 45: 311–316.

Jenkinson, D.S., R.H. Fox, and J.H. Rayner. 1985. Interactions between fertilizer nitrogen and soil nitrogen—the so-called "priming" effect. *J. Soil Sci.* 36: 425–444.

John, P.S., R.J. Buresh, R. Prasad, and R.K. Pandey. 1989. Nitrogen gas (N_2 + N_2O) flux from urea applied to lowland rice as affected by green manure. *Plant Soil* 119:7–13.

Jury, W.A., J. Letey, and T. Collins. 1982. Analyses of chamber methods used for measuring nitrous oxide production in the field. *Soil Sci. Soc. Am. J.* 46: 250–256.

Katyal, J.C., Bijay-Singh, V.K. Sharma, and E.T. Craswell. 1985. Fate and efficiency of nitrogen fertilizers applied to wetland rice. II. *Punjab, India. Fert. Res.* 6: 279–290.

Keeney, D.R., I.R. Fillery, and G.P. Max. 1979. Effect of temperature on the gaseous nitrogen products of denitrification in a silt-loam soil. *Soil Sci. Soc. Am. J.* 43: 1125–1128.

Keeney, D.R., J.S. Schepers, J.E. Blodgett, G.R. Hallberg, and P.F. Pratt. 1987. Nitrate contributions to groundwater by agricultural practices. Proc. 1986 Soil Sci. Soc. Am. Workshop. New Orleans, La. Soil Sci. Soc. Am., Madison, Wis.

Khan, M.F.A., and A.W. Moore. 1968. Denitrifying capacity of some Alberta soils. *Can. J. Soil Sci.* 48: 89–91.

Kieft, T.L., E. Soroker, and M.K. Firestone. 1987. Microbial biomass response to a rapid increase in water potential when a dry soil is wetted. *Soil Biol. Biochem.* 19: 119–126.

Klemedtsson, L., B.H. Svensson, T. Lindberg, and T. Rosswall. 1977. The use of acetylene inhibition of nitrous oxide reductase in quantifying denitrification in soils. *Swedish J. Agric. Res.* 7: 179–185.

Knowles, R. 1981. Denitrification, pp. 323–369. In: Paul, E.A., Ladd, J. (eds.) *Soil Biochemistry*, vol. 5. Marcel Dekker, New York.

Knowles, R. 1982. Denitrification. *Microbiol. Rev.* 46: 43–70.

Kohl, D.H., F. Vithayathil, P. Whitlow, G. Shearer, and S.H. Chien. 1976. Denitrification kinetics in soil systems: the significance of good fits of data to mathematical forms. *Soil Sci. Soc. Am. J.* 40: 249–253.

Koskinen, W.C., and D.R. Keeney. 1982. Effect of pH on the rate of gaseous products of denitrification in a silt loam soil. *Soil Sci. Soc. Am. J.* 46: 1165–1167.

Krishnappa, A.M., and J.E. Shinde. 1980. Fate of ^{15}N-labeled urea fertilizer under conditions of tropical flooded culture, pp. 127–144. *Soil Nitrogen as Fertilizer or Pollutant.* IAEA, Vienna, Austria.

Lance, J.C., R.C. Rice, and F.D. Whisler. 1978. Effects of vegetation on denitrification and phosphate movement during rapid infiltration of soil columns. *J. Water Pollut. Control Fed.* 50: 2183–2188.

Lijinsky, W. 1977. How nitrosamines cause cancer. *New Sci.* 27: 216–217.

Lindau, C.W., W.H. Patrick, Jr., R.D. DeLaune, K.R. Reddy, and P.K. Bollich. 1988. Entrapment of nitrogen-15 dinitrogen during soil denitrification. *Soil Sci. Soc. Am. J.* 52: 538–540.

Lindstrom, M.J., W.B. Voorhees, and G.W. Randall. 1981. Long-term tillage effects on interrow runoff and infiltration. *Soil Sci. Soc. Am. J.* 45: 945–948.

Linn, D.M., and J.W. Doran. 1984a. Aerobic and anaerobic microbial populations in no-till and plowed soils. *Soil Sci. Soc. Am. J.* 48: 794–799.

Linn, D.M., and J.W. Doran. 1984b. Effect of water-filled pore space on carbon dioxide and nitrous oxide production in tilled and nontilled soils. *Soil Sci. Soc. Am. J.* 48: 1267–1272.

Lund, L.J., D.C. Adriano, and P.F. Pratt. 1974. Nitrate concentrations in deep soil cores as related to soil profile characteristics. *J. Environ. Qual.* 3: 78–82.

Mann, L.D., D.D. Focht, H.A. Joseph, and L.H. Stolzy. 1972. Increased denitrification in soil by additions of sulphur as an energy source. *J. Environ. Qual.* 1: 329–332.

Matthias, A.D., A.M. Blackmer, and J.M. Bremner. 1980. A simple chamber technique for field measurement of emissions of nitrous oxide from soils. *J. Environ. Qual.* 9: 251–256.

McConnaughey, P.K., and J.M. Duxbury. 1986. Introduction of acetylene into soil for measurement of denitrification. *Soil Sci. Soc. Am. J.* 50: 260–263.

McElroy, M.B., S.C. Wofsy, and Y.L. Yung. 1977. Nitrogen cycle perturbations due to man and their impact on atmospheric N_2O and O_3. *Philos. Trans. R. Sci. Lond.* B277: 159–181.

McGarity, J.W. 1961. Denitrification studies on some South Australian soils. *Plant Soil* 14: 1–21.

Meek, B.D., L.B. Grass, and A.J. MacKenzie. 1969. Applied nitrogen losses in relation to oxygen status of soils. *Soil Sci. Soc. Am. Proc.* 33: 575–578.

Melen, J., and H. Nömmik. 1983. Denitrification measurements in intact soil cores. *Acta Agric. Scand.* 33: 145–152.

Melillo, J.M., J.D. Aber, P.A. Steudler, and J.P. Schimel. 1983. Denitrification potentials in a successional sequence of northern hardwood forest stands. In: Hallberg, R. (ed.) Environmental Biogeochemistry. *Ecol. Bull. (Stockholm)* 35: 217–228.

Meyer, M.L., P.R. Bloom, and J. Grava. 1989. Transformation and losses of applied nitrogen-15 labeled ammonium in a flooded organic soil. *Soil Sci. Soc. Am. J.* 53: 79–85.

Mielke, L.N., J.W. Doran, and K.A. Richards. 1986. Physical environment near the surface of plowed and no-tilled soils. *Soil Tillage Res.* 7: 355–366.

Miller, R.D., and D.D. Johnson. 1964. The effect of soil moisture tension on CO_2 evolution, nitrification, and nitrogen mineralization. *Soil Sci. Soc. Am. Proc.* 28: 644–647.

Minami, R., T. Shibuya, Y. Ogawa, and S. Fukushi. 1990. Effect of nitrification inhibitors on emission of nitrous oxide from soils. *Trans 14th congress of Soil Science* 2: 267–271.

Mitsui, S., I. Watanabe, M. Honma, and S. Honda. 1964. The effect of pesticides on the denitrification in paddy soil. *Soil Sci. Plant Nutr.* 10: 15–23.

Mohanty, S.K., and A.R. Mosier. 1990. Nitrification-denitrification in flooded rice soils. *Trans. 14th International Congress Soil Science* 4: 326–331.

Mosier, A.R. 1980. Acetylene inhibition of ammonium oxidation in soil. *Soil Biol. Biochem.* 12: 443–444.

Mosier, A.R. 1989. Chamber and isotope techniques, pp. 175–187. In: Andreae, M.O., Schimel, D.S. (eds.) *Exchange of Trace Gases Between Terrestrial Ecosystems and the Atmosphere*. Wiley, Chichester.

Mosier, A.R., S.L. Chapman, and J.R. Freney. 1989. Determination of dinitrogen emission and retention in floodwater and porewater of a lowland rice field fertilized with ^{15}N-urea. *Fert. Res.* 19: 127–136.

Mosier, A.R., W.D. Guenzi, and E.E. Schweizer. 1986a. Soil losses of dinitrogen and nitrous oxide from irrigated crops in north-eastern Colorado. *Soil Sci. Soc. Am. J.* 50: 344–348.

Mosier, A.R., W.D. Guenzi, and E.E. Schweizer. 1986b. Field denitrification estimation by nitrogen-15 and acetylene inhibition techniques. *Soil Sci. Soc. Am. J.* 50: 831–833.

Mosier, A.R., K. Haider, and O. Heinemeyer. 1988. Field and phytotron assessment of the effect of crop plants on denitrification, pp. 271–284. In:

Giovannozzi-Sermanni, G. Nannipieri, P. (eds.) *Current Perspectives in Environmental Biogeochemistry*. CNR-IPRA Rome, Italy.
Mosier, A.R., and O. Heinemeyer. 1985. Current methods used to estimate N_2O and N_2 emissions from field soils, pp. 79–99. In: Golterman, H.I. (ed.) *Denitrification in the Nitrogen Cycle*. Plenum, New York.
Mosier, A.R., O. Heinemeyer, and K. Haider. 1990a. Measurement of denitrification in agricultural fields. In: Proc. Internat. Denitrification Workshop, Giessen, FRG. March 1989. *Mitt. Dtsch. Boden Ges.* 60: 13–18.
Mosier, A.R., S.K. Mohanty, A. Bhadrachalm, and S.P. Chakravorti. 1990b. Evolution of dinitrogen and nitrous oxide from the soil to the atmosphere through rice plants. *Biol. Fertil. Soils* 9: 31–36.
Mosier, A.R., and W.J. Parton. 1985. Denitrification in a shortgrass prairie: a modeling approach, pp. 441–452. In: Caldwell, D.E., Brierley, J.A., Brierley, C.L. (eds.) *Planetary Ecology*. Van Nostrand Reinhold, New York.
Mosier, A.R., W.J. Parton, and G.L. Hutchinson. 1983. Modeling nitrous oxide evolution from cropped and native soils. In: Hallberg, R. (ed.) Environmental Biogeochemistry. *Ecol. Bull. (Stockholm)* 35: 229–241.
Mosier, A.R., and D.S. Schimel. 1992. Nitrification and denitrification in soil system. In: Knowles, R., Blackburn, T.H. (eds.) *Nitrogen Isotope Techniques in Plant, Soil, and Aquatic Biology*. Academic Press, New York (in press).
Mosier, A.R., M. Stillwell, W.J. Parton, and R.G. Woodmansee. 1981. Nitrous oxide emissions from a native shortgrass prairie. *Soil Sci. Soc. Am. J.* 45: 617–619.
Mulvaney, R.L. 1984. Determination of 15-N labeled dinitrogen and nitrous oxide with triple-collector mass spectrometers. *Soil Sci. Soc. Am. J.* 48: 690–692.
Mulvaney, R.L. 1988. Evaluation of nitrogen-15 tracer techniques for direct measurement of denitrification in soil. III. Laboratory Studies. *Soil Sci. Soc. Am. J.* 52: 1327–1332.
Mulvaney, R.L., and C.W. Boast. 1986. Equations for determination of nitrogen-15 labeled dinitrogen and nitrous oxide by mass spectrometry. *Soil Sci. Soc. Am. J.* 50: 360–363.
Mulvaney, R.L., and L.T. Kurtz. 1982. A new method for determination of 15-N labeled nitrous oxide. *Soil Sci. Soc. Am. J.* 46: 1178–1184.
Mulvaney, R.L., and R.M. Vanden Heuvel. 1988. Evaluation of nitrogen-15 tracer techniques for direct measurement of denitrification in soil: IV. Field studies. *Soil Sci. Soc. Am. J.* 52: 1332–1337.
Müller, M.M., V. Sundman, and J.J. Skujins. 1980. Denitrification in low pH Spodosols and peats determined with the acetylene inhibition method. *Appl. Environ. Microbiol.* 40: 235–239.
Myers, R.J.K., and J.W. McGarity. 1971. Factors influencing high denitrifying activity in the subsoil of solodized solonetz. *Plant Soil* 35: 145–160.
Myrold, D.D. 1988. Denitrification in ryegrass and winter wheat cropping systems of western Oregon. *Soil Sci. Soc. Am. J.* 52: 412–416.
Myrold, D.D., and J.M. Tiedje. 1985. Establishment of denitrification capacity in soil: effects of carbon, nitrate and moisture. *Soil Biol. Biochem.* 17: 819–822.

Nömmik, H. 1956. Investigations on denitrification in soil. *Acta Agric. Scand.* 6: 195–228.
Ott, S., and W.L. Hargrove. 1989. Profits and risks of using crimson clover and hairy vetch cover crops in no-till corn production. *Am. J. Alternat. Agric.* 4: 65–70.
Parker, D.T., and W.E. Larson. 1962. Nitrification as affected by temperature and moisture content of mulched soils. *Soil Sci. Soc. Am. Proc.* 26: 238–242.
Parkin, T.B. 1987. Soil microsites as a source of denitrification variability. *Soil Sci. Soc. Am. J.* 51: 1194–1199.
Parkin, T.B., S.T. Chester, and J.A. Robinson. 1990. Calculating confidence intervals for the mean of a lognormally distributed variable. *Soil Sci. Soc. Am. J.* 54: 321–326.
Parkin, T.B., H.F. Kaspar, A.J. Sexstone, and J.M. Tiedje. 1984. A gas-flow soil core method to measure field denitrification rates. *Soil Biol. Biochem.* 16: 323–330.
Parkin, T.B., and J.J. Meisinger. 1989. Denitrification below the crop rooting zone as influenced by surface tillage. *J. Environ. Qual.* 18: 12–16.
Parkin, T.B., J.J. Meisinger, S.T. Chester, J.L. Starr, and J.A. Robinson. 1988. Evaluation of statistical estimation methods for lognormally distributed variables. *Soil Sci. Soc. Am. J.* 52: 323–329.
Parkin, T.B., and J.A. Robinson. 1989. Stochastic models of soil denitrification. *Appl. Environ. Microbiol.* 55: 72–77.
Parkin, T.B., A.J. Sexstone, and J.M. Tiedje. 1985b. Comparison of field denitrification rates by acetylene-based soil core and nitrogen-15 methods. *Soil Sci. Soc. Am. J.* 49: 94–99.
Parkin, T.B., A.L. Sexstone, and J.M. Tiedje. 1985a. Adaption of denitrifying populations to low soil pH. *Appl. Environ. Microbiol.* 49: 1053–1056.
Parton, W.J., A.R. Mosier, and D.S. Schimel. 1988. Rates and pathways of nitrous oxide production in a shortgrass steppe. *Biogeochemistry* 6: 45–58.
Patten, D.K., J.M. Bremner, and A.M. Blackmer. 1980. Effects of drying and air-dry storage of soils on their capacity for denitrification of nitrate. *Soil Sci. Soc. Am. J.* 44: 67–70.
Paul, E.A., and F.E. Clark. 1989. *Soil Microbiology and Biochemistry*. Academic Press, New York.
Paul, E.A., and R.J.K. Myers. 1971. Effect of soil moisture strees on uptake and recovery of tagged nitrogen by wheat. *Can. J. Soil Sci.* 51: 37–43.
Payne, W.J. 1981. Denitrification. Wiley, New York.
Poth, M., and D.D. Focht. 1985. ^{15}N kinetic analysis of N_2O production by *Nitrosomonas europea*: an examination of nitrifier denitrification. *Appl. Environ. Microbiol.* 49: 1134–1141.
Power, J.F., and J.S. Schepers. 1989. Nitrate contamination of groundwater in North America. *Agri. Ecosys. Environ.* 26: 165–187.
Prade, K., and G. Trolldenier. 1990. Denitrification in the rhizosphere of plants with inherently different aerenchyma formation: wheat (*Triticum aestivum*) and rice (*Oryza sativa*). *Biol. Fertil. Soils* 9: 215–219.
Reddy, K.R., E. D'Angelo, C. Lindau, and W.H. Patrick, Jr. 1990. Urea losses in flooded soils with established oxidized and reduced soil layers. *Biol. Fertil. Soils* 9: 283–287.

Reddy, K.R., R. Khaleel, and M.R. Overcash. 1980a. Nitrogen, phosphorus and carbon transformations in a coastal plain soil treated with animal manures. *Agric. Wastes* 2: 225–238.
Reddy, K.R., and W.H. Patrick, Jr. 1975. Effect of alternative aerobic and anaerobic conditions on redox potential, organic matter decomposition and nitrogen loss in a flooded soil. Soil *Biol. Biochem.* 7: 87–94.
Reddy, K.R., and W.H. Patrick, Jr. 1986a. Fate of fertilizer nitrogen in the rice root zone. *Soil Sci. Soc. Am. J.* 50: 649–651.
Reddy, K.R., and W.H. Patrick, Jr. 1986b. Denitrification losses in flooded rice fields. *Fert. Res.* 9: 99–116.
Reddy, K.R., P.D. Sacco, and D.A. Graetz. 1980b. Nitrate reduction in an organic soil-water system. *J. Environ. Qual.* 9: 283–288.
Rekhi, R.S., O.P. Meelu, and R.K. Gupta. 1982. Lysimeter studies on recovery of ^{15}N-labeled urea in wetland rice. *Plant Soil* 66: 57–67.
Rice, C.W., and M.S. Smith. 1982. Denitrification in no-till and plowed soils. *Soil Sci. Soc. Am. J.* 46: 1168–1173.
Robertson, G.P., M.A. Huston, F.C. Evans, and J.M. Tiedje. 1988. Spacial variability in a successional plant community: patterns of nitrogen availability. Ecology 69: 1517–1524.
Robertson, G.P., and J.M. Tiedje. 1984. Denitrification and nitrous oxide production in successional and old growth Michigan forests. *Soil Sci. Soc. Am. J.* 48: 383–389.
Robertson, G.P., P.M. Vitousek, P.A. Matson, and J.M. Tiedje. 1987. Denitrification in a clearcut loblolly pine (*Pinus taeda* L.) plantation in the southeastern U.S. *Plant Soil* 97: 119–129.
Rolston, D.E., A.N. Sharpley, D.W. Toy, and F.E. Broadbent. 1982. Field measurement of denitrification. III. Rates during irrigation cycles. *Soil Sci. Soc. Am. J.* 46: 289–296.
Rolston, D.E., M. Fried, and D.A. Goldhamer. 1976. Denitrification measured directly from nitrogen and nitrous oxide gase fluxes. *Soil Sci. Soc. Am. J.* 40: 256–266.
Rönner, U. 1983. Distribution, production and consumption of nitrous oxide in the Baltic Sea. *Geochim. Cosmochim.* Acta 47: 2179–2188.
Ryden, J.C. 1981. N_2O exchange between a grassland soil and the atmosphere. *Nature* 292: 235–237.
Ryden, J.C. 1983. Denitrification loss from a grassland soil in the field receiving different rates of nitrogen as ammonium nitrate. *J. Soil Sci.* 34: 355–365.
Ryden, J.C., L.J. Lund, J. Letey, and D.D. Focht. 1979. Direct measurement of denitrification loss from soils. II. Development and application of field methods. *Soil Sci. Soc. Am. J.* 43: 110–118.
Ryden, J.C., and L.J. Lund. 1980. Nature and extent of directly measured denitrification losses from some irrigated vegetable crop production units. *Soil Sci. Soc. Am. J.* 44: 505–511.
Ryden, J.C., J.H. Skiner, and D.J. Nixon. 1987. Soil core incubation system for the field measurement of denitrification using acetylene-inhibition. *Soil Biol. Biochem.* 19: 753–757.
Sacks, L.E., and H.A. Barker. 1952. Substrate oxidation and nitrous oxide utilization in denitrification. *J. Bacteriol.* 64: 247–252.

Sahrawat, K.L., and D.R. Keeney. 1986. Nitrous oxide emission from soils. *Adv. Soil Sci.* 4: 103–148.
Samson, M.I., R.J. Buresh, and S.K. DeDatta. 1990. Evolution and soil entrapment of nitrogen gases formed by denitrification in flooded soil. *Soil Sci. Plant Nutr.* 36: 299–307.
Savant, N.K., and S.K. DeDatta. 1982. Nitrogen transformation in wetland rice soils. *Adv. Agron.* 35: 241–302.
Sebacher, D.I., R.C. Harris, and K.B. Bertlett. 1985. Methane emissions to the atmosphere through aquatic plants. *J. Environ. Qual.* 14: 40–46.
Seiler, W., A. Holzapfel-Pschorn, R. Conrad, and D. Scharite. 1984. Methane emission from rice paddies. *J. Atom. Chem.* 1: 241–268.
Sexstone, A.J., T.B. Parkin, and J.M. Tiedje. 1985a. Temporal response of soil denitrification rates to rainfall and irrigation. *Soil Sci. Soc. Am. J.* 49: 99–103.
Sexstone, A.J., T.B. Parkin, and J.M. Tiedje. 1988. Denitrification response to soil wetting in aggregated and unaggregated soil. *Soil Biol. Biochem.* 20: 767–769.
Sexstone, A.J., N.P. Revsbech, T.B. Parkin, and J.M. Tiedje. 1985b. Direct measurement of oxygen profiles and denitrification rates in soil aggregates. *Soil Sci. Soc. Am. J.* 49: 645–651.
Shea, C.P. 1988. Protecting life on earth: steps to save the ozone layer. *World Watch Paper 87.* World Watch Institute, Washington, D.C.
Siegel, R.S., R.D. Hauck, and L.T. Kurtz. 1982. Determination of $^{30}N_2$, and application to measurement of N_2 evolution during denitrification. *Soil Sci. Soc. Am. J.* 46: 68–74.
Skujins, J.J. 1976. Extracellular enzymes in soil. *C.R.C. Critical Rev. Micribiol.* 4: 383–421.
Smid, A.E., and E.G. Beauchamp. 1976. Effects of temperature and organic matter on denitrification in soil. *Can. J. Soil Sci.* 56: 385–391.
Smith, C.J. 1987. Denitrification in the field, pp. 387–398. In: Wilson, J.R. (ed.) *Advances in Nitrogen Cycling in Agricultural Ecosystems.* C.A.B. International, Wallingford, Oxon, U.K.
Smith, M.S., and J.M. Tiedje. 1979. The effects of roots on soil denitrification. *Soil Sci. Soc. Am. J.* 43: 951–955.
Sommers, L.E., C.M. Gilmour, R.E. Wildung, and S.M. Beck. 1981. The effect of water potential on decomposition processes in soils, pp. 97–117. In: Parr, J.F., Gardner, W.R. Elliott, L.F. (eds.) *Water Potential Relations in Soil Microbiology.* Spec. Pub. 9, Soil Science Society of America, Madison, Wis.
Stanford, G., and E. Epstein. 1974. Nitrogen mineralization-water relations in soils. *Soil Sci. Soc. Am. Proc.* 38: 103–107.
Stanford, G., S. Dzienia, and R.A. VanderPol. 1975a. Effects of temperature on denitrification rate in soils. *Soil Sci. Soc. Am. Proc.* 39: 867–870.
Stanford, G., R.A. VanderPol, and S. Dzienia. 1975b. Denitrification rates in relation to total and extractable soil carbon. *Soil Sci. Soc. Am. Proc.* 39: 284–289.
Stanier, R.Y., E.A. Adelberg, and J. Ingraham. 1976. *The Microbial World.* Prentice-Hall, Englewood Cliffs, N.J.
Starr, J.L., and J.Y. Parlange. 1975. Nonlinear denitrification kinetics with continuous flow in soil columns. *Soil Sci. Soc. Am. Proc.* 39: 875–880.

Stefanson, R.C. 1972. Soil denitrification in sealed soil-plant systems. 1. Effect of plants, soil water content and soil organic matter content. *Plant Soil* 33: 113–127.

Stevenson, F.J. 1982. Origin and distribution of nitrogen in soil, pp. 1–42. In: Stevenson, F.J. (ed.) *Nitrogen in Agricultural Soils*. Agronomy Monograph no. 22. American Society of Agronomy, Madison, Wis.

Storrier, R.R. 1966. The pretreatment and storage of soil samples for nitrogen analysis. *J. Aust. Inst. Agric. Sci.* 32: 106–113.

Strong, W.M., E.R. Austin, L.S. Holz, and R.J. Buresh. 1987. Determination of the combined nitrogen-15 content of dinitrogen and nitrous oxide in air. *Soil Sci. Soc. Am. J.* 51: 1344–1350.

Terry, R.E., and J.M. Duxbury. 1985. Acetylene decomposition in soils. *Soil Sci. Soc. Am. J.* 49: 90–94.

Terry, R.E., E.N. Jellen, and D.P. Breakwell. 1986. Effect of irrigation methods and acetylene exposure on field denitrification measurements. *Soil Sci. Soc. Am. J.* 50: 115–120.

Tiedje, J.M. 1988. Ecology of denitrification and dissimilatory nitrate reduction to ammonium, pp. 179–243. In: Zehnder, A.J.B. (ed.) *Biology of Anaerobic Micro-organisms*. Wiley, New York.

Tiedje, J.M., S. Simkins, and P.M. Groffman. 1989. Perspectives on measurement of denitrification in the field including recommended protocols for acetylene based methods. *Plant Soil* 115: 261–284.

Tiedje, J.M., J. Sorensen, and Y.Y.L. Chang. 1981. Assimilatory and dissimilatory nitrate reduction: perspectives and methodology for simultaneous measurement of several nitrogen cycle processes. In: Clark, F.E., Rosswall, T. (eds.) Terrestrial Nitrogen Cycles. *Ecol. Bull. (Stockholm)* 33: 331–342.

van Raalte, M.H. 1941. On the oxygen supply of rice plants. *Ann. Bot. Gard. Buitenzorg*. 51: 43–57.

van der Staay, R.L., and D.D. Focht. 1977. Effects of surface area upon bacterial denitrification rates. *Soil Sci.* 123: 18–24.

Vlek, P.L.G., and B.H. Byrnes. 1986. The Efficacy and loss of fertilizer N in lowland rice, pp. 131–147. In: DeDatta, S.K., Patrick, W.H., Jr. (eds.) *Nitrogen Economy of Flooded Rice Soils*. Martinus Nijhoff, Dordrecht, The Netherlands.

Volz, M.G., M.S. Ardakani, R.K. Schulz, L.H. Stolzy, and A.D. McLaren. 1976. Soil nitrate loss during irrigation: enhancement by plant roots. *Agron. J.* 68: 621–627.

Walter, H.M., D.R. Keeney, and I.R. Fillery. 1979. Inhibition of nitrification by acetylene. *Soil Sci. Soc. Am. J.* 43: 195–196.

Wang, W.C., Y.L. Yung, A.L. Lacis, T. Mo, and J.E. Hanson. 1976. Greenhouse effects due to man-made perturbations of trace gases. *Science* 194: 685–689.

Waring, S.A., and J.W. Gilliam. 1983. The effect of acidity on nitrate reduction and denitrification in lower coastal plain soils. *Soil Sci. Soc. Am. J.* 47: 246–251.

Watanabe, I., S.K. DeDatta, and P.A. Roger. 1987. Nitrogen cycling in wetland rice soils, pp. 239–256. In: Wilson, J.R. (ed.) *Advances in Nitrogen Cycling in Agricultural Ecosystems*. C.A.B. International, Wallingford, Oxon, U.K.

Watanabe, I., B.C. Padre, Jr., and S.T. Santiago. 1981. Quantitative study on nitrification in flooded rice soil. *Soil Sci. Plant Nutr.* 27: 373–382.

Webster, C.P., and R.J. Dowdell. 1982. Nitrous oxide emission from permanent grass swards. *J. Sci. Food Agric.* 33: 227–230.

Webster, C.P., and K.W.T. Goulding. 1989. Influence of soil carbon content on denitrification from fallow land during autumn. *J. Sci. Food Agric.* 49: 131–142.

Weier, K.L., and J.W. Doran. 1987. The potential for biolgoical denitrification in a brigalow clay soil under different management systems, pp. 327–339. In: Bacon, P.E., Evans, J., Storrier, R.R., Taylor, A.C. (eds.) *Nitrogen Cycling in Temperate Agricultural Systems*. Australian Society of Soil Science, Riverina Branch.

Weier, K.L., and J.W. Gilliam. 1986. Effect of acidity on denitrification and nitrous oxide evolution from Atlantic coastal plain soils. *Soil Sci. Soc. Am. J.* 50: 1202–1205.

Yeomans, J.C., and E.G. Beauchamp. 1982. Acetylene as a possible substrate in the denitrification process. *Can. J. Soil Sci.* 62: 139–144.

Yoshinari, T., R. Hynes, and R. Knowles. 1977. Acetylene inhibition of nitrous oxide reduction and measurement of denitrification and nitrogen fixation in soil. *Soil Biol. Biochem.* 9: 177–183.

Yoshinari, T., and R. Knowles. 1977. Acetylene inhibition of nitrous oxide reduction by denitrifying bacteria. *Biochem. Biophys. Res. Commun.* 69: 705–710.

Freezing and Its Effect on Chemical and Biological Properties of Soil

A.C. Edwards and M.S. Cresser

I. Introduction

The effects of subzero temperatures on soil are varied, influencing a wide range of physical, biological, and chemical processes. Since the formation of ice in soil requires an understanding of thermal gradients and water movement, the major thrust of recent work has been from a physically based approach. The visible effects of freezing and thawing, such as changes in soil structure and damage to building and road foundations, have also received considerable attention. There is far less information available concerning possible effects on chemical and biological soil reactions. Our aim, therefore, is to present an overview of the freeze/

Advances in Soil Science, Volume 18

thaw cycle, placing particular emphasis on factors relating to the less well-documented and understood influences on nutrient availability.

II. Factors Influencing the Degree of Frost Penetration

Heat transfer between the soil surface and surrounding atmosphere can occur through a variety of mechanisms, which include radiation, convection, and conduction. The relative importance of each will vary with location and seasonally, in response to changing conditions at the same site.

Factors affecting the degree of frost penetration can be conveniently grouped as those related to the heat flux away from the freezing front and those related to water supply from the unfrozen soil (Larson and Allmaras, 1971).

The long-term likelihood of a soil experiencing freezing will depend ultimately upon geographic location and broad-scale climatic conditions. Many soils in the Arctic/Antarctic region are permanently frozen, experiencing only a limited annual thaw of their upper horizons (active layer). A greater land area is subject to periodic, short-term freezing and thawing cycles. In general the severity of climatic conditions and consequently the number of freezing days (where arithmetic mean of the maximum and minimum temperatures for the day is below 0° C) increases sharply inland and also with altitude. This relationship is highlighted for north-east Scotland in Fig. 1.

This general pattern is, however, subject to modification by many local factors, the most significant of which are discussed below.

A. Ground Cover

The amount of insulation afforded by vegetation and snow cover is particularly important in determining not only the rate and depth of freezing, but also the thawing characteristics. Initial work by Post and Dreibelbis (1942) showed that frost penetration increased as vegetation type changed, from woodland through meadow pasture to wheat (*Triticum aestivum*), for sites in Ohio, USA. A related pattern was also observed by Harris (1972), who compared coniferous, deciduous, and open field sites at thaw and reported that the coniferous plantation remained frozen longer.

Numerous studies have shown that frost penetration is inversely linked with the amount of litter and hence degree of insulation. In an agricultural situation it is therefore possible to modify the freezing characteristics through specific management strategies. Kohnke and Werkhoven (1963) compared the effects of bare (fallow) soil and a mulch treatment ($3.8\,t\,ha^{-1}$) of surface applied wheat straw. They noted that during the

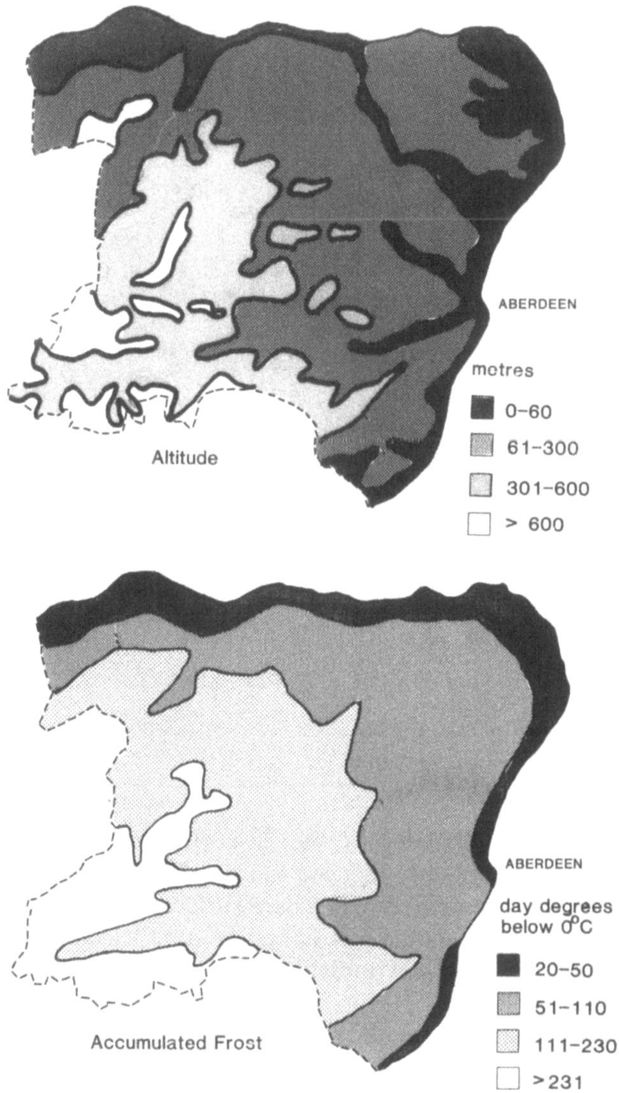

Figure 1. Relationship between altitude and the number of freezing days for northeast Scotland (modified from Birse and Robertson, 1970).

winter the near-surface soil of a mulched area remained warmer by approximately 1°C, while the opposite was true for the spring period. The daily temperature fluctuation at a soil depth of 25 mm was twice as great in bare soil than where the mulch had been applied. This had a considerable influence on the frequency of freeze/thaw cycles experienced by the surface soil layers (Fig. 2).

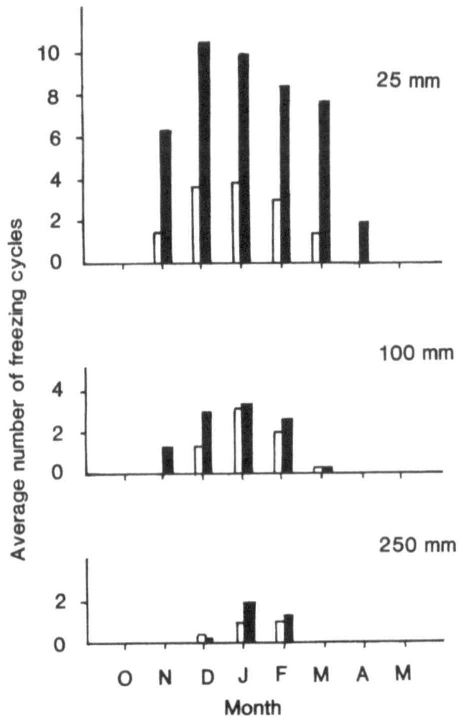

Figure 2. Frequency of freezing and thawing at three soil depths for bare (shaded) and mulched (open) soil (Kohnke and Werkhoven, 1963).

A 4-year study by Pikul et al. (1986) in northeast Oregon showed that standing stubble consistently reduced the depth of frost penetration, by an average of 35%, compared with a bare surface. The persistence of soil frost during thawing weather was linked to both surface residue cover through its influence on net radiation and soil heat flux and climatic conditions during the thaw.

Snow depth and persistence will also be modified considerably by vegetation cover. The degree of insulation afforded by snow can be considerable. Removal of snow cover in a Minnesota oak stand, for example (Table 1), caused deeper and more persistent soil freezing than did either litter removal or soil compaction (Thorud and Duncan, 1972). It was also apparent that the thawing process was not constant over the whole experimental plot with some areas remaining frozen longer than others. This has important consequences as regards infiltration and soil hydraulic conductivity. A 30- to 60-cm reduction in frost penetration was reported by Crawford and Legget (1957) for each extra 30 cm of snow cover. Reducing snow blow loss on exposed sites may also be modified through land use and management options favoring snow entrapment.

Table 1. The difference in average depth (cm) and persistence of frozen soil between various experimental treatments and an undisturbed plot for a Minnesota oak stand. (Modified from Thorud and Duncan, 1972)

	Difference in average depth (cm)				
Treatment	Jan.[a] 10	Jan. 30	Feb. 27	Mar. 26	Apr. 3
Litter removed vs undisturbed	18*	25*	14*	16*	24*
Snow removed[b] vs undisturbed	5.0	35*	33*	28*	38*
Compacted soil vs undisturbed	8.0*	8.0	6.0	4.0	7.0

	Percentage of sample points with frozen soil			
Date	Undisturbed	Litter	Snow	Compacted
Mar. 26	100	100	100	100
Apr. 3	80	80	100	75
Apr. 9	60	25	100	50
Apr. 15	10	10	95	5
Apr. 20	0	0	95	0
Apr. 24	0	0	0	0

[a] Snowpack of between 5 and 13 cm depth was continuously present.
[b] Snow removed after every storm.
* Significantly different.

B. Influence of Moisture Content

Dry soils in the autumn have been shown to freeze faster and to greater depth than wet soils (Willis et al., 1961). The higher specific heat of moist soils requires a greater energy loss for cooling and subsequent freezing. Ice rarely forms within soil at 0° C, the greatest temperature at which ice can exist under a given set of circumstances depending on the freezing point depression (Miller, 1980), which in turn depends on the nature and concentration of soil solution solute. The latter is a function of moisture content. Thermal conductivities of soils depend on the nature of the soil, as well as its moisture content. Mineral soils have been found to have thermal conductivities roughly an order of magnitude greater than dry peat (Brown and Williams, 1972). This results in the potential for a greater annual range in soil temperatures for mineral soils. The influence of soil moisture on the rate and degree of soil freezing is, therefore, difficult to quantify on a field scale, because of the complex set of interactive parameters involved.

The direct proportional relationship between thermal conductivity and moisture content for a given soil means that seasonally high water tables can play a very significant role in determining the degree of frost penetration. This is well shown in a detailed investigation by Price (1983) of a Saskatchewan catchment area. A range of terrain types (including both organic and inorganic soils) and different hydrological conditions were monitored. The circulation of water through near-surface soil horizons reduced the rate of freezing, but caused a more rapid cooling in deeper layers. Where little flow occurred, the cooling of lower soil layers lagged considerably, resulting in a more rapid frost peneration in upper layers. The surface topography was also found to interact with water table depth causing further local modifications. The freezing of sphagnum hummocks, therefore, proceeded more rapidly than in adjacent depressions. Increased exposure in this situation would also be an important consideration.

C. Movement of Water in Soil Undergoing Freezing

It has long been recognised that water can undergo a considerable amount of redistribution in a freezing soil. Water can be drawn towards a developing freezing front or point and Bouyoucos and McCool (1915), using sealed soil columns, showed that water moved from areas of high to areas of low temperature (with freezing conditions not being necessary). There followed considerable debate in the literature as to the actual mechanisms involved and their relative importance. Water can move in either the liquid or the vapor phase, with potential arising from various gradients (including osmotic, matric, or thermal) which exist in the soil (Hoeckstra, 1966; Oliphant et al., 1983).

In a laboratory study, using soil columns which were initially at uniform temperatures and water contents, Dirksen and Miller (1966) showed, by lowering the temperature at one end of the column, that the water (ice) content of the freezing zone increased at the expense of the unfrozen section. The moisture profiles as freezing progressed were as shown in Fig. 3. Extrapolation from their laboratory column studies to the natural situation is complicated by the "closed" nature of their system, where the only source of water was that which was present initially. This situation is, therefore, unlike that which is usually found in the field where there may well be a supply of water to the freezing zone from deeper soil horizons and additional sources, such as precipitation. New techniques which include time domain reflectometry (Hayhoe and Bailey, 1985) offer potential for improving our understanding of water migration within the field situation.

The water content of the frozen soil increases with time (Fig. 3) and can reach levels greater than the saturation value of the unfrozen soil. This gradual build-up suggests that water is relatively mobile within the ice zone at temperatures well below 0° C. The proportion of water that remains unfrozen within the frozen zone will decrease as the temperature drops. The exact location of liquid in soil is difficult to predict. The

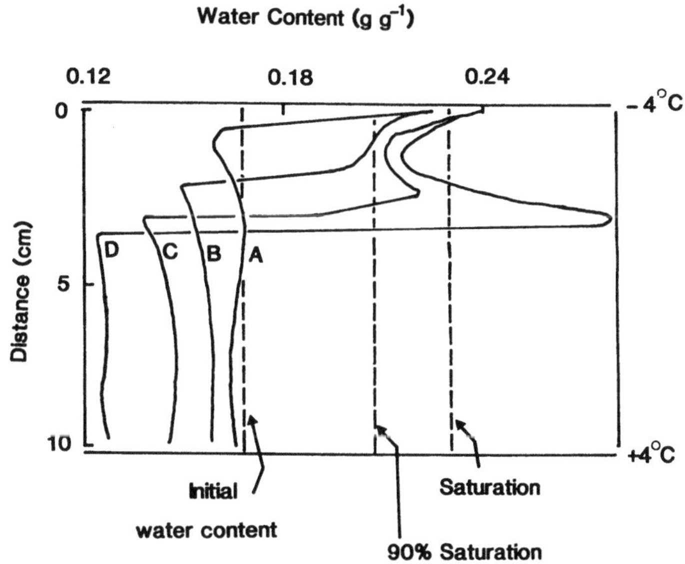

Figure 3. Freezing induced redistribution of water in a New Hampshire silt, water content includes ice (after Dirksen and Miller, 1966).

thickness of films on mineral grain/ice boundaries for a given soil has been shown to depend mainly on temperature (Anderson and Hoekstra, 1965). Relatively high rates of ion diffusion were found in frozen Na-Wyoming bentonite pastes when compared with unfrozen ones (Murrman et al., 1968) by using a radioactive tracer (^{22}Na). The researchers suggested that any reduction in ion mobilities that did occur at the lower temperature were due to a decreased thickness of unfrozen water film. The thickness of the liquid films decreased from 50 Å or more at 0° C to about 9 Å at −5° C (Anderson, 1970). This has important consequences with regard to exchange and sorption reactions which will be discussed later. Flow of water and solutions through the frozen zone will occur along these unfrozen liquid channels or films. The continuity of these films and any factors which influence it are, therefore, particularly important (Fig. 4). For example, with the increased formation of ice, soil pore enlargement may occur, which results in the soil particles being pushed apart (Loch and Miller, 1975) and a break-up of the continuous liquid films reducing the transmission of water (Benz et al., 1968).

The importance of water movement in the vapor phase has been difficult to assess (see Gurr et al., 1952). Dirksen and Miller (1966) noted that the transport mechanism in their soil columns exceeded, by several orders of magnitude, that which could be accounted for by vapor movement, suggesting the liquid phase is of primary importance. In some instances vapor phase transport may be significant (Cary and Mayland,

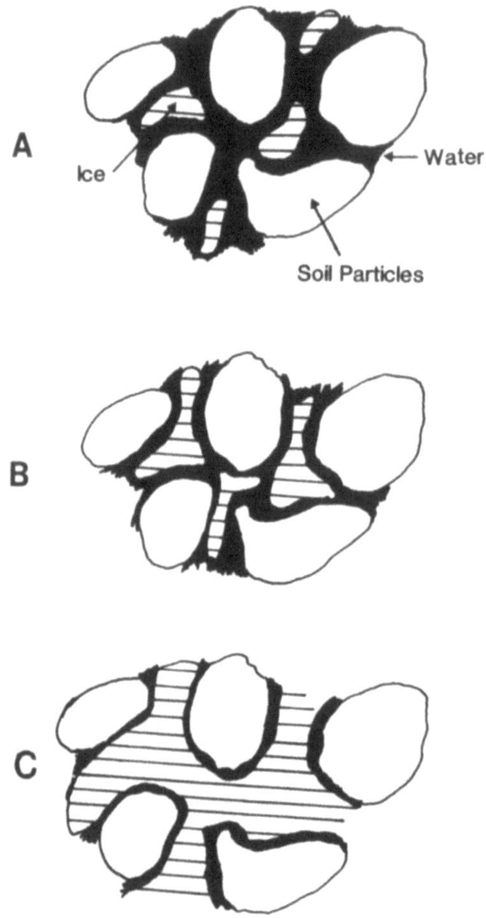

Figure 4. Schematic representation of the relationship between ice formation and unfrozen water films. The amount of ice present increases from A to C reducing the water film thickness and its continuity.

1972). In the present context, however, links between water and solutes are of more interest.

It has also been suggested that ice is not immobile under conditions in which steady liquid transport is to be expected (Miller et al., 1975). A continuous transformation of ice to film water occurs at one side while at the opposite edge film water is constantly transformed into ice.

D. Solute Concentration

The presence of solutes in the soil water significantly lowers the temperature range over which the freeze/thaw zone develops (Fuchs et al., 1978). Both composition and concentration of solutes are important (Fig. 5). Accumulation of solutes within the freezing zone may also modify the freezing point of the water within this region with the result that a leap-

Figure 5. Expected reduction in freezing points for varying solution concentrations of (a) sea water (b) sodium chloride.

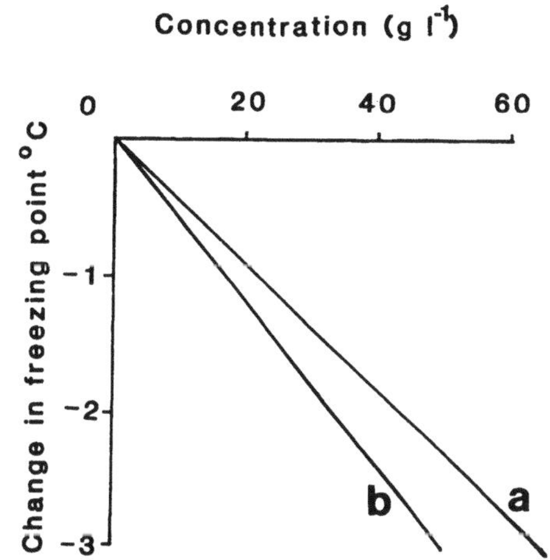

frog type (ice starts forming at a point beyond this high solute concentration zone) progression of ice formation occurs (Kay and Groenevelt, 1983; Miller, 1980).

III. Influence of Rate and Degree of Freezing on the Structure of Ice

An important aspect of freezing which influences subsequent effects is the rate at which it occurs. Nucleation of ice crystals begins in larger pores where interfacial forces are lower because of the greater distances between surfaces. Lower temperatures will be required to freeze water in smaller pores (Larson and Allmaras, 1971). Rapid freezing results in the development of numerous ice centers and overall smaller crystal sizes. At slower freezing rates the number of freezing centers decreases, due to the formation of larger crystals and greater physical pressures exerted on the soil matrix. The soil mineral and organic solids impose a restriction on the freezing of soil moisture through interaction (i.e., surface/water interactions) with the liquid phase and structural limitations imposed on growth of the ice phase (Miller, 1980).

Post and Dreibelbis (1942) identified a number of ice structures which appeared to be related to the depth of freezing, moisture content, and type of vegetation. Shallow freezing occurred particularly on meadow and pasture soils and often developed more open structures, while deep freezing was associated with a typically massive "concrete" structure (characterized by small crystals and a dense matrix). These differences in ice structure can greatly influence the hydraulic characteristics of soil and effects on the physical properties of soil during the period of thawing.

IV. Effects of Freezing on Individual Physical Characteristics of Soil

The development of ice can dramatically alter soil water storage and drainage characteristics, not only in the short term during snowmelt but also in the longer term as a consequence of structural modifications. Laboratory studies of upward movement of water towards the frozen soil zone have been mentioned earlier. Field observations also indicate such upward movement. Willis et al. (1964) showed that localized lowering of the water table appeared to be related to increased depth and amount of ice in the zone above. Water from depth is, however, not necessarily the only source. Sartz (1969) noted that surface-derived (precipitation and melt) water percolated through more than 60 cm of hard frozen ground before freezing. An associated decline in permeability occurred with time, causing reduced infiltration rates. This resulted in a build-up of water in the frozen zone during the spring melt period and the potential for abnormal quantities of runoff.

Burt and Williams (1976) studied the hydraulic conductivity of a frozen soil and showed a decline in conductivity below 0° C. They noted the importance soil type has on freezing and conductivity, through the dependence of moisture content on pore size distribution, void ratio, particle size, and surface area for frozen soils. For sandy soils, where the majority of water freezes close to 0° C, very low hydraulic conductivities (10^{-8} cm s^{-1}) were found compared with those for unfrozen sand (10^{-4} cm s^{-1}). Silts and clays, however, can contain a relatively large proportion of unfrozen water below 0° C (e.g., 30% at −1° C) and maintain relatively high conductivity rates compared with their unfrozen values, which although low, do allow considerable flow over a long period of time (Burt and Williams, 1976).

Vegetation cover has been shown to influence infiltration rates of frozen ground through its effect on the dominant ice structure (Augustine, 1941; Stoeckeler and Weitzman, 1960). Forest soils appeared to maintain almost "normal" infiltration when frozen to a depth of 10 cm, while corn (*Zea mays*) and pasture areas had greatly reduced rates when frozen to similar depths. Penman and Schofield (1941) noted that ground freezing often resulted in a period of reduced drainage. Both ice formation and snow accumulation may result in the temporary storage of substantial volumes of water, which may be released rapidly under thawing conditions. When snow fell on previously frozen areas, the eventual drainage depended on the rate at which soil thawed, as well as the rate at which snow melted (Penman and Schofield, 1941).

The effects of freezing and thawing can be either beneficial or harmful to soil structure (Larson and Allmaras, 1971). During the winter disruption of clods produced by autumn cultivation is desirable, but excessive disruption may well increase the susceptibility of a soil to wind and water erosion. Hinman and Frederick (1968) noted in a laboratory study that the percentage of aggregates 1 mm or more in diameter may be increased, decreased, or unaffected by freezing and thawing. Similar variability in observed effects has been noted in the field situation and the end result is dependent on a number of factors such as moisture content (Bisal and Nielsen, 1967), rate of freezing or melting (Zuzel et al., 1982), autumn tillage operation (e.g., Kay et al., 1985), and number of freeze/thaw cycles. Reducing the number of freeze/thaw events can reduce the time a soil is susceptible to erosion. With time and conditions favoring evaporation and drainage, soil strength and erosion resistance increase (Formanek et al., 1983). Edwards and Burney (1989) compared the extent of soil loss and runoff from freeze/thaw soil with and without subsoil compaction and ground cover. Soil compaction resulted in a 45% increase in sediment loss, compared with that from noncompacted controls.

The greater penetration of ice in compacted soils has been previously demonstrated by Barnett (1937), who also noted increased thawing rates in the spring as a result of greater thermal conductivity of compacted soil. The work by Edwards and Burney (1989) clearly showed the experimental difficulties associated with this general subject area. After thawing, continuously frozen soil profiles gave much greater sediment and runoff values than the equivalent profiles subjected to alternate freeze/thaw cycles. This was found to be partly due to the incomplete thawing of the continuously frozen soil, which resulted in saturated conditions near the surface and highly favorable conditions of runoff and erosion.

The change in soil volume associated with freezing is called frost heave. The degree of heaving can vary considerably with such factors as texture and moisture content. An excellent discussion of heaving is given by Miller (1980). In general coarse, sandy soils would not be expected to show signs of heaving (except perhaps under saturated conditions). In finer textured, clay soils, water migrates towards the freezing front from adjacent unfrozen soil, which undergoes some degree of shrinkage. The net overall expansion corresponds to the volume change on freezing of this redistributed water. The most favorable conditions which allow substantial heaving to occur are when there is a readily available supply of water in silt-textured soils.

V. Chemical and Biological Effects of Soil Freezing

The chemical and biological aspects of soil freezing are poorly understood. Initial interest in the subject was prompted by observations that soils collected in the spring contained appreciably more exchangeable nutrients than those collected the previous autumn. The exact nature of effects of freeze/thaw cycles is however very difficult to separate. Both increased and decreased availabilities of nutrients have been reported, but many of the studies have not been designed to test for specific effects. The following discussion is, of necessity, rather speculative, but attempts to explain why storing soil samples in a frozen condition prior to analysis is not advisable. Any effects are also likely to be highly soil and element specific, which further complicates the issue. In addition, the factors which influence key processes such as freezing rate (ice crystal size), moisture content, and frequency of freeze/thaw cycles (as discussed earlier) will vary considerably from site to site.

A. Effects of Soil Freezing on Chemical Reactions of Soil

From a knowledge of the physical aspects relating to freezing it is possible to suggest a number of situations which may arise. The division will be based upon whether the dominant factor is purely physical, or physicochemical.

1. Physical Disruption of Soil

The structural disintegration of soil, associated with the volumetric changes of water upon freezing, could result in the exposure of fresh reactive surfaces. This would not be limited to mineral soils, as the physical disruption of organic matter has also been reported. Both reversible and irreversible changes in the structure of soil-derived organic matter can occur (Forsyth and Fraser, 1947; Dergacheva and Dedkov, 1977; Giesy and Briese, 1978). The outcome would be a provision of new sites for ion fixation or the exposure of what used to be "unavailable" supplies of ions. This may well help to explain some of the apparently conflicting changes reported in nutrient availability after freezing. For example, the impact of repeated freeze/thaw cycles was investigated by Hinman (1970) for some Canadian soils. Numerous freeze/thaw treatments caused additional increases, over that produced by a single cycle, for extractable ammonium and phosphorus, while showing little effect upon nitrate. At the same time Hinman (1970) found a reduction in potassium availability, suggesting some type of fixation. The importance of physical disruption was also suggested by Allen and Grimshaw (1962), who prevented the formation of ice crystals (at $-15°$C) in a soil by adding glycerol. There were negligible changes in extractable ammonium or phosphorus following this treatment, compared with a water only control. Allen and Grimshaw (1962) also compared the response to freezing of mineral and organic soils. They noted that by far the greatest effects of freezing on nutrient availability were associated with highly organic soils (Fig. 6). For example, for the peat soil approximately one-eighth of the total phosphorus present was released during a single period of freezing.

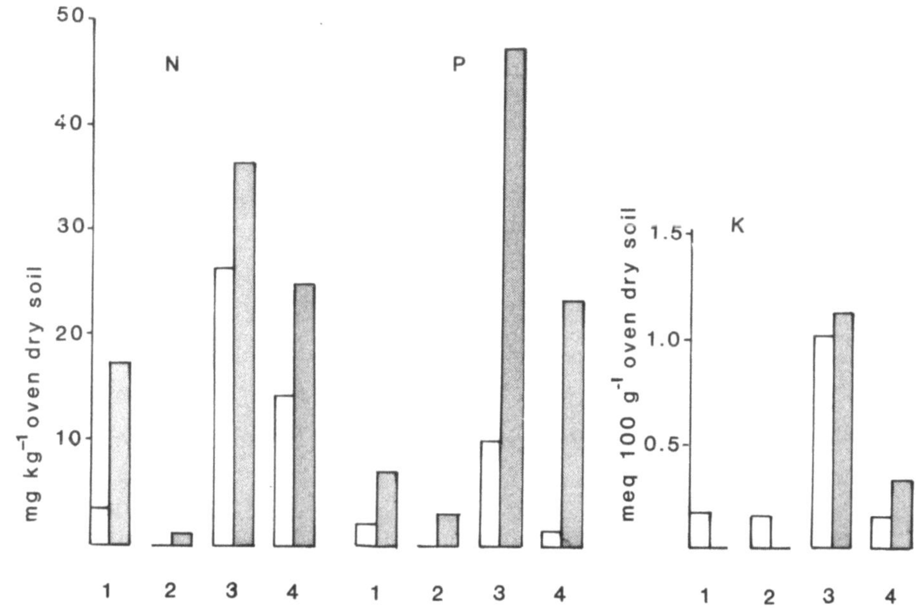

Figure 6. A comparison of the extractable nitrogen, phosphorus, and potassium between unfrozen (open) and frozen (shaded) treatments for a range of four soil types: (1) calcareous, (2) slate, (3) litter, and (4) peat (modified from Allen and Grimshaw, 1962).

2. Physicochemical Reactions

Locally high solute concentrations can arise from either the exclusion of ions from active freezing centers, or as a consequence of the water migration from unfrozen soil to the freezing front. An important consequence of this is the possibility of transient high concentrations of ions in the surviving solution films and in the unfrozen soil, below the freezing front. These ions will be in intimate contact with soil surface exchange/sorption sites. Precipiation reactions could occur at these increased solute levels. It is known that, for example, freezing as a method of sample preservation is not advised when subsequent analysis may include iron, aluminum, and silicon determinations, due to irreversible hydrolysis and polymerization of hydrous oxides of these elements (Florence, 1982). The relevance of these precipitation reactions to soil mineralogy and certain pedogenic processes may also be significant. Silicon readily precipitates from solution (Mahler and Bechtold, 1980; Wada and Nagasto, 1983; Edwards, 1984) and could be implicated in processes such as induration development which is thought to be periglacial in origin (FitzPatrick, 1956). Factors controlling the extent of precipitation are, however, com-

Table 2. The effect of freezing on the determination of silicate in various water samples

	Detectable silicon ($\mu g\,ml^{-1}$)		
Sample type	Before freezing	After freezing	Loss (%)
Soil solution	3.2	ND	100
River water (1)	7.5	3.9	52
River water (2)	7.6	1.3	83
Standard solution[a]	6.0	5.2	13

Relatively high (1) and low (2) total ionic contents; ND, not detectable.
[a] Standards made from sodium fluorosilicate.

plicated (Table 2) and appear to be related to the total solution ion content and composition.

Other possible consequences of solute concentration following freezing include secondary cementation, clay mineral formation, crystallization of amorphous material, and possibly even iron pan formation. It has been suggested that various aluminosilicates and silicates may be formed as a result of freezing (Mahler and Bechtold, 1980; Wada and Nagasto, 1983; Inoue and Huang, 1984). It is not inconceivable to think of a freezing front (with its associated higher ionic concentration) regularly reaching similar depths (particularly under periglacial conditions) resulting in a distinct area of concentrated precipitation.

B. Freezing and Soil Biology

Freezing and thawing also have effects on the soil's microbial population. Again, however, there are conflicting reports as to the likely effects (Jager, 1967). Soulides and Allison (1961) were able to show a decrease in bacterial population, after wet soil was frozen at −11°C for 24h. Sudden, severe frosts have been found to eliminate most of the bacteria in exposed soils (Logsdail and Webber, 1959). Upon thawing, large flushes in respiration measured by CO_2 production (Soulides and Allison, 1961) or O_2 uptake (Wynn-Williams, 1982), similar to those caused by drying soil, have been noted. This burst of activity is related to the presence of a substantial supply of readily available nutrients (especially soluble carbon compounds) released through the mechanisms discussed earlier and possibly also through rupture of cell membranes (as discussed later). For example, concentrations of free amino acids in soils have been shown to display a marked increase as a result of freezing to −14°C (Ivarson and Sowden, 1966).

Increased rates of nitrogen mineralization have frequently been reported after soil freezing (Soulides and Allison, 1961; Mack, 1963; Jager, 1967). Rates of denitrification can also be significantly increased

(McGarity, 1961; Christianson and Cho, 1983). Bremner et al. (1980) estimated that 6% to 21% of the annual N_2O flux occurred during the period when topsoil thawed in spring. In an attempt to assess the origin of N_2O emissions from thawing soils, Goodroad and Keeney (1984) added $CHCl_3$ to soil cores prior to freezing. A slower rate of N_2O emissions occurred in the treated soil, but denitrification did not cease, which indicated some of the N_2O flux resulted from a physicochemical release, but with an additional source from biological denitrification processes. Edwards and Killham (1986) studied the effect of freeze/thaw on gaseous nitrogen losses from urea-amended and untreated upland soils under either heather or improved pasture. Rates of both denitrification and NH_3-volatilization were increased following the freeze/thaw cycle, particularly when the intact soil blocks had been previously fertilized with urea. A series of laboratory and field experiments showed that N_2O production was two orders of magnitude higher at thaw in the spring than at any other time during the rest of the year and represents a significant source of atmospheric N_2O in temperate regions (Christensen and Tiedje, 1990). Interestingly soil thaw in midwinter did not result in any increase in N_2O flux. Factors influencing N_2O emission are of particular interest at the present time because it functions as a greenhouse gas.

An additional source of solutes could also arise from the disruption of decomposing plant cells and soil microfauna, as intracellular fluid expands upon freezing. White (1973) reported increased leaching of a wide range of elements from frozen prairie vegetation. He showed that the degree of leaching was related to the stage of growth or decomposition of the vegetation and that rupturing of plant cells was involved.

VI. Consequences for Soil Fertility

The literature concerned with the consequences of freeze/thaw on soil fertility status is limited and often contradictory. Various factors have been suggested to influence the overall response of a soil. For example, Fine et al. (1940) reported a potential liberation of potassium, although their results were by no means clear-cut. Eight of the twelve soils they sampled showed an increase in extractable potassium (up to $168\,kg\,ha^{-1}$), while three showed a marked decrease. Potassium status was an important factor, the majority of soils displaying a positive effect, having low or moderate potassium levels, while two out of the three soils displaying a negative effect had high potassium status. A further complication arose when a liming factor was included. Lime on its own tended to inhibit the liberation of potassium, regardless of the soil's initial potassium status. However, with both lime and potassium added, extra potassium was released. In contrast, Sinclair (1979) noted no change in available potassium as a result of freezing and thawing.

In a study, using highly organic soils from three British mires, Summerfield and Rieley (1973) found the acid-extractable (0.05 *M* HCl) concentrations of sodium, potassium, calcium, magnesium, and iron removed from treated (i.e., subjected to freeze/thaw) samples were greater (up to 5% of the original concentration) than for the untreated controls. The largest effects were observed at the most exposed site and these authors suggest that the more severe climate of such upland sites during the winter months could result in additional inputs of available nutrients. They also stressed the importance of the other environmental factors which can influence the water chemistry of mires, such as wetting and drying cycles.

The study of the impact of repeated freeze/thaw cycles by Hinman (1970) for some Canadian soils has already been mentioned. Successive freeze/thaw treatments caused additional increases, in excess of that produced by a single treatment, for ammonium and phosphorus, while having no effect on nitrate and even decreasing potasssium availability. It appears, therefore, that both the physical and chemical effects associated with ice formation are important in determining the amounts of these nutrients released for subsequent plant uptake.

VII. Possible Effects on Natural Surface Water

The previous discussion on solute releases should be thought of in conjunction with the modifications in hydrology mentioned earlier, which significantly alter the "normal" water pathways. It is, therefore, appropriate to consider here possible environmental consequences of freeze/thaw cycles. Young and Mutchler (1976) highlighted an important aspect of excessive runoff periods through the increased potential for pollution episodes during the spring, as a consequence of manure spreading on frozen ground. Soil loss through erosion can also be considerable.

Table 3. Effect of freezing on the amount of solutes leached from replicated repacked soil columns (upper organic rich horizons)

	Unfrozen (μg)	Frozen (μg)	Difference between means
Sodium	183	357	NS
Potassium	123	536.5	***
Calcium	44	41	NS
Magnesium	23.5	45.5	*
Nitrate-N	0.6	2.9	*
Silicon	63.5	68	NS
Aluminum	6.7	63	**
Total organic carbon	327	8565	**

*, $P < 0.05$; **, $P < 0.01$; ***, $P < 0.001$; NS, not significant.

Detailed long-term monitoring of river and drainage water chemisry has highlighted peaks in nitrate levels during the spring (e.g., Likens et al., 1977). These have been attributed to increased nitrification following freeze/thaw cycles and the consequent mobilization of nitrate to drainage waters (Arefyeva and Kolesnikof, 1964; McGarity and Rajaratnam, 1973). Such losses of nitrate may be an important factor contributing to eutrophication, particularly as they occur at a time when plant demand for nitrogen is low, Reduced river pH values during the spring have also been attributed to preferential thawing of the snow pack, with large initial releases of pollutants (Johannessen and Henriksen, 1978; Johannessen et al., 1980; Christopherson et al., 1984). This theory, however, does not seem adequate to explain the high aluminum levels reported during this period, which is a particularly important and sensitive time in the development of young fish. The present authors have highlighted a possible second (perhaps complementary) mechanism for these increased solute concentrations involving freeze/thaw (Edwards et al., 1986) based on work at their two catchment areas in northeast Scotland. Significantly increased levels of many elements (Table 3) were obtained from leaching repacked columns of highly organic soil from their sites, after a short period of freezing at $-12°C$. Large and significant increases in aluminum, potassium, and organic carbon were obvious (Table 3).

VIII. Conclusions

Freezing and thawing can have a significant influence on many important biological, chemical, and physical soil processes. The effects of solute exclusion from freezing centers and the substantial local increases in concentrations of elements that can develop are particularly interesting. Large-scale changes in land use and its management (e.g., afforestation) could have direct consequences for snow accumulation and the degree of frost penetration. The importance of freeze/thaw cycles within annual nutrient budgets and the maintenance of a site's long-term fertility have received little attention.

There is a considerable need to increase our awareness through a multidisciplined approach of the possible consequences resulting from freeze/thaw.

The freezing process may play a role, in hitherto little investigated ways in:

1. Pedogenesis
2. Modification of microbial processes and populations
3. Modification of soil physicochemical parameters and hence soil fertility
4. Modification of catchment hydrological pathways
5. Changing the fate of pollutants

Acknowledgment. Malcolm S. Cresser is grateful to the UK Department of the Environment for financial support for this work.

References

Allen, S.E., and H.M. Grimshaw. 1962. Effect of low-temperature storage on the extractable nutrient ions in soils. *J. Sci. Food Agric.* 13: 525–529.

Anderson, D.M. 1970. Phase boundary water in frozen soils. *U.S. Army Corps of Engineers, Cold Regions Research and Engineering Laboratory Research Report* 274.

Anderson, D.M., and P. Hoekstra. 1965. Migration of interlamellar water during freezing and thawing of Wyoming bentonite. *Soil Sci. Soc. Proc.* 29: 498–504.

Arefyeva, Z.N., and B.D. Kolesnikof. 1964. Chemistry and biochemistry dynamics of ammonia and nitrate nitrogen in forest soils of the Transurals at high and low temperature. *Soviet Soil Sci.* 3: 246–260.

Augustine, M.T. 1941. Infiltration runs on frozen ground. *Soil Sci. Soc. Proc.* 6: 435.

Barnett, R.J. 1937. Ground cover affects frost penetration. *Kansas. Acad. Sci. Trans.* 4: 203–207.

Bisal, F., and K.F. Nielsen. 1967. Effect of frost action on the size of soil aggregates. *Soil Sci.* 104: 268–272.

Benz, L.C., W.O. Willis, F.M. Sandoval, and R.H. Mickelson. 1968. Soil water translocation in a high water table area. *Water Resources Res.* 4: 95–101.

Birse, E.L., and L. Robertson. 1970. Assessment of climatic conditions in Scotland. 2. Based on exposure and accumulated frost (Soil Survey of Scotland, Aberdeen).

Bouyoucos, G.J., and M.M. McCool. 1915. The freezing point method as a new means of measuring the concentration of the soil solution directly in the soil. *Mich. Agric. College Exper. Station Tech. Bull. No. 24.*

Bremner, J.M., S.G. Robbins, and A.M. Blackmer. 1980. Seasonal variability in emission of nitrous oxide from soil. *Geophys. Res. Lett.* 7: 641–644.

Brown, R.J.E., and G.P. Williams. 1972. The freezing of peatland. *N.R.C. Technical Paper 381, N.R.C. 12881.*

Burt, T.P., and P.J. Williams. 1976. Hydraulic conductivity in frozen soils. *Earth Surf. Process.* 1: 349–360.

Cary, J.W., and H.F. Mayland. 1972. Salt and water movement in unsaturated frozen soil. *Soil. Sci. Soc. Proc.* 36: 549–555.

Christensen, S., and J.M. Tiedje. 1990. Brief and vigorous N_2 production by soil at spring thaw. *J. Soil Sci.* 41: 1–4.

Christianson, C.B., and C.M. Cho. 1983. Chemical denitrification of nitrite in frozen soils. *Soil Sci. Soc. Proc.* 47: 39–42.

Christopherson, N., S. Rustad, and H.M. Seip. 1984. Modelling streamwater chemistry with snowmelt. *Phil. Trans. Royal. Sco., Lond.* B305: 427–438.

Crawford, C.B., and R.F. Legget. 1957. Ground temperature investigation in Canada. *Engin. J.* 40: 1–8.

Dergacheva, M.I., and V.S. Dedkov. 1977. Influence of freezing and thawing on soil organic matter in OB forest tundra. *Soviet J. Ecol.* 8: 111–118.

Dirksen, C., and R.D. Miller. 1966. Closed-system freezing of unsaturated soil. *Soil Sci. Soc. Proc.* 30: 168–173.

Edwards, A.C. 1984. Some factors influencing elemental mobilities in an upland catchment in the Grampian Region. PhD thesis, University of Aberdeen.

Edwards, A.C., J. Creasey, and M.S. Cresser. 1986. Soil freezing effects on upland stream solute chemistry. *Wat. Res.* 20: 832–834.

Edwards, A.C., and K. Killham. 1986. The effect of freeze/thaw on gaseous nitrogen loss from upland soils. *Soil Use Manag.* 2: 86–91.

Edwards, L.M., and J.R. Burney. 1989. The effect of antecedent freeze-thaw frequency on runoff and soil loss from frozen soil with and without subsoil compaction and ground cover. *Can. J. Soil Sci.* 69: 799–811.

Fine, L.O., T.A. Bailey, and E. Truog. 1940. Availability of fixed potassium as influenced by freezing and thawing. *Soil Sci. Soc. Proc.* 5: 183–186.

FitzPatrick, E.A. 1956. An indurated soil horizon formed by permafrost. *J. Soil Sci.* 7: 248–254.

Florence, T.M. 1982. The speciation of trace elements in waters. *Talanta* 29: 345–364.

Formanek, G.E., D.K. McCool, and R.I. Papendick. 1983. Effect of freeze-thaw cycles on erosion in the Palouse. Am. Soc. Agric. Engineers, St. Joseph, Mich. paper No. 83–2069.

Forsyth, W.G.C., and G.K. Fraser. 1947. Freezing as an aid to the drying and purification of humus and allied materials. *Nature Lond.* 160: 607.

Fuchs, M., G.S. Campbell, and R.I. Papendick. 1978. An analysis of sensible and latent heat flow in a partially frozen unsaturated soil. *Soil Sci. Soc. Proc.* 42: 379–385.

Giesy, J.P., and L.A. Briese. 1978. Particulate formation due to freezing humic waters. *Water Resources Res.* 14: 542–544.

Goodroad, L.L., and D.R. Keeney. 1984. Nitrogen oxide emissions from soils during thawing. *Can. J. Soil Sci.* 64: 187–194.

Gurr, C.G., T.J. Marshall, and J.T. Hutton. 1952. Movement of water in soil due to a temperature gradient. *Soil Sci.* 74: 335–345.

Harris, A.F. 1972. Infiltration rate as affected by soil freezing under three cover types. *Soil Sci. Soc. Proc.* 36: 489–492.

Hayhoe, H.N., and N.G. Bailey. 1985. Monitoring changes in total and unfrozen water content in seasonally frozen soil using time domain reflectometry and neutron moderation techniques. *Water Resources Res.* 21: 1077–1084.

Hinman, W.G. 1970. Effects of freezing and thawing on some chemical properties of three soils. *Can. J. Soil Sci.* 50: 179–182.

Hinman, W.C., and B. Frederick. 1968. Alterations of soil structure upon freezing and thawing and subsequent drying. *Can. J. Soil Sci.* 48: 193–197.

Hoeckstra, P. 1966. Moisture movement in soils under temperature gradients with the cold-side temperature below freezing. *Water Resources Res.* 2: 241–250.

Inoue, K., and P.M. Huang. 1984. Influence of citric acid on the natural formation of imogolite. *Nature Lond.* 308: 58–60.

Ivarson, K.C., and F.J. Sowden. 1966. Effect of freezing on the free amino acids in soil. *Can. J. Soil Sci.* 46: 115–120.

Jager, G. 1967. Changes in the activity of soil microorganisms influenced by physical factors (drying-remoistening freezing-thawing). In: Graff, O., Satchell, J.E. (eds.) *Progress in Soil Biology*, pp. 178–191, North-Holland, Amsterdam.

Johannessen, M., and A. Henriksen. 1978. Chemistry of snowmelt water: changes in concentrations during snowmelt. *Water Resources Res.* 14: 615–619.

Johannessen, M., A. Skartveit, and R.F. Wright. 1980. Streamwater chemistry before, during and after snowmelt. Proceedings of the International Conference on the Ecological Impact of Acid Precipitation, pp. 224–225, Norway.

Kay, B.D., and P.H. Groenevelt. 1983. The redistribution of solutes in freezing soil: exclusion of solutes. *Permafrost: 4th International Conference Proceedings.* pp. 584–588, National Academy Press, Washington.

Kay, B.D., C.D. Grant, and P.H. Groenevelt. 1985. Significance of ground freezing on soil bulk density under zero tillage. *Soil Sci. Soc. Proc.* 49: 973–978.

Kohnke, H., and C.H. Werkhoven. 1963. Soil temperature and soil freezing as affected by an organic mulch. *Soil Sci. Soc. Proc.* 27: 13–17.

Larson, W.E., and R.R. Allmaras. 1971. Management and natural force as related to compaction. In: Barnes, K.K., Carleton, W.M., Taylor, H.M., Throckmorton, R.I., Vanden Berg, G. (eds) *Compaction of Agricultural Soils.* Am. Soc. Agric. Engineers, pp. 367–427, St. Joseph, Mich.

Likens, G.E., F.H. Bormann, R.S. Pierce, J.S. Eaton, and N.M. Johnson. 1977. *Biogeochemistry of a Forested Ecosystem.* Springer-Verlag, New York.

Loch, J.P.G., and R.D. Miller. 1975. Tests of the concept of secondary frost heaving. *Soil Sci. Soc. Proc.* 39: 1036–1041.

Logsdail, D.E., and L.R. Webber. 1959. Effect of frost action on structure of Haldimand clay. *Can. J. Soil Sci.* 39: 103–106.

Mack, A.R. 1963. Biological activity and mineralization of nitrogen in three soils as influenced by freezing and drying. *Can. J. Soil Sci.* 43: 316–324.

Mahler, W., and M.F. Bechtold. 1980. Freeze-formed silica fibers. *Nature Lond.* 285: 27–28.

McGarity, J.W. 1961. Denitrification studies in some south Australian soils. *Plant Soil* 14: 1–21.

McGarity, J.W., and J.A. Rajaratnam. 1973. Apparatus for the measurement of losses of nitrogen as gas from the field and simulated field environments. *Soil Biol. Bioch.* 5: 121–131.

Miller, R.D. 1980. Freezing phenomena in soils. In: Hillel, D. (ed). *Applications of Soils Physics.* pp. 254–299, Academic Press, New York.

Miller, R.D., J.P.G. Loch, and E. Brester. 1975. Transport of water and heat in a frozen permeameter. *Soil Sci. Soc. Proc.* 39: 1029–1036.

Murrmann, R.P., P. Hoekstra, and R.C. Bialkowski. 1968. Self-diffusion of sodium ions in frozen Wyoming bentonite-water paste. *Soil Sci. Soc. Proc.* 32: 502–506.

Oliphant, J.L., A.R. Tice, and Y. Nakano. 1983. Water migration due to a temperature gradient in frozen soil. *Permafrost: 4th International Conference Proceedings.* pp. 951–956, National Academy Press, Washington.

Penman, H.L., and R.K. Schofield. 1941. Drainage and evaporation from fallow soil at Rothamsted. *J. Agric. Sci.* 31: 74–109.

Pikul, Jr., J.L., J.F. Zuzel, and R.N. Greenwalt. 1986. Formation of soil frost as influenced by tillage and residue management. *J. Soil Water Cons.* 41: 196–199.

Post, F.A., and F.R. Dreibelbis. 1942. Some influence of frost penetration and microclimate on the water relationships of woodland, pasture and cultivated soils. *Soil Sci. Soc. Proc.* 7: 95–104.

Price, J.S. 1983. The effect of hydrology on ground freezing in a watershed with organic terrain, pp. 95–104. *Permafrost: 4th International Conference Proceedings*. National Academy Press, Washington.

Sartz, R.S. 1969. Soil water movement as affected by deep freezing. *Soil Sci. Soc. Proc.* 33: 333–337.

Sinclair, A.H. 1979. Availability of potassium to ryegrass from Scottish soils. I. Effects of intensive cropping on potassium parameters. *J. Soil Sci.* 30: 757–773.

Soulides, D.A., and F.E. Allison. 1961. Effect of drying and freezing soil on carbon dioxide production, available mineral nutrients, aggregation and bacterial population. *Soil Sci.* 91: 291–298.

Stoeckeler, J.H., and S. Weitzman. 1960. Infiltration rates in frozen soils in Northern Minnesota. *Soil Sci. Soc. Proc.* 24: 137–139.

Summerfield, R.J., and J.O. Rieley. 1973. Substrate freezing and thawing as a factor in the mineral nutrient status of mire ecosystems. *Plant Soil* 38: 557–566.

Thorud, D.B., and D.P. Duncan. 1972. Effects of snow removal, litter removal and soil compaction on soil freezing and thawing in a Minnesota Oak Stand. *Soil Sci. Soc. Proc.* 36: 153–157.

Wada, S., and A. Nagasto. 1983. Formation of silica microplates by freezing dilute silicic acid solutions. *Soil Sci. Plant Nutr.* 29: 93–95.

White, E.M. 1973. Water-leachable nutrients from frozen or dried prairie vegetation. *J. Environ. Qual.* 2: 104–109.

Willis, W.O., C.W. Carlson, J. Alessi, and H.J. Haas. 1961. Depth of freezing and spring run off as related to full soil-moisture levels. *Can. J. Soil Sci.* 41: 115–123.

Willis, W.O., H.L. Parkinson, C.W. Carlson, and H.J. Haass. 1964. Water table changes and soil moisture loss under frozen conditions. *Soil Sci.* 98: 244–248.

Wynn-Williams, D.D. 1982. Simulation of seasonal changes in microbial activity of maritime antarctic peat. *Soil Biol. Biochem.* 14: 1–12.

Young, R.A., and C.K. Mutchler. 1976. Potential pollution of manure spread on frozen ground. *J. Environ. Qual.* 5: 174–179.

Zuzel, J.F., R.R. Allmaras, and R. Greenwalt. 1982. Runoff and soil erosion on frozen soils in northeastern Oregon. *J. Soil Water Cons.* 37: 351–354.

Soil and Fertilizer Phosphorus and Crop Responses in the Dryland Mediterranean Zone

A. Matar, J. Torrent, and J. Ryan

Advances in Soil Science, Volume 18

I. Introduction

As we approach the end of the century, the disparity between the rich and poor nations of the world is startling. While poverty knows no climatic or geographical boundary, a disproportionate number of the less well-off are concentrated in dryland or rainfed areas of the world. There, poverty has many bedfellows; low literacy levels, underemployment, small land holdings, limited and poor quality resources, and, consequently, low agricultural output. Because of over-stocking and poor grazing management, soil degradation is an inevitable consequence and one that undermines a country's agricultural production capacity. This phenomenon is apparent in virtually all dryland areas of the world (Majeed, 1986). While soil degradation is evident in such places as diverse as Brazil and China, most countries of Africa are affected. Though the sub-Saharan Sahel has attracted attention because of disastrous famine in the past two decades, the North Africa region is just as much threatened. Indeed, the problem is common in West Asian countries as well. It is ironic that water, a scarce resource, which dictates the course of man's fortunes in dryland areas, can also cause a depletion of the soil resource, if not managed properly.

The challenge to reverse the trends that have occurred in dryland agriculture, and improve productivity in a sustainable fashion, is immense. Recognition of the constraints to doing so is the first step in this process (Steiner et al., 1988). Looming large in the litany of constraints are institutional and infrastructural ones, such as availability of credit, fertilizers, and pesticides; inadequate marketing and distribution systems; and weak research and educational capability. Socioeconomic factors such as fragmented holdings, inappropriate ownership norms, common grazing land, and a host of other factors, exacerbated by rapid population growth, also impinge on land use. Physical and technological obstacles to dryland farming are even more daunting. Chief among these are fragile impoverished soils and precarious rainfall.

This review deals with an aspect of soils, i.e., fertility with respect to soil phosphorus (P), in relation to crop growth. A brief description of dryland soils, climates, and cropping systems, with specific reference to those of the Mediterranean zone, is pertinent. Subsequently, aspects of soil P chemistry and its relationship to P fertilizer use by crops are presented. Where appropriate, reference is also made to relevant work from similar environments. While the review is heavily biased toward the authors' institutions in Syria, Morocco, and Spain, it nevertheless high-

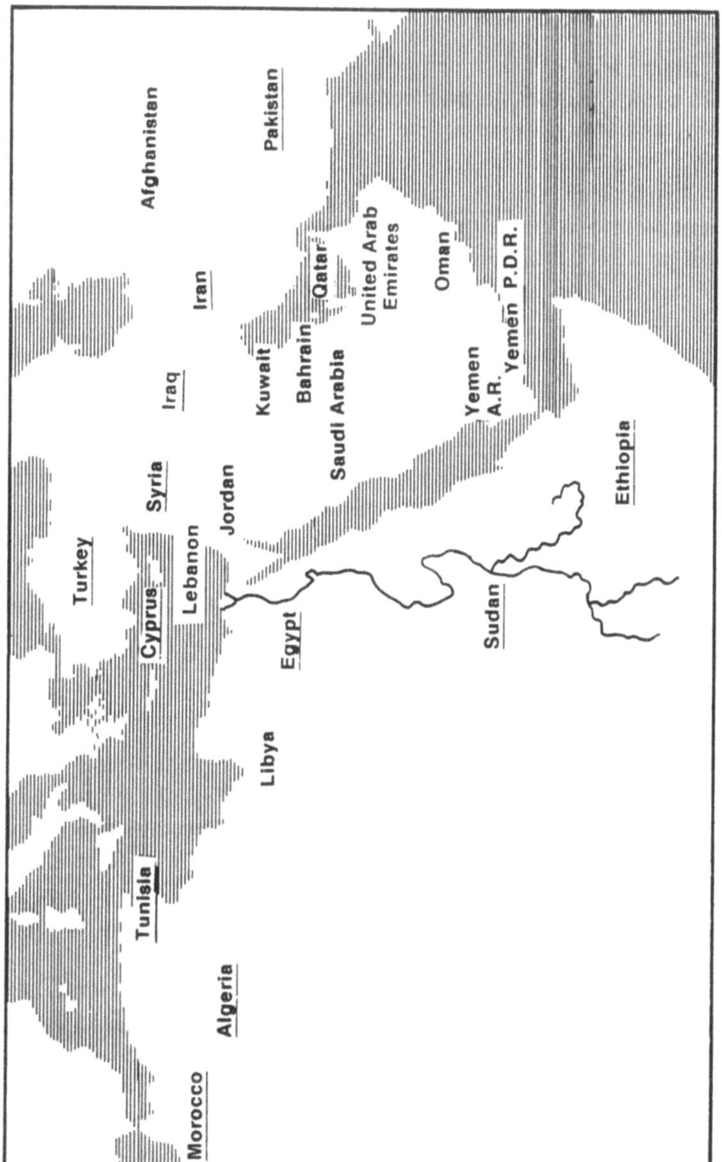

Figure 1. Schematic map of the Mediterranean region. Underlining indicates major food producer (absolute or per capita basis).

lights relationships that probably hold in other countries of the whole West Asia-North Africa (WANA) region and the northern shores of the Mediterranean (Fig. 1) as well. Though documented resources of P research varied from country to country, the absence of such reports does not imply that P is not significant to the agricultures of those countries.

The lands bordering the Mediterranean are redolent with the history of man's origins and struggles to survive and flourish. However, the "cradle of civilization" has been cultivated for millennia, and worn threadbare in the process. In a way, science and technology seem to have by-passed these ancient lands. While a general picture of the region's agriculture (Clawson et al., 1971) and of its climate and soil resources (Kassam, 1981) has emerged, the issue of P in the region's agriculture, except for some brief localized reviews (Ryan, 1983; Ryan, 1990), has not been addressed. Indeed, symposia on Mediterranean areas are few; the regional meeting in Amman, Jordan (FAO, 1980) addressed broad issues related to dryland agriculture with scant reference to P from a soil or fertilizer perspective. However, this meeting set in motion a process to cooperatively tackle these issues. In an assessment of dryland Mediterranean agriculture in southern Italy, Paris (1986), while stressing the importance of fertilizers, pointed to the need to consider soil, crop, and management factors in determining fertilization needs and rates. We hope that the synthesis presented here will be a catalyst for improving this important aspect of the region's agriculture.

A. Soil Resources

Because of unique circumstances surrounding their genesis, soils of arid regions broadly differ from those of temperate or tropical regions. Nonetheless, at least half of the world's soil orders are represented in dryland areas: Entisols, Inceptisols, Alfisols, Mollisols, Vertisols, and in the extremely dry areas, Aridisols. Data provided by Kassam (1981) for semiarid cropped soils of the Mediterranean basin are illustrated (Table 1) for the North African belt of Morocco, Algeria, Libya, Tunisia, and Egypt and West Asia. Cropland in North Africa is dominated by Inceptisols (20%), Lithosols or shallow soils (15.5%), Entisols (11%), and Aridisols (16.2%). In West Asia, the landscape is dominated by shallow soils (Lithosols) and Aridisols. Mollisols, with relatively high levels of organic matter in the soil profile, are not widespread, but are important locally in countries such as Morocco. As the FAO classification used by Kassam (1981) does not correspond exactly to the US Soil Taxonomy, it is apparent that wide diversity exists at lower levels of classification.

While broad soil variation can be expected as a result of wide climatic variation, a diversity of soils can occur over a small range. For instance, three experimental sites in northern Syria varied in soils at the order

Table 1. Distribution of major soil groups in North Africa and West Asia with a 75- to 210-day growing period (10^6 ha). Adapted from Kassam (1981) using soil units from FAO/UNESCO soil map

Soil Group	North Africa	West Asia
Cambisols	7.0	10.2
Lithosols	5.4	20.6
Luvisols	3.8	7.7
Fluvisols	2.1	1.9
Regosols	2.4	7.9
Vertisols	1.2	4.1
Xerosols	5.6	22.0
Yermosols	3.1	6.3
Total	34.7	89.8

level, i.e., higher level of classification, over a 190-mm rainfall range (Matar and Brown, 1989a), i.e., Breda (Typic Calciorthid, 281 mm), Tel Hadya (Vertic Rhodoxeralf, 328 mm), and Jindiress (Palexerollic Chromoxerert, 471 mm). Even more extreme variation was noted in Morocco (Shroyer et al., 1990). In the environs of Settat, large areas of Petrocalcic Palexerolls, Typic Chromoxererts, and Vertic Calcixerolls occur.

Though not as widely studied as soils of temperate regions, some reviews (Dregne, 1976) permit broad generalizations about dryland soils. While ranging in texture from sands to clays, most are shallow and have serious inherent or external drawbacks. Limitation on depth, in turn, limits the water-holding capacity of the soil—a major factor, since with infrequent rains most rainfed crops survive on residual soil moisture. Also, shallow soils are particularly vulnerable to soil erosion. While clay soils, i.e., Vertisols, are deep and inherently productive with high water-holding capacity, the range at which they can be tilled is limited. Frequently, they are either too wet or too dry to cultivate. Extensive areas of Vertisols are found from Morocco to Syria. Crop yields from these soils are usually more than double those of shallower soils in the same rainfall zone (Abdel Monem et al., 1990a). In North Africa, 1.2 million ha is under cultivation, while 4.1 million ha is cultivated in West Asia. Vertisols are difficult to hand-cultivated (Kassam, 1981), and are usually worked mechanically.

Dryland soils are usually low in organic matter (OM), which, in turn, limits soil structure and chemical fertility. Arid soils usually contain from 0.1% to 1% OM, while semiarid soils range from 1% to 3% (Hagin and Tucker, 1982). For instance, Jackson et al., (1983) cited long-term cropping studies from the Pacific Northwest of the USA, which showed that, with decreasing rainfall at 500, 350, and 200 mm/year, representative OM values were 3% to 4%, 2.0%, and 1.3%, respectively. Organic matter serves as a nutrient reserve, particularly for nitrogen (N) and, to a lesser extent, P. With cultivation and intensification of agriculture, declines in OM invariably occur. As a consequence, P behavior in dryland soils is dominated by inorganic soil compounds. As most dryland soils are calcareous, solubility relationships dictated by high pH and $CaCO_3$ combine to produce low levels of soluble P in soils. As a result, most dryland soils which have not been fertilized are P-deficient. For normal cropping, use of soil P—indeed any nutrient element—for crop production is subordinate to the overwhelming influence of climate.

B. Mediterranean Climatic Environments

Classification of global climates is based on variations in rainfall and temperature. Considerable deviation from climatic norms exists in any one climatic zone due to varying latitude and altitude. Thus, any thermal climate usually has arid and humid variants, and a whole range of "environments." Mediterranean-type climates, based on rainfall during the colder winter period and dry hot summers, while dominating the WANA belt, do occur in other hemispheres, i.e., west coast of the USA in the northern hemisphere and in Chile, South Africa, and Australia in the southern hemisphere. Typical Mediterranean climates have rainfall ranging from 200 to 600 mm/year and cool (5° to 18° C) or cold (<5° C) winter with hot (>25° C) summers.

Mediterranean climates dominate the vast area from Morocco to Afghanistan, which is under the mandate of the International Center for Agricultural Research in Dry Areas (ICARDA) at Aleppo, Syria, whose primary mission is improvement of the region's rainfed agriculture. This area merges with temperate climates to the north in Turkey, subtropical summer rainfall and tropical Sahelian climates to the south, and to tropical monsoon climates to the east in Pakistan, northern India, and southwest Nepal. Two major climatic zones in the winter rainfall area are recognized; "lowland" areas less than 1200 m in altitude, and "plateau" areas above that. This distinction is based on the observation that at between 1000 and 1500 m altitude, moisture and thermal regimes are visibly different. Such differences are deemed to dictate cropping land use patterns. Based on calculations of Kassam (1981), dryland cereal cropping is feasible on 53% (or 460×10^6 ha) of the land area in North Africa and on 61% (or 416×10^6 ha) of that in West Asia.

Even within any relatively small areas, considerable climatic variation exists. Some examples of such variation are presented as they are relevant to the many ICARDA studies subsequently cited in this review. For instance, in the northern area of Syria, three subzones are recognized. Data from Brown et al. (1987) show long-term rainfall averages of 476 mm at Jindiress and 286 mm at Breda. At both sites, monthly rainfall was highest in January, i.e., 102 mm at Jindiress and 47.2 mm at Breda. The number of frost days at these sites were 33 and 63, respectively. At Jindiress, average temperatures ranged from a low of 6.6° C in January to 27.6° C in July; data for the Breda site were comparable. As cropping and water loss are dependent on the extent of evaporation, which is an outcome of temperature and rainfall, evaporation data are generally recorded. Monthly class A pan evaporation was in the order of from 34 to 37 mm (December, January, February) to 355 to 492 mm in June and July, with little difference between sites. Tel Hayda, ICARDA's main station outside Aleppo, with 328 mm/year is intermediate between the wetter Jindiress and drier Breda.

As several crop-P response studies are also cited from Morocco, it is pertinent to present parallel climatic and soil data. Rainfall generally decreases to the south and the east. The northern area (Meknes, Fez, Tangier) has relatively high rainfall (500–600 mm/year), while at Merchouche, about 100 km south, it declines to 460 mm/year. Further south at Settat in the Chaouia region, rainfall averages 386 mm. Another 150 km south in Abda Province, rainfall is marginal and can be as low as 200 mm/year. Frost is rare in Morocco's rainfed cereal belt, in contrast to the more continental colder conditions in Syria. While agroecological zones are not as clearly delineated as in Syria, climatic differences are considered in field experimentation and crop management.

While few references are made in this review to areas north of the Mediterranean Sea, an overview of southern Italy (Paris, 1986) illustrates similar climatic settings and cropping concerns as in North Africa and West Asia. While rainfall is generally higher than in the latter areas, it is concentrated in the October to March period. Regional and seasonal variability is a consistent feature. With temperatures favorable throughout the cropping season, frost days are essentially absent. The primary constraint to dryland southern Italian agriculture is high terrain, with associated runoff and soil workability problems in addition to colder temperatures which impede growth at higher elevations.

C. Cropping Systems

Most of the cultivated area of winter rainfall in the Mediterranean zone is under rainfed wheat (*Triticum* spp.) (Kassam, 1981). Barley (*Hordeum vulgare* L.) tends to be grown in drier areas and in association with livestock production (Keatinge et al., 1986). Indeed, the comparative advantage of improved barley cultivars over landraces is obscured by low rainfall conditions (Cooper et al., 1987). The relative importance of bread (T. *aestivum* L.) and durum (T. *durum* L.) wheat and barley depends on rainfall, and varies from country to country. Cereals are rotated with weedy fallow or a food legume. Only where the growing period—defined by temperatures above 5° C and precipitation exceeding 0.5 potential evapotranspiration (PET)—is greater than 150 days is wheat normally cropped without fallow (Kassam, 1981). When the growing period is lower, i.e., less and more erratic rainfall, cropping occurs once every 2 or 3 years. While designed to conserve moisture, the benefits of fallowing are not always apparent. A growing season with a humid period, i.e., precipitation exceeds PET, is considered normal for the dryland rainfed Mediterranean area. In extremely dry seasons precipitation never exceeds 0.5 PET.

The cropping period is normally from October–November, at the onset of fall rains, to May–June, when harvesting occurs. As the soil has little or no residual moisture after the dry summer and early fall months, adequate rainfall is crucial at the crop establishment phase; at this stage PET greatly exceeds precipitation. After decreasing in the December–January–February period, vapor pressure deficits increase again. Several studies at ICARDA on crop water-use efficiency (WUE) (Cooper, 1983; Cooper et al., 1987; Gregory et al., 1986) stressed the importance of fertilization on phenological development and transpiration efficiency during the colder period with more favorable PET/precipitation ratios. Fertilizer P not only increased dry matter accumulation, but also hastened anthesis and maturity. Thus, while differences in WUE exist between varieties, the only feasible way it can be increased by most of the region's farmers is by fertilization. Fertilizer P is indispensable to this goal.

Tillage systems vary throughout the region. While mechanization is being used to an increased extent, much primary and secondary cultivation is done by animal traction. In Morocco, the ground is prepared by an offset disc harrow, followed by hand-broadcasting of seed and fertilizer, if used, and subsequent incorporation by a second run of the harrow. While drills are more efficient for seed and fertilizer application, they are frequently not available for most farmers. Despite the potential for soil erosion, the concept of conservation tillage is new—soil loss represents depletion of soil and fertilizer applied P as well.

Despite moisture deficits and limited growing seasons, attempts have been made to manipulate crops to obviate these constraints. The use of barley, which has a shorter growing season and is more drought resistant than wheat in more stressed areas, has evolved naturally. A major objective of the region's breeders is to incorporate early maturity into improved cultivars. The past decade has witnessed the introduction of triticale (*X triticosecale*) to the region (Mergoum et al., 1990; Ryan et al., 1991a). Being relatively disease and drought tolerant, and with high yield potential, it may have a major complementary role to wheat in dryland areas of the Mediterranean zone.

Notwithstanding efforts to improve rainfed cereals in the WANA region, several generalizations are still applicable. Production is generally characterized as subsistence, capital intensity is low, uncosted family labor is high, fertilizer and chemical inputs are low, while local rather than improved cultivars are used. To add to these woes, market organization and extension-research services are weak. While many of these constraints are practically immutable for the forseeable future, the most feasible means of readily impacting on production is by using fertilizers—a technology that can be adapted to even the most traditional forms of agriculture. While the consistent impact of N has been clearly demonstrated from field trials in the region (e.g., Abdel Monem et al., 1990a; Harmsen et al., 1983; Paris, 1986), the other element of major importance, P, is far more complex. Field crop response is dependent on several factors; soil test levels, soil type, mineralization, and rainfall and temperature. Prior to considering these factors, it is important to understand the basic soil factors which govern the chemical behavior of P in the soil.

II. Factors Governing the Behavior of Phosphorus in Mediterranean Soils

Though P occurs in comparatively small amounts in soils, the chemistry of soil P, either inorganic or organic, is exceedingly complex. Its behavior with respect to the plant—reflecting its solubility relationships—is influenced by virtually all physical and chemical components of the dynamic heterogeneous medium called soil. Nevertheless, there are some constituents that exert an undue influence on soil P. Though such components do not exist separately in soils, their unique behaviors with respect to soil P are highlighted separately here.

A. Soil Mineral Constituents

Phosphorus in soils is part of organic and inorganic soil phases—and it reacts extensively with the latter. Of primary concern in this regard are the clay mineral suites, complex crystalline and amorphous iron (Fe) and aluminum (AL) oxides, and calcium carbonate ($CaCO_3$), which dominates the chemistry of calcareous soils.

1. Clay Mineral Suite

The dynamics of P in soil is greatly affected by the mineralogical composition of the clay fraction. It is pertinent, therefore, to examine the distinctive properties of this fraction in the soils of Mediterranean regions (henceforth referred to as "Mediterranean soils"); such soils predominate in the semiarid dryland farming countries of North Africa and the Middle East. Most of the clay fraction is usually made up of clay minerals, which originate by inheritance from the soil parent material, by transformation from structurally similar minerals (micas), or by neoformation from dissolved or amorphous materials (Millot, 1970). In the xeric moisture regime typical of Mediterranean soils, the degree of weathering and the removal of the weathering products are limited. This has the following consequences: (i) Inherited minerals can constitute a substantial portion of soil clay, and (ii) transformation and neoformation processes take place in solutions having medium to high concentration of silicic acid and basic cations, giving rise, essentially, to 2:1 minerals.

Illite is the most abundant clay mineral in Mediterranean soils (Gradusov, 1974), being inherited or resulting from the transformation of muscovite. Smectite is present in many soils and can be dominant in soils having a nonleaching regime, where silica and divalent cation concentrations are high (e.g., Vertisols); in this case, much of that mineral can be neoformed. In many soils derived from Quaternary or Tertiary sediments, much of the smectite can, however, be inherited. Vermiculite is much less abundant than the former minerals, although it may be a common transformation product of biotite weathering (e.g., Nettleton et al., 1970). Chlorite is absent or appears in small amounts, unless inherited from chlorite-rich parent materials. In contrast with soils of more humid, temperate areas, Mediterranean soils do not have significant amounts of hydroxy-Al interlayered vermiculite and smectite, since the necessary active weathering and low pH conditions (Rich, 1968) do not usually exist. Interstratified minerals are less frequent in Mediterranean soils that in soils of humid temperate areas, although biotite-vermiculite (Parra et al., 1983) and illite-smectite (Bech et al., 1982), among others, are occasionally reported.

Although generally appearing in small amounts, kaolinite is almost universal in Mediterranean soils. The ubiquity of this mineral probably stems from the fact that many soil solutions lie within its stability field (Bohn et al., 1979). Neoformation of kaolinite has been reported in soils derived from igneous rocks (Barshad, 1966), and it probably forms in many well-drained soils. Halloysite has been reported in some soils from volcanic deposits and acid igneous rocks (e.g., Torrent and Benayas, 1977), but it occurs far less frequently than kaolinite. Paligorskite, also reported in some Mediterranean soils, is found in some calcareous sedimentary rocks (Gal et al., 1974), but does not seem to be stable in pedogenic environments, except for some calcareous soil horizons (Millot et al., 1969).

A recent study on 114 soils from the Mediterranean part of Spain (Peña, 1990) substantiates these generalizations regarding mineralogical composition of Mediterranean soils. These were mainly Entisols, Inceptisols, Alfisols, and Vertisols developed from widely different parent materials and under different climatic conditions (mean annual temperature, 10° to 18° C; mean annual rainfall, 250 to 750 mm). The percentages (and ranges) of illite, smectite and kaolinite in the carbonate and Fe oxide-free clay fraction were, 55 (6–97), 22 (0–87), and 20 (1–87), respectively; in about 20% of the samples, other minerals, such as illite-smectite, chlorite or vermiculite, were also present. Oxalate acid-ammonium oxalate-extractable Al (pH 3.0), which is an estimate of allophane and imogolite in soils (Wada, 1977), ranged from 0.2 to only 3.7 (mean = 1.0) $g\,kg^{-1}$ soil. This is consistent with the general observation that significant amounts of noncrystalline clay minerals are absent in most Mediterranean soils.

2. Iron and Aluminum oxides

Iron oxides (a term used, for brevity, to designate both oxides and oxyhydroxides of Fe) are usually concentrated in the clay fraction. In Mediterranean soils, they usually constitute from 1% to 10%, depending on the nature of the parent material, degree of weathering, and other factors. The most common Fe oxides are goethite (α-FeOOH) and hematite (α-Fe_2O_3), which may be inherited from the parent material or of pedogenic origin, and magnetite (Fe_3O_4), which is inherited. Occasionally, pedogenic maghemite (γ-Fe_2O_3) has also been reported (Abreu and Robert, 1985). Two other oxides, lepidocrocite (γ-FeOOH) and ferrihydrite (poorly crystalline $HFe_5O_8.H_2O$) are found in some environments, e.g., hydromorphic soils.

The total amount of free Fe oxides is determined in the laboratory by reductive dissolution which citrate-bicarbonate-dithionite (CBD) (Mehra and Jackson, 1960). Treatment with oxalic acid-ammonium oxalate (pH 3.0) extracts the poorly and noncrystalline Fe, i.e., ferrihydrite and organically bound Fe (Schwertmann, 1964). Thus, the CBD-extractable Fe (Fe_d) minus the oxalate-extractable Fe (Fe_o) gives a measure of the crystalline Fe oxides.

In many Mediterranean soils, the ratio Fe_o/Fe_d, i.e., the noncrystalline/crystalline oxide ratio, is usually less than 0.2 (Singer, 1978; Torrent et al., 1980; Ryan et al., 1985a), apparently because, in xeric soils, ferrihydrite formed in the course of weathering transforms rapidly into crystalline Fe oxides (Schwertmann, 1985). The relative proportion of the two common crystalline oxides, goethite and hematite, also depends on pedoenvironmental factors; with increasing soil temperature and dryness, and decreasing organic matter contents, more hematite is formed compared with goethite (Schwertmann, 1985). Pedogenic hematite formation, referred to as rubification because of the red color imparted by hematite, is common in Mediterranean areas. Soils with only goethite are frequent and show brown hues due to the yellow color of this mineral. Little information is available on the properties of the crystalline Fe oxides in Mediterranean soils, largely because they occur in such low concentrations in the clay fraction. Differential X-ray diffraction (DXRD) (Schulze, 1981) can be used to overcome this difficulty (e.g., Boero and Schwertmann, 1989). To lower detection limits, this technique can be combined with previous concentration of Fe oxides in the clay fraction by dissolution of the clay minerals with $5M$ KOH, density gradient separation, or high gradient magnetic separation (Schulze, 1988). Even with these refined techniques, identification is difficult at low concentration, i.e., 1% to 2%, in the clay fraction.

From DXRD patterns, Peña and Torrent (1984) calculated that, in 26 Alfisols from a river terrace sequence in Spain, goethite and hematite had similar crystal size, being of the order of up to 100 nm. Calculated specific surface areas ranged from 65 to $104\,m^2\,g^{-1}$ for goethite and from 44 to $113\,m^2\,g^{-1}$ for hematite. In 48 Terra Rossa soils from different Mediterranean region sites, Boero and Schwertmann (1989) found that the mean calculated goethite and hematite sizes were also similar (about 30 nm) and differed little from one soil to another. The calculated specific surface areas were of the order of $100\,m^2\,g^{-1}$. Consequently, Fe oxides appear to have specific surfaces similar in magnitude to the external specific surfaces of the clay minerals common in Mediterranean soils. Gibbsite (α-$Al(OH)_3$) is the only crystalline Al oxide reported in Mediterranean soils. It appears almost exclusively in well-drained soils developed on granitic saprolites (Macías Vázquez, 1981), where it is present in both the silt and clay fractions.

3. Carbonates

Calcite, dolomite, and Mg-calcite are the most common soil carbonates. Dolomite is of lithogenic origin, being much less frequent than calcite, which can be partly or entirely pedogenic. Calcite is common and often abundant in Mediterranean soils, and can be present in any size fraction. When it is pedogenic, it occurs in different forms, such as nodules, calcans, neocalcans, and crystallaria (Drees and Wilding, 1987). The fine-grained particles (micrite, $<4\,\mu m$) can constitute a substantial part of the carbonate present in soil.

The determination of total $CaCO_3$ equivalent (CCE) is routinely performed by acidifying the soil and measuring the CO_2 evolved. The CCE is, however, a poor indicator of $CaCO_3$ reactivity. For that reason, "active" $CaCO_3$ (ACCE), i.e., the fine-sized $CaCO_3$ which reacts with ammonium oxalate (Drouineau, 1942), or the clay-sized $CaCO_3$ is determined. In 36 soils from different locations in Spain, Borrero et al. (1988) found a good correlation between ACCE and clay-sized $CaCO_3$ (clay-sized $CaCO_3 = -0.5 + 0.72$ ACCE; $r = 0.98$; $P < 0.001$). The $CaCO_3$ content of the clay fraction may, in practice, be estimated from the CCE values (Fig. 2), since good correlation exists between soil CCE % and $CaCO_3$ % in the clay fraction (Gal et al., 1974).

Active or clay-sized $CaCO_3$ may sometimes be a poor indicator of $CaCO_3$ reactivity because the specific surface area of this component can be widely different for particles differing in size within the clay fraction. Determination of the specific surface area therefore appears desirable.

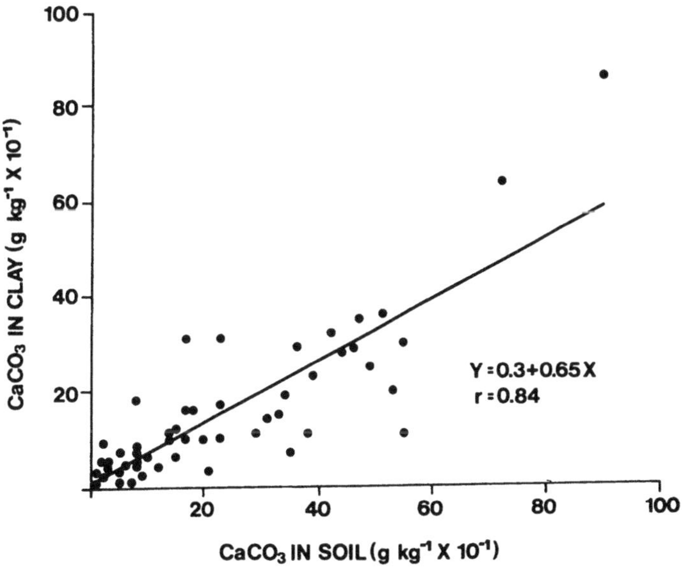

Figure 2. Relationship between the $CaCO_3$ content in clay and the $CaCO_3$ content in soil. Data taken from Gal et al. (1974) and Borrero et al. (1988).

This determination involves the measurement of surface area before and after dissolving the $CaCO_3$ with an acid buffer, the difference in surface being attributed to $CaCO_3$ (Talibudeen and Arambarri, 1964). Nevertheless, treatment with an acid buffer can modify the surfaces of the soil sample or expose new surfaces and therefore need to be interpreted with caution.

Little information is available on the specific surface areas of soil carbonates in Mediterranean areas. Borrero et al. (1988) found widely different values, but, on a soil basis, the $CaCO_3$ surface areas were lower than $12\,m^2\,g^{-1}$ soil. In calcareous soils of Azerbaijan, Abedi and Talibudeen (1974) found that the specific surface area of $CaCO_3$ ranged between 3 and $373\,m^2\,g^{-1}$ and was inversely correlated with total $CaCO_3$ in the soil; total surface area of $CaCO_3$ ranged from 0.5 to $4\,m^2\,g^{-1}$ soil. The inverse relationship between specific surface area of $CaCO_3$ and total $CaCO_3$ was also observed by Holford and Mattingly (1975a). Although these authors studied soils in which $CaCO_3$ resulted from the dissolution of lithogenic carbonate under a humid climate, total surface area of carbonate varied little, i.e., 4 to $8\,m^2\,g^{-1}$ soil. More information on the reactive surface of $CaCO_3$ in soils will accumulate if simple, rapid, routine procedures to measure it are successfully developed. Precipitation of calcium oxalate on $CaCO_3$ surfaces (Kaushansky et al., 1984) or steady-state acid dissolution (Moore et al., 1990) are some of the procedures which may prove successful in the future.

B. Forms of Soil Phosphorus

The total amount of P in soils of Mediterranean regions ranges widely, depending, among other factors, on parent material and soil genesis. Values as high as 2500 μg P g^{-1} (Hasan, 1980) or as low as 100 μg P g^{-1} (Eleizalde, 1977) have been reported. The most important primary P-bearing mineral is apatite, usually fluorapatite ($Ca_5(PO_4)_3F$), which appears as crystals of fine sand and silt size in many rocks and soils. Apatite can weather with relative ease, even in the xeric regime of Mediterranean soils (Torrent, 1976), P released during weathering being reprecipitated in diverse secondary phosphates, incorporated into the organic matter, and adsorbed by, or occluded in, soil minerals.

Organic P constitutes only a small portion of total P in soils of Mediterranean and other semiarid areas. This is due to the generally low content (0.5% to 1.0%) in organic matter of these soils, especially when they have been cultivated for a long time. Organic P/inorganic P ratios usually range from 0.05 to 0.15 (Ryan and Zghard, 1980; Meixner and Singer, 1985; Solís and Torrent, 1989a). Most of the secondary inorganic forms of P occur in the fine silt and clay fractions of the soil; their positive identification by usual mineralogical and chemical methods is rather difficult. Indirect methods to identify and quantify P forms include the study of solubility relationship and "fractionation" of different P forms using various extractants. The fractionation scheme of Change and Jackson (1957) considered Al-, Fe-, and Ca-bound forms and reductant-soluble (in Fe oxides) P. This original scheme has been modified in calcareous soils by several workers. For instance, Syers et al. (1972) suggested that the Fe- and Al-bound P should be removed together. Due to these different fractionation methods the results obtained by different authors are not strictly comparable, but they nevertheless provide an insight into the P forms and their evolution in Mediterranean soils.

With increasing soil age, Ca-bound P in the form of apatite decreases, whereas organic P, nonoccluded Fe- and Al-bound P, secondary Ca-bound P, and P occluded in the Fe oxides increase. In later stages of soil development, nonoccluded P forms decrease while occluded forms increase (Walker and Syers, 1976). Even in relatively young soils or similar parent material changes in P inorganic fractions are evident with time, as shown by Ryan and Zghard (1980) for a chronosequence of Pakistani soils. These general trends are followed by soils of Mediterranean areas, although when soils are formed on previously weathered materials, no definite clear-cut age-P form relationships are found (Meixner and Singer, 1985). The relative amounts of the different forms depends not only on soil age, but on parent material, pedoclimate, fertilization history, and other factors. In general, a direct relationship is found between the Ca-P/(Fe + Al)-P ratio and pH (e.g., Hernando et al., 1968). Many Mediterranean soils are calcareous or have pH values around neutrality. This dictates that Fe- and Al-P forms are usually less abundant than secondary Ca-forms (Sayegh and Majid, 1969, Eleizalde, 1976; Eleizalde, 1983; Ryan and Zghard, 1980; Luque and Arambarri, 1983; Solís and Torrent, 1989a). While providing a reasonable basis for comparing soils. P fractionation schemes do not give adequate information about the exact nature of the P compounds extracted in each fractionation step. For this reason, their usefulness appears, at present, to be limited.

III. Phosphate Reactions in Soils

Orthophosphate ions strongly react with various soil components, essentially by processes of adsorption and formation of solid phosphates. The term P is used here when no particular form of P is implied. The nature and practical implications of all the processes leading to what is commonly called P "fixation" or "retention" have been reviewed by several authors (Larsen, 1967; Parfitt, 1978; Beek and van Riemsdijk, 1979; Sample et al., 1980; White, 1981; Barrow, 1987). Phosphate retention is currently considered as a complex summation of physical adsorption, chemisorption, precipitation, and hypothetical penetration into soil particles. In general, at P solution concentrations up to the millimolar range, adsorption reactions predominate. In contrast, precipitation can be dominant at higher concentrations such as, for instance, those produced by the addition of a fertilizer granule to the soil. In the following sections the focus will be on aspects of P reactions specifically applicable to Mediterranean soils.

A. Adsorption by Soil Components

The adsorption of P has been usually characterized by adsorption curves, also called adsorption isotherms, and, more properly, sorption curves or quantity/intensity (Q/I) plots. Sorption will be the term preferred here when no evidence for lack of precipitation exists. Sorption curves are constructed by plotting, for different soil suspensions in which P has been added, the amount of P sorbed against the P concentration in the equilibrium solution. These curves have been described by different "isotherms" (e.g., Langmuir or Freundlich) which, rather than an insight into the sorption mechanisms, provide a convenient way to describe the process (Barrow, 1978). The shape of the sorption curves depends on many factors, such as soil: solution-ratio, supporting electrolyte, pH, and time of equilibration, making comparison between the results of different authors difficult. These constraints are taken into account in the following evaluation of the effect of soil components on P sorption by Mediterranean soils.

Only a few studies report the use of isotherms in short-term P sorption. In Lebanon, Ryan et al. (1985a) found that the Langmuir isotherm was appropriate for describing P sorption, while in Iraq, Berigari et al. (1985) found that their data conformed to linear plots of the Freundlich, Langmuir, and Temkin equations. Employing parameters from the Langmuir equation, Pavel and Toma (1972) concluded that most Iraqi soils had a high P adsorption maximum but also a low bonding energy. Limited work with one soil from Jordan (Khader and Abu Sharar, 1979) showed that P adsorption at both 20° and 40° C conformed to the Langmuir equation; the partial molar free energy decreased with increasing adsorption but increased with temperature. In the following sections we will consider, first, rapid sorption, usually evaluated at 1 to 6 days, focusing on the soil constituents involved in the process.

1. Iron Oxides

Phosphate is specifically adsorbed on surfaces of variable charge such as those of Fe and Al oxides where H^+ and OH^- determine the surface charge and potential, and those of calcite, where Ca^{2+} and $CO_3{}^{2-}$ are the potential-determining ions. The Fe and Al hydroxilated surfaces, where orthophosphate is adsorbed by a ligand exchange mechanism (Parfitt, 1978), are the most abundant P-reactive surfaces of soils. The Fe-OH surface groups are almost exclusively on Fe oxides, whereas the Al-OH groups appear on both Al oxides and clay minerals.

Although the important role of Fe oxides in P sorption is well documented in other soils, only in the past 15 years has evidence accumulated to show that these minerals have a dominant influence on the P sorption by Mediterranean soils (Singer, 1978; Eleizalde, 1978; Peña and Torrent, 1984; Ryan et al., 1985a; Torrent, 1987; Borrero et al., 1988; Solís and Torrent, 1989b; Peña, 1990). This evidence comes from regression analyses showing the correlation between Fe oxides and P sorbed either at a fixed equilibrium concentration (e.g., $1\,\mu g\,P\,mL^{-1}$) or from a fixed amount of added P, or the maximum P sorption capacity from the fitted Langmuir isotherm. Good correlations are often found for soils of common origin, such as those from a river terrace sequence in southern Spain (Peña and Torrent, 1984), as illustrated in Fig. 3. The fact that the Fe oxides are mainly crystalline explains why the correlation between P sorption and Fe_o is often poor. In a study of 114 widely different soils from the Mediterranean part of Spain (Peña, 1990), P sorption was well correlated with Fe_d ($r = 0.71$; $P < 0.001$), but less with Fe_o ($r = 0.39$; $P < 0.001$). However, in other soils, where the amount of noncrystalline Fe oxides is not negligible, the correlation between P sorption and Fe_o is higher (Ryan et al., 1985a).

It is obvious that the relative contribution of crystalline and noncrystalline Fe oxides to P sorption must depend not only on their relative

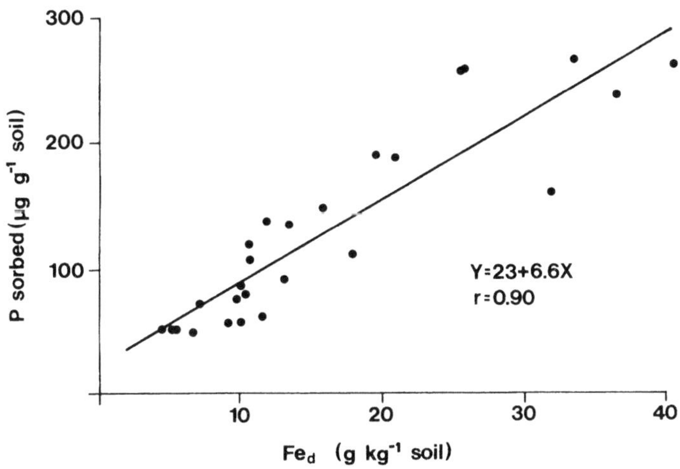

Figure 3. Phosphate sorbed at an equilibrium P concentration of $0.3\,\mu g\,P\,mL^{-1}$ against Fe_d for a group of Alfisols of a river terrace sequence in southern Spain (from Peña and Torrent, 1984).

amounts but on their relative specific surface and sorption capacities. Surface area of noncrystalline Fe oxides is of the order of $400\,m^2\,g^{-1}$ (Borggaard, 1983a), whereas crystalline Fe oxides of Mediterranean soils may have values of about $100\,m^2\,g^{-1}$. Consequently, noncrystalline Fe oxides might have much more weight in P sorption than that simply deduced from the Fe_o/Fe_d ratio, thus explaining some of the correlations referred to earlier.

Studies on synthetic samples have shown that sorption capacities of several Fe oxides (as given, for instance, by the maximum of the Langmuir equation fitted to low concentration values) are of the order of $2.5\,\mu mol\,P\,m^{-2}$ (Borggaard, 1983b). For actual soils, sorption of a similar order of magnitude was demonstrated (Peña and Torrent, 1984). These authors compared P sorption capacity of these soils with the Fe oxides' surface area (calculated from XRD data), and found an average value of $2.1\,\mu mol\,P\,m^{-2}$. The possibility exists, however, that different Fe oxides have widely different P sorption capacities that vary according to their different specific surfaces and their crystalline characteristics since not all crystal faces are able to adsorb similar P amounts. Experiments have shown that synthetic hematite (Barrón et al., 1988) can adsorb only as little as $0.2\,\mu mol\,P\,m^{-2}$ when it has a platy habit with predominant (001), (101), and (104) faces; in contrast, synthetic goethites, prepared in different ways and with predominant (110) faces, always adsorb around $2.5\,\mu mol\,P\,m^{-2}$ (Torrent et al., 1990). The extent to which hematitic Mediterranean soils can be less P-sorptive than the goethitic ones is not known. However, Peña and Torrent (1984) found no significant difference between the two groups. In contrast, V. Barron and J. Torrent (unpublished data) found that, per unit weight of Fe oxides, the yellow (goethitic) mottles of an Acrisol from southern Spain sorbed more P than the contiguous red (hematitic) mottles. More work is needed to clarify these discrepancies.

Regression analyses provide only indirect evidence of the quantitative importance of Fe oxides in P sorption. More direct evidence could be obtained if Fe oxides could be effectively isolated from remaining soil components. As indicated above, this is not feasible at present. Alternatively, the sorbing capacity of Fe oxides could be calculated by measuring the difference in sorption capacity before and after their removal by CBD. Using this procedure, Ryan et al. (1985a), Torrent (1987), and Peña (1990) found inconsistent results, i.e., sorption could increase, decrease, or remain similar after CBD, without any relationship with the nature and amount of the Fe oxides. These results are probably due to the effect of CBD on soil minerals, either by changing the nature of the non-Fe oxide surfaces or by exposing surfaces formerly covered by the Fe oxides.

2. Crystalline Clay Minerals

Crystalline clay minerals are able to specifically adsorb P through a ligand exchange reaction with the (OH)H groups coordinated with the Al ion on the edge of the crystal (Muljadi et al., 1966; Pissarides et al., 1968; Beek and van Riemsdijk, 1979; White, 1981). Thus, the adsorption capacity depends, among other factors, on the proportion of the surface area occupied by edge faces and the number of reactive Al-OH(H) groups per unit area of edge face. According to this, kaolinite usually adsorbs more P per unit area than 2:1 clay minerals.

Data from different studies on P adsorption capacity of clay minerals are not strictly comparable, since they have been obtained under different experimental conditions (pH, exchangeable cations, ionic strength, etc.) and often without measuring specific surface area. Early data were also invalidated by the use of high (millimolar to molar) P concentrations which caused crystal disruption. Muljadi et al. (1966) reported highly specific adsorption values of about 1.5 μmol P m^{-2} edge face for kaolinite at an equilibrium P concentration of 100 μM at pH 7, equivalent to 5 μmol P g^{-1}. Chen et al. (1973) found a value of 4 μmol P g^{-1} for K-kaolinite at pH 7. Bar-Yosef et al. (1988) reported values of about 5 and 3 μmol P g^{-1} for Ca-saturated kaolinite and montmorillonite, respectively. Santa Cruz et al. (1982) studied CBD-treated soil clays for several Spanish soils; at an equilibrium concentration of 100 μM they found values of about 0.5, 4.5, and 5 μmol P g^{-1} for K-kaolinite, illite, and montmorillonite, respectively.

The former data provide a basis for comparing sorption by clays with sorption by Fe oxides in Mediterranean soils if one considers the relative amounts in which these two types of minerals may occur. Peña (1990) found that the average Fe_d/clay ratio in a group of 114 Mediterranean soils was 0.06. If the Fe oxides are estimated by the $Fe_d \times 1.5$ product, have $100\,m^2\,g^{-1}$, sorb $2\,\mu mol\,P\,m^2$ at $100\,mM$ of P, and the clay sorbs an average of $3\,\mu mol$ P g^{-1}, the relative sorption can be calculated as follows:

$$\frac{\textit{Fe oxide P sorption}}{\textit{Clay P sorption}} = \frac{6\,g\,Fe_d}{100\,g\ clay} \times \frac{6 \times 1.5\,g\ Fe\ oxides}{6\,g\,Fe_d} \times \frac{100\,m^2\ Fe\ oxides}{1\,g\ Fe\ oxides} \times \frac{2\,\mu mol\,P}{m^2\ Fe\ oxides} \times \frac{1\,g\ clay}{3\,\mu mol\,P} = 6 \quad (1)$$

This value, for the reasons explained above, must be taken very cautiously, but it conveys, nevertheless, the idea that the Fe oxides dominate sorption, at least in noncalcareous Mediterranean soils. Good correlations between P sorption and clay content of Mediterranean soils have been found by several authors (Turan et al., 1976; Eleizalde, 1978; Kishk and Lashin, 1978; Peña and Torrent, 1984; Ryan et al., 1985a; Oskay, 1986; Nakos, 1987). From these results, it is difficult to infer the contribution of clay minerals to sorption, inasmuch as Fe_d and clay show a high covariance for many groups of soils. In some cases Fe_d was not determined and, consequently, the correlation between Fe_d and clay is not known.

3. Soil Carbonates

Frequent and contradictory reports on the effect of $CaCO_3$ on P sorption have appeared in the literature. Part of the contrast between the results of different authors stems from the fact that experimental conditions varied widely, especially equilibrium P concentration. Phosphate concentrations above about $100\,\mu M$ can lead to precipitation of Ca-phosphates (Griffin and Jurinak, 1973; Freeman and Rowell, 1981; Hundal, 1988); we will consider, therefore, data obtained near or below this limit.

At the calcite surface, P ions are specifically adsorbed by replacing water molecules, bicarbonate ions, or hydroxyl ions that fill the vacant coordinate positions of exposed Ca^{2+} ions. In synthetic calcites, Griffin and Jurinak (1973) found that, below and equilibrium P concentration of $20\,\mu M$, about 5% of the total calcite surface was covered with specifically adsorbed P ions; above this limit cluster formation took place prior to crystal growth. The 5% coverage corresponds to an adsorption values of $0.3\,\mu mol\,P\,m^{-2}$.

In soils, calcite is mixed with other minerals and no direct measurement can be made on the P adsorption by that mineral. On removal of $CaCO_3$

from a group of 36 Mediterranean Spanish soils, Borrero et al. (1988) found an average P sorption capacity of 0.13 μmol P m^{-2} (equilibrium P concentration = 20 μ*M*), a value not very different from that of synthetic calcites, and about one order of magnitude less than that of Fe oxides. The contribution of $CaCO_3$ to P sorption in soils must be closely correlated with its total surface area, since, as the above results suggest, P sorption per unit area has a relatively well-known value. Although not much information is available on the specific surface of $CaCO_3$ in Mediterranean soils, we can take, as a simple calculation exercise, the average values of 3 m^2 $CaCO_3$ g^{-1} soil given by Borrerc et al. (1988). In the soils studied by these authors the average Fe oxide content was 14 mg g^{-1} soil. If the values of Eq. 1 are taken, the comparison between the P sorbed by Fe oxides and $CaCO_3$ gives:

$$\frac{\text{Fe oxide-sorbed P}}{\text{CaCO}_3\text{-sorbed P}} = \frac{14\,\text{mg Fe oxides}}{4\,\text{m}^2\,\text{CaCO}_3} \times \frac{0.1\,\text{m}^2\,\text{Fe oxides}}{1\,\text{mg Fe oxides}} \times \frac{2\,\mu\text{mol P}}{1\,\text{m}^2\,\text{Fe oxides}} \times \frac{1\,\text{m}^2\,\text{CaCO}_3}{0.2\,\mu\text{mol P}} = 3.5 \qquad (2)$$

The above value, although merely indicative, suggests that $CaCO_3$ may play only a secondary role in P sorption provided that the Fe oxide contents are not very low. Regression analyses on different groups of soils support this idea (Ryan et al., 1985a; Borrero et al., 1988; Peña, 1990). In other studies on soils of Mediterranean or semiarid regions, P sorption has been found to be either unrelated (e.g., Turan et al., 1976; Santa Cruz et al., 1981) or related to $CaCO_3$ content (e.g., Kacar, 1967; Kishk and Lashin, 1978; Lajtha and Bloomer, 1988). Many of these results are difficult to reconcile and compare, again because of different experimental conditions and because Fe oxides and clay were not always determined. In some Jordanian work, Taimeh and Hattar (1988) observed $CaCO_3$ phases such as exchangeable Ca, $CaCO_3$ in the clay fraction, and "active" $CaCO_3$ to be equally well correlated with P sorption in some calcareous Vertisols. What is clear from the above studies, however, is the influence of $CaCO_3$ in soils where Fe_d and clay contents are very low (e.g., Eleizalde, 1978; Aggarwal et al., 1987).

Comparisons between P-sorbing properties of noncalcareous and calcareous soils also elucidate the role played by $CaCO_3$. With 114 Spanish soils, Peña (1990) found average values for Fe_d and clay to be 20 and 260 mg g^{-1}, respectively, for the noncalcareous group (n = 71), and 13 and 274 mg g^{-1}, respectively, for the calcareous group. The average $CaCO_3$ content of the latter was 139 mg g^{-1}. The calcareous soils sorbed, on average, less than the noncalcareous soils (73 vs 97 μg P g^{-1}, at an equilibrium P concentration of 7 μ*M*). This difference points to the limited role played by $CaCO_3$ and the dominant role played by the Fe oxides, which occur in higher amounts in the noncalcareous group. It is not discounted, however, that in highly calcareous soils with low Fe oxide content, P sorption may be controlled by $CaCO_3$. A final point to consider is the apparently higher P adsorption energy of the Fe oxides with respect to $CaCO_3$. Some support for this notion comes from, among others, the work of Holford and Mattingly (1975b), who, with a group of calcareous soils, showed a correlation between Fe oxides and high-energy bonding sites, and between $CaCO_3$ and low-energy bonding sites.

4. Other Soil Components

Several other components which affect P sorption in soils of other regions are of little importance in Mediterranean areas. This is the case with gibbsite and amorphous aluminosilicates, which are usually absent or occur only in small amounts. The presence of gypsum has been reported to have little effect on P sorption (Pavel and Toma, 1972), probably because it is usually present as relatively large crystals.

Organic matter does not adsorb P unless it is associated with cations such as Fe^{3+}, Al^{3+} and Ca^{2+} (White, 1981). In Mediterranean soils, which have pH values usually above 5.5, little Al and Fe can be expected to be complexed by organic matter, but the Ca associated with organic components may contribute to P sorption. However, organic compounds may block the adsorption sites, either directly or through their effect on soil structure formation, thus decreasing sorption. The net outcome of these competing processes can vary. In general, P sorption is negatively correlated, or not correlated at all, with organic matter for many groups of Mediterranean soils. Positive correlations have, nevertheless, been reported (Nychas and Kosmas, 1984; Nakos, 1987; Aggarwal et al., 1987).

The limited work on other soil organic components from the region has been inconclusive (Karim et al., 1989). These authors showed that citric acid reduced P sorption due to the formation of more stable citrate-metal chelates compared with the corresponding metal-P precipitates. Both oxalic and tartaric acids, however, enhanced P adsorption. The effects of these acids were dependent on the P concentration. The practical significance of organic acids in relation to P is still unclear.

B. Sorption and Soil-Solution Characteristics

Phosphate adsorption is controlled by numerous soil-solution factors, including P concentration, pH, ionic strength, type of cation, and competing anions. Models that describe adsorption and take these factors into account have been developed in the past 20 years, mostly on oxide surfaces (Barrow, 1987), and also in soils (e.g., Goldberg and Sposito, 1984) and clays (e.g., Bar-Yosef et al., 1988), thus helping to interpret data. In practice, however, the effect of several chemical properties on P sorption by Mediterranean soils may not be evident from regression analyses when one considers a group of soils having different mineralogical properties. For instance, within the usual pH range found in Mediterranean soils (pH 5.5 to 8.5), P sorption is known to decrease with increasing pH (e.g., Goldberg and Sposito, 1984), but a relationship between the two variables is rarely found for a diverse group of soils. In some cases, pH appears as a secondary explicative variable in stepwise regression analyses (Peña, 1990).

Phosphate adsorption is also markedly affected by the nature and concentration of background electrolyte. For nonsaline soils with Ca^{2+} as the dominant saturating cation, which represent the vast majority of Mediterranean soils, little difference in P sorption can be expected for that reason. According to some models, saline soils show higher sorption values than the equivalent nonsaline soils (e.g., Eleizalde and Fernández, 1982). In contrast, sodic soils sorb less P than similar Ca-saturated soils (Russell et al., 1988).

1. Estimating P Sorption

At low P concentration and short periods of soil-solution contact, P sorption can be estimated in Mediterranean soils by considering the mineralogical components which seem to be relevant for this process: Fe oxides, clay minerals, and $CaCO_3$. Multiple regression equations, usually obtained using a stepwise procedure, are used. Table 2 gives examples for

Table 2. Regression equations relating P sorption and soil properties in Mediterranean soils

Soil type	Sample No.	Supporting electrolyte	Equil. P conc. for sorption	Equation	r^2
			$\mu g\ mL^{-1}$		
Alfisols, mainly noncalcareous, in a chronosequence[a]	26	0.1 *M* KCl	0.3	$Y = 6 + 3.5\ Fe_d + 0.22$ clay	0.90
Calcareous soils of different origin[b]	36	0.1 *M* KCl	0.62	$Y = 22 + 2.8\ Fe_d + 0.13$ clay $+ 0.37$ active $CaCO_3$	0.63
Noncalcareous soils of different origin[c]	71	0.01 *M* $CaCl_2$	0.2	$Y = 10 + 2.1\ Fe_d + 0.18$ clay	0.71
Calcareous soils of different origin[c]	43	0.01 *M* $CaCl_2$	0.2	$Y = 48 + 1.9\ Fe_d$	0.53
Calcareous Vertisols and Inceptisols[d]	24	0.01 *M* $CaCl_2$	1.0	$Y = 23 + 8.5\ Fe_d + 0.17$ clay $+ 0.22$ active $CaCO_3$	0.71

Y = sorption in $\mu g\ P\ g^{-1}$ soil; Fe_d, clay and $CaCO_3$ are expressed in $mg\ g^{-1}$ soil.

[a] Peña and Torrent (1984); [b] Borrero et al. (1988); [c] Peña (1990); [d] Solis and Torrent (1989b).

several soil groups. Although not necessarily implying a cause-effect relationship, the larger regression coefficients of Fe_d vs clay and $CaCO_3$ are compatible with the suggested dominant role of Fe oxides. Equations of high predictive value are obtained when the soil group is relatively homogeneous (e.g., soils having similar parent material and developed in the same climatic area).

2. Slow (Long-Term) Sorption

Phosphate sorption is rapid at first (Ryan et al., 1985a). After a few days, it becomes slower, continuing usually for a long time (Ryan et al., 1985b; Hooker et al., 1980). At low P concentrations (10 to $100\,\mu M$), where precipitation processes do not take place or are extremely slow, the continuous loss of P in solution has been attributed to microbial uptake and immobilization, diffusive penetration into the bulk of soil particles (Barrow, 1987), change of a mono- to a binuclear surface complex (Kafkafi et al., 1967), diffusion to adsorption sites in interparticle pores, of diffusion into surface defects and pores (Parfitt, 1989).

Slow P sorption due to the presence of surface pores has been reported for synthetic Fe oxides (Madrid and Arambarri, 1985). This might hold for natural Fe oxides. Indeed, in Mediterranean soils, Fe oxides seem to be involved in long-term P sorption. Ryan et al. (1985b) found, for six Lebanese soils, that the decrease in P extractability with time was related to Fe_d. Torrent (1987) found that the slow sorption for a group of 20 clays of Mediterranean soils could be described by a modified Elovich equation ($C = Co - qt$), the slope of which, q, was correlated with Fe_d. For 24 calcareous Vertisols and Inceptisols of Spain, Solís and Torrent (1989b) found that both active $CaCO_3$ and Fe_d were the best variables to explain slow sorption. In his study of 114 soils, Peña (1990) also found that the q value of the Elovich equation was weakly, but significantly, correlated with Fe_d. More studies are needed, however, to establish possible contribution of different components and factors to the slow P sorption.

At concentrations in the micromolar range, such as those usually found in soils which have not been recently fertilized, slow sorption appears, in most cases, to be of little quantitative importance. Peña (1990) found that slow sorption, i.e., between 6 and 126 days after P addition averaged only about 5% of the rapid, initial 6-day sorption. In this study, soil suspensions were periodically shaken; under natural soil conditions, P diffusion is probably much slower and the slow/rapid sorption ratio accordingly wider.

C. Precipitation Products and Phosphate Potentials

Precipitation of many different phosphates is active at P concentrations in the millimolar to the molar range, as those occurring around a fertilizer granule. Talibudeen (1981) pointed out four steps in the precipitation process, when dissolved P reacts with soil: (i) formation of a surface-adsorbed P complex; (ii) dissolution of soil minerals which increases the P-reactive metal ion concentration in solution; (iii) slow desorption of surface-adsorbed P compounds, and (iv) slow nucleation, crystallization, and recrystallization of P compounds. Reviews on the reactions and products resulting from the addition of P fertilizers to soils have been made by Olsen and Flowerday (1971), Lindsay and Vlek (1977), and Sample et al. (1980).

The nature of precipitated P compounds is influenced by characteristics of the reacting P solution as well as of the soil. Concentrated P solutions may cause degradation of clay mineral structures and subsequent precipitation of Al phosphates. Particles of Fe and Al oxides may become coated by Fe- or Al-P films when exposed to those solutions (e.g., Moreno et al., 1960). Recent experiments by Martin et al. (1988) have shown that precipitation of a Fe-P can take place on the surface of goethite at P concentrations as low as 1 m*M* and pH 4.5. Local soil conditions leading to the precipitation of Al and Fe phosphates (high P concentration, low pH, presence of some ions, etc.) usually change with time; in the long run, the stability of these minerals will mostly depend on "average" soil properties.

For Mediterranean soils, in which pH is usually above 6, even the most stable Fe and Al phosphates, i.e., strengite ($FePO_4.H_2O$) and variscite ($AlPO_4.H_2O$), are expected to be unstable with respect to Ca phosphates (Lindsay, 1979), but they can probably remain metastable for a long time. For example, Roca and Pomares (1987) found that more than half of the P applied as superphosphate, i.e., monocalcium phosphate (MCP), for 7 years in a calcareous soil from Spain was recovered in the Al-P and Fe-P fractions extracted by Chang and Jackson (1958) procedure. In calcareous soils from the Midwestern USA which were fertilized with superphosphate for 5 years, Hooker et al. (1980) found that the Fe-Al-P fraction increased to reach a steady state, but this state was cyclic, indicating a possible shift to less soluble forms, i.e., Ca-P and reductant-soluble P. However, in Spain, only small amounts of Fe-Al-P compounds were formed by incubation of powdered MCP with calcareous Vertisols and Inceptisols (Solis and Torrent, 1989a). More work is evidently needed to study the solubility and persistence of Fe and Al phosphates in Mediterranean soils, including both calcareous and noncalcareous groups.

Precipitation of Ca phosphates and reaction of soluble P with $CaCO_3$ and calcareous soils has attracted much interest. At low concentrations, P ions are specifically adosrbed onto the calcite surface, as previously discussed. Griffin and Jurinak (1973) described adsorption on synthetic calcite by a two-surface Langmuir isotherm, in which the second linear portion starts at a concentration of about 20 μM corresponding, according to these authors, to the division between OCP or octacalcium phosphate ($Ca_8H_2(PO_4)_6.5H_2O$) and HA or hydroxyapatite ($Ca_5(OH)(PO_4)_3$). In this region, heterogeneous nucleation processes, leading ultimately to the precipitation of Ca phosphates, would take place. The nuclei are presumably amorphous Ca phosphate, which appears to be transformed, in many cases, to dicalcium phosphate dihydrate ($CaHPO_4.2H_2O$ (DCDP)) (Cole et al., 1953; Freeman and Rowell, 1981). This DCDP can gradually change to less soluble, more basic forms, according to the solubility sequence: OCP > β-TCP (tricalcium phosphate, $Ca_3(PO_4)_2$) > HA. Alternatively, HA or OCP nuclei can form onto the $CaCO_3$ surface depending, probably, on the Ca/P ratio of the system. These observations were generally confirmed by researchers in the Mediterranean area.

The precipitation of Ca phosphates induced by $CaCO_3$ is influenced by serveral factors. High specific surface areas of $CaCO_3$ give more rapid formation of Ca phosphates (Stumm and Leckie, 1971) while magnesium has been shown to prevent adosorption of phosphate onto calcite and to make difficult or retard nucleation (Yadav et al., 1984). The presence of small amounts of pyrophosphate inhibits or retards nucleation and/or crystal growth of DCDP in either pure $CaCO_3$ or calcareous soils (El-Zahaby and Chien, 1982; Amer et al., 1985). Phosphate concentration, especially at low supersaturation, affects the kinetics of precipitation and the nature of the precipitate (Al-Khateeb et al., 1986; Shaviv et al., 1989) since it affects the Ca:P activity ratio. Hundal (1988) contended that P sorption in an alkaline soil followed a four-region Langmuir isotherm due to the successive formation of HA, β-TCP, OCP, and DCDP. In many soils, it is to be expected, however, that precipitation of Ca phosphates takes place according to very complex patterns and in competition with adsorption phenomena. The effect of $CaCO_3$ on precipitation of Ca phosphates is less marked at high P concentrations (high supersaturation), such as those occurring near a fertilizer granule. The formation of Ca phosphates under such conditions has been extensively dealt with in several reviews, as previously indicated.

Except for those experimental settings in which high P concentrations are reacted with small soil volumes, the products resulting from the P-soil reactions, because of their small size and low concentrations, cannot be identified by conventional X-ray diffraction, optical microscopy, or infrared spectroscopy. For this reason, P-solubility characteristics have been frequently used to gain information, albeit indirect, on those reaction compounds. The P-solubility relationships for slightly acid to calcareous soils are best expressed by the monocalcium phosphate potential (0.5 pCa + pH_2PO_4) (Aslyng, 1954), plotted frequently against the lime potential (pH − 0.5 pCa). The technique has been used to identify the Ca phosphates controlling the P concentration in the soil solution by comparing the soil solution characteristics with the solubility data coined for different minerals (Lindsay, 1979).

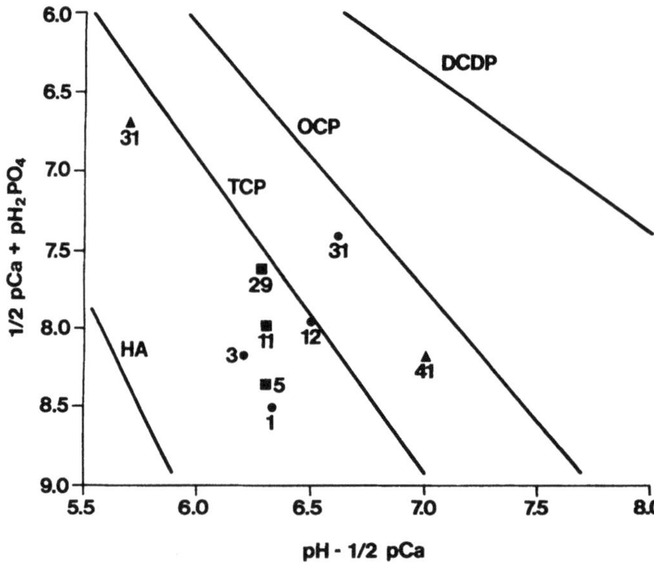

Figure 4. Monocalcium phosphate and lime potentials for some selected calcareous soils. Data from Fixen et al. (1983) (△), Adepoju et al. (1986) (◇), and Solís (1988) (□). Figures by the symbols are Olsen P-values ($\mu g \, P \, g^{-1}$).

Phosphate potentials are useful to monitor time changes in the soil P compounds (Fixen et al., 1983; Havlin and Westfall, 1984; Adepoju et al., 1986). They cannot indicate, however, if phases less soluble than the one controlling the potential are present. Neither do they indicate the amount of the compound that supplies the phosphate to the solution. In addition, solubility data are available only for well-characterized, crystalline compounds; phosphates that form in soils are likely to be poorly crystalline or to have foreign ions.

Figure 4 shows solubility data for different calcareous soils. Several authors (e.g., Fixen et al., 1983; Havlin and Westfall, 1984) have found that in highly fertilized soils OCP seems to control solution P levels. For soils testing medium to low in available P, there seems to be considerable buffering on or near the TCP (whitlockite) isotherm (Fixen et al., 1983), although this mineral has never been identified in soils. It therefore appears that some definite Ca-P phases are likely to influence P in solution in many calcareous soils.

D. Kinetics and Reversibility of Phosphate-Soil Reactions

The reactions discussed above take place at very different rates. Adsorption, *sensu stricto*, is an endothermic reaction that is essentially completed in minutes. Precipitation is usually much slower, its rate depending on the presence of nuclei, temperature, degree of supersaturation, and presence of foreign compounds. For all these reasons, and other causes not pertaining strictly to the nature of the P-soil reaction (mixing and diffusion, immobilization in organic compounds, penetration into crystal pores or defects, etc.), there is always a slow sorption component. The mechanisms and practical implications of the slow sorption have been discussed by Barrow (1980). The same author (Barrow, 1987) also showed some of the possible ways to model slow sorption by oxides.

The fact that adsorption processes are rapid and dominate at low concentrations, and that precipitation is slower and more active at higher concentrations, helps to explain the apparent contradictions found in the literature when the effect of soil components on P sorption is evaluated by means of Q/I (sorption) curves. For instance, it has been previously indicated that, at low ($<100\,\mu M$) concentrations, sorption in Mediterranean soils appears to be affected essentially by Fe oxides. In contrast, for high concentrations and/or long times, $CaCO_3$ can contribute substantially to, or control sorption (Kacar, 1967; Kishk and Lashin, 1978; Oskay, 1984; Lajtha and Bloomer, 1988; Solís and Torrent, 1989b).

The Q/I curves provide information about the quantity of P required to raise the solution P concentration to a "critical" level that is considered adequate for plant growth. The concentration of P in the soil solution (i.e., the "intensity" factor, I) determines the instantaneous rate of P absorption by plants but total P uptake depends essentially on the reserve (i.e., the "quantity" factor, Q) capable of replenishing P loss in the soil solution. The Q factor *significant for plants*, as well as its variation with I (dQ/dI, or "buffer capacity") could be obtained directly from the Q/I curves constructed in the laboratory, provided these curves were reversible. This is certainly not the case, since desorption curves obtained with different methods in the laboratory differ, often markedly, from sorption curves. However, Barrow (1983) stated that "sorption is reversible in the sense that a continuous curve of sorbed and desorbed phosphate is obtained when these are measured in opposite directions by increasing and decreasing the solution P concentration. However, because dynamic processes are involved, an earlier portion of a plot of sorbed P against concentration is not desorbed when the concentration is changed." What it is clear is that even though individual sorption processes may be reversible, the rate of some of them may be very slow. It is not surprising, therefore, that desorption measured over a short period in the laboratory is incomplete. In addition, some true irreversible processes may take place.

Different methods have been used to study P desorption. This can be carried out, for instance, by measuring the rate or release of phosphate to an anion-exchange resin (Amer et al., 1955; Elrashidi et al., 1975). This method provides the possibility of comparing soils and establishing the existence of a true irreversible process. Resins provide, at the same time, a measure of the "labile" P (Larsen, 1967), i.e., the P that readily exchanges with solution P. Resins extract similar amounts to those obtained by the isotopic exchange reference method for labile P (Bache and Ireland, 1980). Laboratory desorption methods cannot obviously simulate the plethora of processes that take place when P is adsorbed by roots under field conditions. In contrast, there is increasing evidence that the slope of the desorption curves under plant uptake conditions is similar to the slope of the sorption curves determined using the usual laboratory methods (Bowman and Olsen, 1985; Holford, 1988; Solís and Torrent, 1989b). These results suggest that the sorption curves may be useful tools for the practical management of the P fertility of soils.

The apparent reversibility of the sorption curves under field conditions deserves careful attention. Future research should be carried out on how the nature of the solid-phase P influences the kinetics of the P release to plants. Soils having similar sorption curves but differing in the components responsible for the sorption processes might have different desorption curves. For Mediterranean soils, in particular, the effect of Fe oxides and $CaCO_3$ should be carefully studied and compared.

IV. Assessing Plant-Available Soil Phosphorus

Phosphorus availability to plants is determined both by factors which affect the soil's ability to supply nutrients and by factors which affect the plant's ability to utilize the supplied nutrients. Knowledge of the various reactions which P undergoes in the soil, and the many soil and environmental factors that affect the plants' ability to absorb P which is present at the root surface, is of utmost importance (Hagin and Tucker, 1982). In the present review, the relevant studies on P availability carried out in the Mediterranean will be summarized and cited.

Table 3. Methods for determining available P currently used in some Mediterranean countries

Method	Extractants	Countries
Olsen	0.5 *M* $NaHCO_3$ pH 8.5	Turkey, Morocco, Tunisia, Syria, Cyprus, Jordan, Spain, Italy, Greece
Bray I	0.025 N HCl + 0.03 *N* NH_4 F	Turkey, Spain, Greece
Bray II	0.10 *N* HCl + 0.03 *N* NH_4 F	Turkey, Iran
Bingham	H_2O	Turkey, Morocco
Dyer	2% citric acid pH 2	Morocco, France
Truog	0.002 *N* H_2SO_4 buffered to pH 3 with $(NH_4)_2SO_4$	Turkey
Morgan	0.5 *N* HOAC + 0.7 *N* NaOAC pH 4.8	Morocco, France

A. Chemical Extractants

Several methods have been used in the WANA region for determining available P in soils; most of these were developed in the USA. The educational background of soil scientists in charge of the region's various soil laboratories was a determining factor in the initial introduction of a particular soil test for determining available P (Table 3). Procedures that had varying degrees of success in dryland calcareous soils were generally adopted. In some countries, several tests are used depending on the institution involved; frequently different tests are used by commercial laboratories as opposed to educational and research institutions. While the Olsen procedure, which involves extraction with 0.5 M $NaHCO_3$ at pH 8.5, has widespread currency in the region, anomalous situations occur where inappropriate or ineffective test methods still exist. As soil testing as a management tool for fertilizer use recommendation is in its infancy in most Middle East countries, the concept of an "official" soil test procedure does not generally exist. A glance at the papers dealing with soil P analyses and crop response reported at a recent regional meeting (Ryan and Matar, 1990) clearly underlines the status of the Olsen test in Tunisia, Yemen, Iraq, and Lebanon in addition to those areas listed in Table 3.

Studies by several authors from the region (Agbani and El Mejahed, 1983; Matar et al., 1988b; Bousselham, L. 1986; Hassan, et al., 1974; Tnani and Kannenberg, 1971; Ryan and Ayubi, 1981) compared, under controlled conditions in pots, various soil P tests with P uptake by ryegrass (*Lolium multiforum* L.), millet (*Panicum ramosum*), or other crops. Ryegrass, through either single or successive cuttings, was generally considered a good indicator of soil P supply, and is used as a reference with which results obtained by other methods could be compared. Representative soil samples were selected from several countries of the region with soil pH ranging between 6.2 and 8.2 and 0% to 40% "free" $CaCO_3$. Determination of available P in these soils by the isotopic exchange (L-value) (Gachon, 1966), desorption by anion resin (Sibbesen, 1978), and Olsen methods (Olsen et al., 1954; Hassan et al., 1974) were best correlated with P uptake by crops during both short- and long-term growth and, as such, were generally satisfactory in estimating available P over a wide range of soil characteristics of the region (Agbani and El Mejahed, 1983; Matar et al., 1988b; Ryan and Ayubi, 1981). Notwithstanding these relationships, a number of practical considerations dictate the ultimate fate of a particular test in terms of widespread acceptance.

Although P^{32} equilibration with soil P (L-value) was highly correlated with crop P availability, this procedure requires a high degree of laboratory sophistication, which militates against its adoption in the WANA region. The P extracted by acid solutions (Joret and Herbert, 1955; Truog, 1930), organic P determination, total P, or P-fixing capacity were much less satisfactory for assessing P availability in soils of the Mediterranean region. Though Kacar (1969) found a good correlation between some inorganic P fractions, e.g., aluminum phosphate and crop P uptake, the use of such fractions is inappropriate for a routine simple soil test procedure. Sharar et al. (1976), determined critical levels for soil P by dividing soils into responsive and nonresponsive groups using four P soil test extractants: lactate, oxalate, as well as $NaHCO_3$, and anion exchange procedures.

In a study on the P status of soils from countries of the Mediterranean region, Amar and Ait Houssa (1990) found that the Olsen and resin methods were the best soil-P tests for soils in general, especially calcareous soils, which correlated best with P uptake by ryegrass. However, they were not reliable for sesquioxide soils. They concluded that the best strategy is to select the most reliable soil P test for each major soil group, i.e., calcareous vs noncalcareous, which should be then calibrated with crop response under field conditions. In another comparison of these procedures, Ryan and Ayubi (1981) found the resin method to be as effective as that of Olsen; however, the former procedure was considered tedious and time-consuming and unlikely for widespread adoption.

Though no soil chemical extractant can reliably estimate P availability in all situations, the quest to improve the Olsen procedure or replace it has, nevertheless, continued. In Pakistan, Rashid et al. (1990) observed good correlations with the $NaHCO_3$-P, NH_4HCO_3-DTPA (Soltanpour, 1985), and Mehlich 3 (Mehlich, 1984) tests; all three were equally well correlated with P concentrations or uptake by wheat, lentil (*Lens culinaris*), chickpea (*Cicer arietinum*), and mustard (*Brassica* spp). With the last two soil tests, one can assess nutrient availability for about ten elements using a single extraction. These have a clear advantage over the $NaHCO_3$ test, especially when micronutrients are limiting to plant growth in the same soils. Multiextractants are more cost-efficient, especially when used with auto-analyzers. In Morocco, the Azzaoui et al. (1989) study of 14 soil types from the Chaouia region showed the Olsen procedure and its DTPA modification were equally satisfactory predictors of available P in calcareous soils.

B. Plant-Tissue Analysis

Plant analysis of nutrient elements is useful in identifying nutrient deficiency or toxicity (Walsh and Beaton, 1973). However, the concentration of P or any other element depends on growth stage, specific plant part, species or cultivar, and other factors. In a list of critical P concentrations in several crops, Hagin and Tucker (1982) indicated a value of 0.30% for wheat—the main crop in the WANA region—when the upper four leaves are sampled at the boot stage. Tissue analysis is an approach, complementary to soil analysis, in crop diagnosis and fertilizer recommendation, albeit a *post mortem* one in the case of P. Where a plant-tissue analysis indicates P deficiency, little can be done about it during the current cropping year because of the difficulty of incorporating P fertilizer into an established crop stand.

Only a few scattered reports from the WANA region deal with tissue analysis of P. While Orphanos and Krentos (1980) found that P fertilizer increased grain and straw yields of durum wheat in Cyprus, the P content of either component was not affected. Later, Rashid et al. (1990) suggested a critical plant P concentration of 0.16% in diagnostic wheat leaves; however, seed P concentration showed no relationship with grain yield. Despite these shortcomings, Ryan et al. (1992) used tissue analyses to assess the nutrient status of over 2000 cereal fields in Morocco; while most N values were below the critical concentration, P values were from 0.22% to 0.69%. These survey data generally corroborated soil analysis and crop response data from the same region.

V. Field-Crop Responses to Phosphorus Fertilizer

The many field trials conducted throughout the WANA zone provide a number of generalizations about crop fertilization. In all except extreme drought situations and where there was sufficient residual N from a previous legume (Abdel Monem et al., 1990a), N is almost invariably deficient as exemplified by recent reports from countries ranging from Morocco (Ryan et al., 1992) to Pakistan (Rice et al., 1990). Though less frequent than N, deficiencies of P are widespread in all countries of the region, with both frequently occurring simultaneously (Michel et al., 1967; Krentos and Orphanos, 1979; Loizides, 1970). Exceptions to this generalization can be largely traced to residual P build-up, which is discussed later. Fortunately, other nutrient deficiencies such as of K are rare for most countries of the region and semiarid areas in general (Dregne, 1976). While deficiencies of micronutrients, i.e., iron and zinc, do occur in crops grown in calcareous soils, they are more frequent in fruit trees and in corn (*Zea mays* L.) and legumes; they are less common with dryland cereals. Specific details are provided from several countries of the region. In view of the dominance of rainfed cereals in the Mediterranean region, most of the research on P has, logically, focused on wheat and barley, and other cereals to a lesser extent. However, with respect to P, noncereal crops, such as legumes, are also of importance.

A. Rainfed Cereals

Results from P trials in Cyprus (Annual Reports, 1963–1978), Syria (FAO, 1970), Jordan (Moore, 1974; NCARTT, 1975–1984), Yemen (Saad, 1990), Morocco (Michel and Bousoubaa, 1980, Ryan et al.,

1991c), and other countries have indicated responses of local wheat and barley cultivars to P fertilization. Increases in the order of 0.5 to 1 ton of grain were generally obtained from application of from 13 to 18 kg P/ha. Most, if not all, soils of the region are calcareous, and naturally low in available P (Harmsen, 1983; Krentos and Orphanos, 1979; Matar, 1976a; Matar et al., 1988a, 1988c; NCARTT. 1975–1984; Abu Rub and Ashab, 1987). But the effect of P fertilizers is less consistent than that of N fertilizers.

Initially, responses to P fertilization were recorded in almost all experiments, but regular dressings of P fertilizers quickly raised available soil P to a level that is adequate to meet crop P requirements. Although the response to P was widespread in the region when fertilizers were first introduced to dryland farming agriculture, further studies indicated little or no P fertilizer response (Moore, 1974; NCARTT, 1975–1984; Loizides, 1980; Matar, 1976; Abdel Monem et al., 1990a). In Syria at present, most of the experiments where there was no P response were conducted in areas where wheat is grown in rotation with irrigated crops such as cotton (*Gossypium hirsutum* L.), vegetables, and other summer crops which are adequately fertilized.

In the past 5 years (1985 to 1990) ICARDA, in collaboration with the Syrian national program involved with soil fertility studies (Soils Directorate), have conducted an extensive research program on assessment of biological and economic effects of N and P fertilizer on barley (Syrian Min. Agric./ICARDA, 1985–88) in the barley-livestock systems across northern Syria. Through multiple-season, multiple-location trials on farmers' fields, different soil and environmental conditions and rotations dominant in barley-growing areas were represented. Data from 75 of these N-P trials showed mean increases of 0.54 ton grain/ha and 0.80 ton straw/ha from 40 kg P/ha over the nonfertilized control. Such responses in the first year of fertilization were not surprising since most barley-growing soils are poor in available P (<4 ppm Olsen P). In most countries of the region, where little fertilizer has been allocated to barley, these results suggest that P fertilization is economical in barley production, even in areas receiving less than 200 mm annual rainfall.

Wheat is the most important staple food crop grown in West Asia and North Africa, where it is the major source of calories and protein in national diets (FAO, 1984). In the ICARDA region of influence, national programs have also focused research on the higher rainfall wheat-based farming systems. Potential crop production is greater in these areas where farmers are financially more secure, showed a high degree of awarenes of fertilizers, and have been using N and P fertilizers for some time. The wheat fertilization program (1986–1989) adopted by ICARDA was similar to that of barley, with due concern for soil type, soil P test levels, rainfall, and previous crop or rotation, with the objective of formulating a comprehensive fertilizer program. The results of 40 N-P on-farm experiments conducted in the first 2 years were published (ICARDA, Soils Directorate, 1989). Responses in terms of grain and straw yields to P fertilization were positive and significant at only ten sites. This limited response to P could be attributed to soil fertility build-up with time due to consistent P fertilization by farmers. Work in Tunisia coincided with the Syrian findings (Gharbi et al., 1990); a fertility survey of 230 sites showed 4% of them to be less than the critical value of 5.2 ppm. In crop trials at nine sites, all seven sites with $NaHCO_3$-P less than 6.0 ppm responded to applied P.

B. Legumes and Forage Crops

In West Asia and North Africa, rainfed food and forage legumes are almost always grown in rotation with cereals. The typically low level of native soil P is accentuated by a higher crop P requirement than for cereals. It is not surprising, therefore, that economic responses to applied P have been observed for lentil, vetch (*Vicia* spp), and faba bean (*Vicia faba* L.) in Syria (FAO, 1970; Matar, 1976a), Libya (Badawy, 1976), Pakistan (Sharar et al., 1976), and other parts of the region.

It was shown that, with the exception of pea (*Pisum sativa* L.), compared with lentil, chickpea, vetch, and faba bean, the average responses to P were much greater and more consistent in a dry year than in a wet one (Table 4). This coincided with results obtained in Jordan (Haddad, 1986a; 1986b) on lentil and chickpea, and was similar to those obtained on wheat and barley (Harmsen et al., 1983; Matar et al., 1987; Orphanos, 1987; Krentos and Orphanos, 1979; Matar, 1977). The response of the pea crop to P application was more consistent than for other legume crops studied. A response of chickpea to P fertilization occurred in only two of the 18 on-farm trials, i.e., fields very low in available P (<3 ppm). This lack of chickpea response to P was noted by other scientists (Saxena, 1984; Itoh, 1987). Several explanations were advanced (Seklani, 1983):

Table 4. Biomass increase in P-fertilized legumes over two seasons at Tel Hadya, Aleppo ($kg\,ha^{-1}$)

1986/87 Season	Pea	Chickpea	Faba bean	Lentil	Vetch	Mean
Residual P	1072	969	683	983	738	889
Applied P	1292	352	553	570	1459	845
Mean P	1182	660	618	776	1098	
1987/88 Season						
Residual P	1604	16	381	290	42	466
Applied P	1244	156	0	656	56	422
Mean P	1442	86	190	473	49	

Rainfall for 1986/87 = 358 mm/year, 1987/88 = 504 mm/year.

the large seed reserve of P in chickpea, thus boosting early growth and development, and the low rhizosphere pH around chickpea roots, which could lead to solubilization of naturally insoluble Ca phosphates, thus making them more available for plant uptake. The role of development and mycorrhizae in P nutrition of chickpea needs to be investigated.

Millions of hectares of marginal land occurs within the WANA region. With increased livestock numbers and consequently grazing intensity, much of this area is susceptible to erosion. Fertilization of natural or introduced forages can help reduce this trend. For instance, in Tunisia, Seklani (1983) applied P fertilizer using shallow tillage to a mixture of alfalfa (*Medicago sativa* L.) and ryegrass with the seed or separately and produced a substantial increase in total dry matter yield, which was dominated by the legume (Fig. 5). Even under the least favorable conditions (i.e., rainfall 220 mm), a low rate of 20 kg P/ha was sufficient to get a

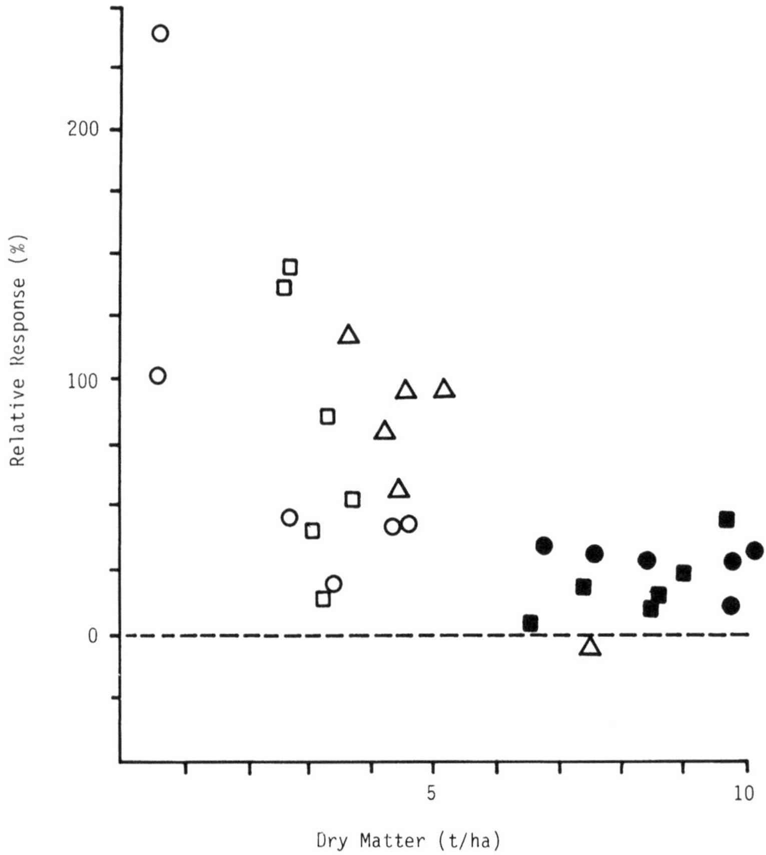

Figure 5. Impact of applied P, alone and with seed, on pasture yield at varying rainfall locations in Tunisia (from Seklani, 1983).

yield increase of 0.5 ton of dry matter. More recent work at ICARDA (Osman et al., 1990) showed that surface P application to natural pastures or marginal grazing areas without tillage increases plant populations and yield of native legume species; in fact, an application of 25 kg P/ha increased legume seed yield from 21 to 99 kg/ha.

Work on forage fertilization is equally encouraging. Greenhouse studies with a range of grasses showed a high degree of responsiveness (Arif et al., 1989). Trials are currently under way in the field with the more adapted and high-yielding species in the dryland zone. A parrallel program has focused on medics (*Medicago* spp.) as a dryland farming system to replace fallow and increase livestock-carrying capacity. In a P-deficient soil, Derkaoui et al. (1990) increased biomass yield by up to threefold; however, one cultivar did not nodulate due, presumably, to the absence of appropriate *Rhizobia*. In the companion greenhouse trial, all eight cultivars showed a wide range of P responses. Differences in yield potential were evident.

VI. Factors Associated with Crop Phosphorus Responses

While crop response data from the field trials within the region provide the basis for P fertilizer recommendations, these recommendations must be tuned or adjusted to consider the many field conditions that affect fertilizer use; some may reduce or eliminate the need for P, while others may accentuate the fertilizer requirement. Some brief comment on these factors is pertinent.

A. Water-Use Efficiency

Under the semiarid Meditterranean climate, water is obviously the most limiting factor that affects plant growth, and one that will continue to dominate the future of the region's agriculture. It is crucial that whatever water is available be used as efficiently as possible. The concept of water-use efficiency (WUE), discussed in detail by Cooper et al. (1988), is a useful guide to improving crop management. It expresses production in terms of yeild per hectare per millimeter of available water. It is defined as:

$$\mathrm{WUE} = \frac{\mathrm{TE}}{(1 + \mathrm{Es/T})}\ \mathrm{Kg/ha/mm} \qquad (3)$$

where TE represents the average seasonal crop transpiration efficiency ($\mathrm{kg\,ha^{-1}\,mm^{-1}}$) and Es the water lost as evaporation from the cropped soil (mm).

It is commonly observed that balanced fertilizer applications give substantial increases in dry matter production and grain yield from essentially

Table 5. Fertilization in relation to barley yield, evapotranspiration, and water-use efficiency (WUE) at Breda, Syria. Adapted from Cooper et al. (1987)

Variables			Measurements		
Season	Rainfall (mm)	Fertilization	Dry Matter (ton ha^{-1})	Evapotranspiration (mm)	WUE (kg ha^{-1} mm^{-1})
1980/81	299	None	3.48	234	16.4
		N, P	7.54	225	33.5
1981/82	324	None	4.54	231	19.7
		N, P	6.13	231	26.5
1982/83	284	None	2.01	224	9.0
		N, P	3.40	235	14.5
1983/84	207	None	1.37	171	7.8
		N, P	2.09	174	12.0

the same quantity of available water. While the WUE increases presented in Table 5 cannot be attributed to P alone, they effectively illustrate how water is used much more efficiently when nutrient deficiencies are corrected. Increases in WUE ranged from 100% in 1980/81 to 35% the following year. However, there was no apparent relationship with total seasonal rainfall.

The interaction between fertilizer and water use is largely due to more rapid growth and canopy development during early growth and the winter period. Solar energy is intercepted by the crop, and less reaches the soil surface. As a result, soil evaporation (Es) is reduced and crop transpiration (T) increases. Fertilizer reduces the Es/T rates and thus leads to greater WUE. When P is applied to deficient soils, it accelerates crop development and can hasten maturity by up to 2 weeks (Shepherd et al., 1987). As fertilizer P also enhances root growth (Cooper et al., 1987), it increases a crop's ability to extract moisture, if available, during the spring and early summer months, when it depends entirely on what is stored in the soil.

The interaction between fertilizer use and weed control is very significant in WANA, and its impact on water-use efficiency is serious, expecially for food legumes. These practices which control weeds, i.e., mechanical, manual, chemical, or use of suitable rotations, are desirable. The impact of P fertilization with or without weed control is illustrated by field trial data with lentils (Table 6); grain yield was greatly enhanced with P fertilization, expecially when weeds were controlled.

B. Rainfall and Soil Moisture

Several scientists in the WANA region had observed relatively larger and more significant yield responses to P application in dry and unproductive years, and/or in shallow soils, while yield responses to N tend to be

Table 6. Phosphate fertilization of lentils; yield and water-use efficiency. Adapted from ICARDA's *Annual Report*, 1983

	Yield ($kg\,ha^{-1}$)		Water use efficiency ($kg\,ha^{-1}\,mm^{-1}$)	
Treatment	Grain	Dry matter	Grain	Dry matter
Weeds				
+P	342	1128	1.3	4.2
−P	275	667	1.1	2.6
No weeds				
+P	1039	2038	4.4	8.6
−P	529	751	2.4	3.4

larger under high rainfall conditions (Harmsen et al., 1983; Krentos and Orphanos, 1979). At the wetter sites, there seemed to be little or no P response, even when the available soil P was in the low range. This could have important consequences, because it would imply that there would be a considerable potential for increasing yields, even in low-rainfall areas, where P fertilizers are not applied at present.

Several theories have been advanced to explain the positive effects of P in improving crop growth in low-rainfall conditions. With the improved stimulation of root growth by P fertilization, the effective availability of nutrients is increased, even under conditions of low fertility, due to increasing the soil volume accessible for water and nutrients by the rooting system. Since the crop takes up most of its P from solution, absorption of P by roots is greatly reduced under conditions of low soil water availability or low-rainfall conditions (Olsen et al., 1961). The thin film of moisture around the soil particles and the small number of contact points between the soluble P and the root surfaces leads to a longer path uptake of P. As a consequence, P application increases the concentration of soluble P in the moisture film, thus compensating for the small number of contact points between soil particles and root surfaces (Matar, 1977).

In a more detailed field study at ICARDA, the interaction between rainfall and available soil P (Olsen test) in relation to barley responses to P application was assessed with 50 on-farm N-P factorial trials (Syr. Min Agric./ICARDA. 1985–1988; Cooper et al., 1988). It was found that when rainfall was greater than a certain amount, the increase in yield due to 27 kg of P/ha levels off and subsequently declines. These differences become more apparent if the same data are expressed as percentage increases in yield over the control (Figs. 6, 7). This positive P response in dry areas has been demonstrated on lentil, wheat, and barley in several countries of the region (Harmsen et al., 1983; Soltanpour et al., 1988; Krentos and Orphanos, 1979; Matar, 1977) and deserves to be considered seriously for increasing food and forage crops in the region.

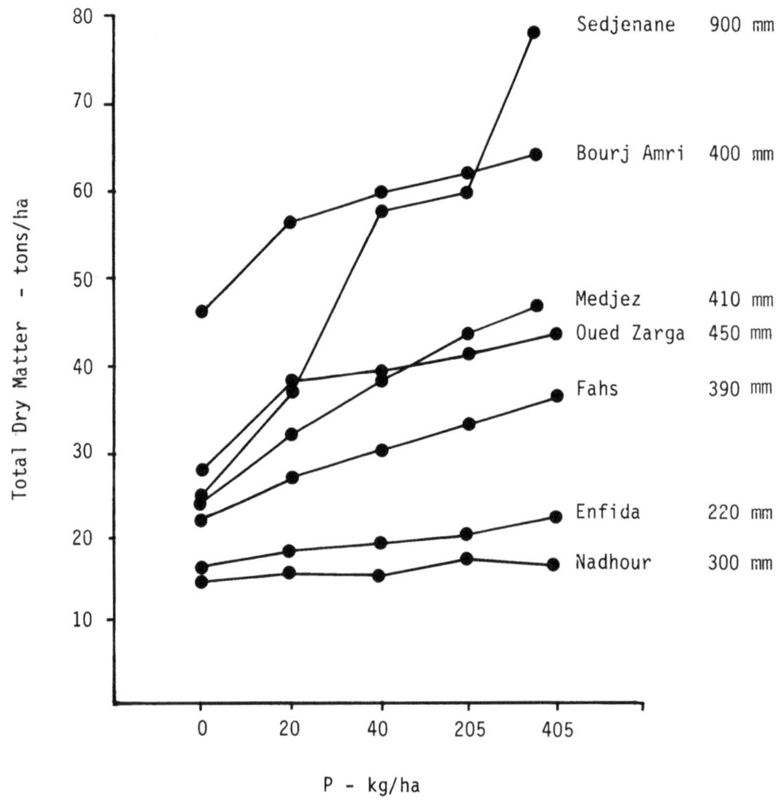

Figure 6. Relative increase in P-fertilized barley at five research stations in Syria (Harmsen et al., 1983). Stations: ○, Khanasser, ●, Tel Hadya; △, Jindiress; □, Breda; ■, Kafr Antoun.

C. Available Soil Phosphorus Levels

The process of determining the crop soil test relationship is commonly referred to as calibration of chemical soil tests. A calibrated soil test value for a particular nutrient will indicate the degree of deficiency of that nutrient and the amount of the fertilizer nutrient to be applied. The work on soil test calibration for rainfed crops started in the mid-1960s in the WANA region. Turkey (Yurtsever, 1965, 1987, 1988; Digdigoglu, 1980) was the first country of the region to calibrate available soil P with wheat and barley responses to P fertilization; the Bray-Mitscherlich equation (Bray, 1958) was generally used. More work started later in Syria (Matar, 1977; Mater et al., 1987), Cyprus (Krentos and Orphanos, 1979), Morocco (Agbani and El Mejahed, 1983; Soltanpour et al., 1989), and Tunisia (Gharbi et al., 1990), and several other countries in the region.

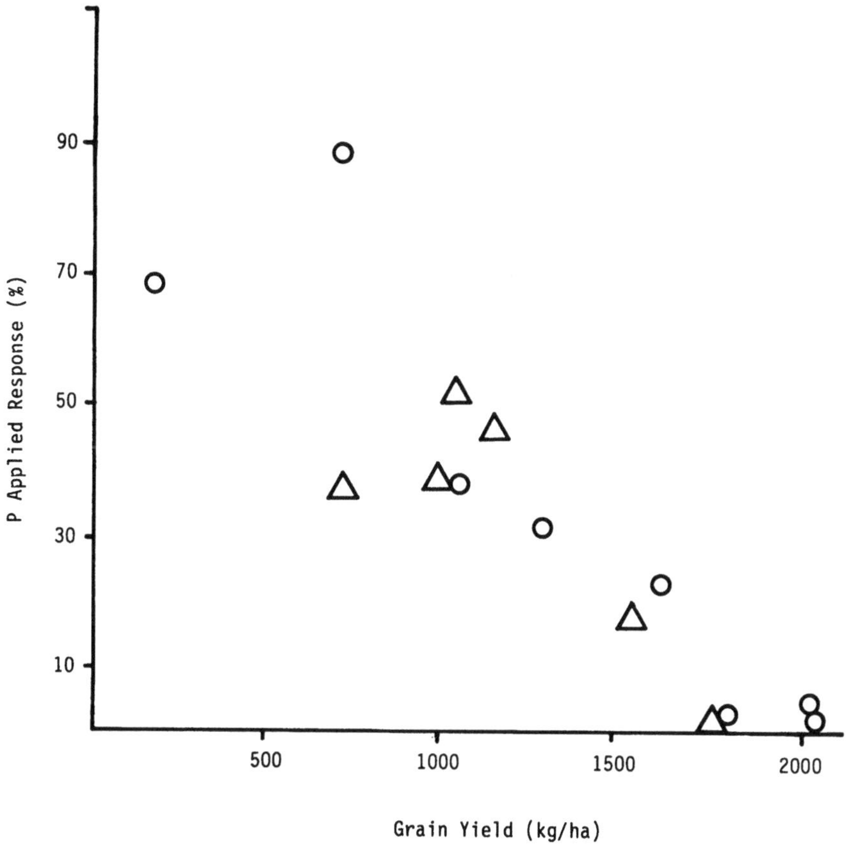

Figure 7. Relationship between average wheat (O) and lentil (Δ) yields with relative P response (from Matar, 1977).

At present, the $NaHCO_3$-extractable P (Olsen-P) test is extensively used in the region as a reliable and most practical test for available P determination in calcareous soils. Not surprisingly, crop response data from the WANA region showed that yield increases from P fertilization were inversely related to the extent of soil P deficiency (Fig. 8). However, in the soil test calibration program, tests such as the Bray I have been included in calibration of field experiments (Yurtsever, 1988). Using the Cate-Nelson graphical method, critical levels of Olsen-P in soils that optimize wheat and barley yields ranged between 5 and 10 ppm, depending upon climate and soil conditions. Higher critical Olsen-P levels occurred under low-rainfall conditions (Harmsen et al., 1983; Krentos and Orphanos, 1979; Matar 1976b). Consequently, the P availability-soil moisture interactions should be considered. Little or no responses of wheat to P application were observed in wet years, even if the available P

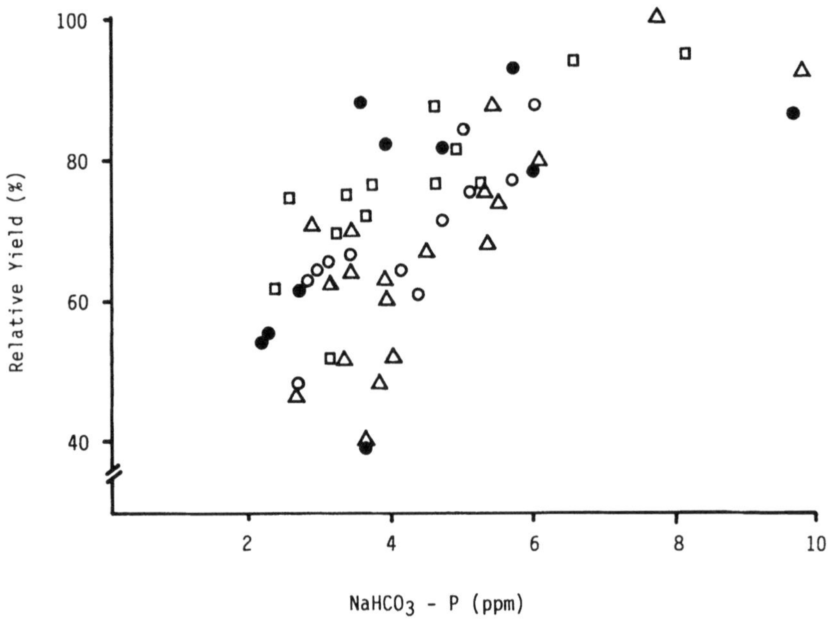

Figure 8. Relationship between available soil P and relative on-farm barley (Arabic Aswad) dry matter yields over four seasons in northern Syria: ●, 1985; ○, 1986; △, 1987; □, 1988.

was around 4 to 5 ppm. However, in dry years (less than 300 mm annual rainfall), the critical level could exceed 8 to 9 ppm (Soltanpour et al., 1988; Matar 1976b; Krentos and Orphanos, 1979). Furthermore, the critical level was well related to soil properties. Higher critical levels were observed with shallow Rendolls (Soltanpour et al., 1988), Gypsiorthids (A. Matar, unpublished data), and sandy soils (Novais and Kamprath, 1978) than with deep Vertisols, Calciorthids, or Xerochrepts. Critical values are higher also under low temperatures. In the low winter temperatures of the Anatolian Plateau in Turkey (Yurtsever, 1987) wheat responses to P application were observed with more than 8 ppm available Olsen-P.

Early work on the relationships of food and forage legumes to soil P availability had found that a level of 4 ppm P was adequate to maximize yields of local lentil (Matar, 1976b), but a linear relationship was observed in dry years between available P (2 to 8 ppm) and lentil grain yield, which reached 0.6 ton/ha. A more detailed study at ICARDA had shown that a level of 8 ppm is adequate to secure near-maximum yields of lentil, chickpea, faba bean, vetch, and pea (Matar et al., 1988a), as illustrated in Fig. 9. Based on results of numerous soil-test calibration trials conducted in the past 4 years in the WANA region, and on the

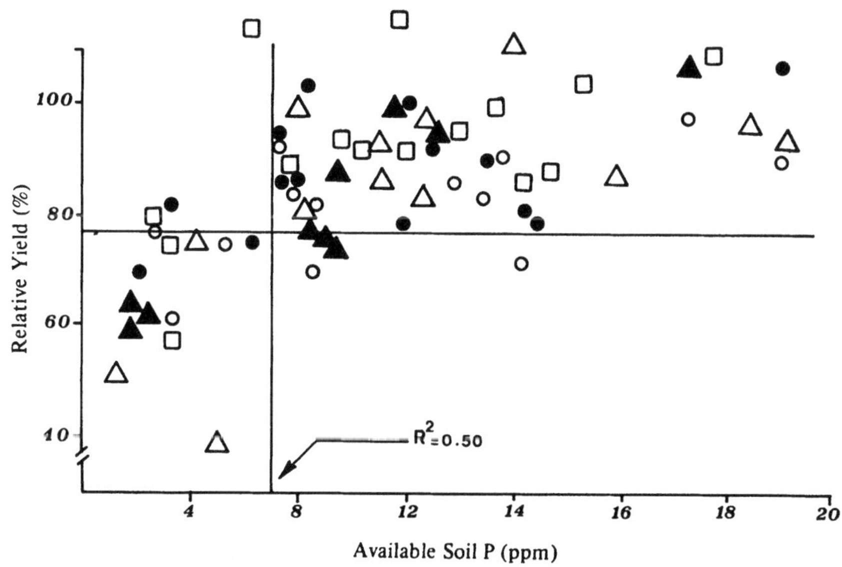

Figure 9. Scatter diagram of relative food legume seed yield and dry matter yield of vetch with available soil P at sowing: ▲, peas; ○, vetch; □, faba bean; ●, lentil; △, chickpea.

limited previous research, it can be concluded that a level of 10 ppm soil-P at sowing of winter cereals or a legume crop should secure an optimum P availability over the range of various soils and climatic conditions of the region.

D. Residual Phosphorus

A common concept in the early days of soil fertility research was that most added P was irreversibly "fixed" in unavailable forms in soil. Much research in the past 2 to 3 decades has shown that this is not strictly true; yearly or large additions of P do indeed build up in the soil, thus reducing or temporarily eliminating the need for P fertilizer. Similar observations were made in the WANA region.

A yearly application of 18 kg P/ha on a Vertisol, with a wheat/lentil rotation, raised Olsen-P from 2 to 8 ppm within 4 years (Matar, 1976a). Similarly, yearly application of 27 kg P/ha for 5 years to a Typic Calciorthid with different barley rotations increased P from 2.5 to 12 ppm (Jones and Matar, 1990). A higher application of 60 kg P/ha on a fine sandy loam soil in Cyprus increased Olsen-P values from 4 to 36 ppm within 5 years. In Tunisia, Gharbi et al. (1990) surveyed available soil P and found more than 84% of the sites contained more than 10 ppm. Apparent recovery fraction for P fertilizers depends on soil type, rate of applied P, type of crops and climatic conditions, and is usually low in the first cropping year, i.e., 5.5% to 10% of applied P (13 to 39 kg P/ha) was recovered by barley and wheat. Thus, despite immobilization reactions which take place between soil constituents and soluble applied P, a large amount of residual P is left in the soils after harvest. Highly significant residual P effects on yields of successive crops were also demonstrated in Cyprus (Orphanos, 1988), Turkey (Kacar, 1972), and Morocco (Abdel Monem et al., 1990c).

Build-up of residual P is particularly noticeable at experimental stations which are normally intensively fertilized. As these are invariably used for purposes such as variety and tillage trials, which do not involve assessment of fertilizer performance, more fertilizer than necessary is usually applied. Work in Lebanon (Ryan et al., 1980) at the American University of Beirut's Bekaa Valley station showed that P had built up to much more than adequate levels in surface horizons; the disparity between P in surface and subsurface horizons was greater in irrigated than in rainfed portions of the station. A recent survey of five stations in Morocco's semiarid zone (250 to 450 mm/year) indicated considerable P build-up in virtually all experimental plots (Ryan et al., 1990b). Fertilizer P trials at such sites lead to invalid conclusions regarding the need for P fertilization.

A knowledge of residual effects and appropriate means of quantifying them is essential for efficient utilization of fertilizer. Phosphate added to soil is subject to chemical transformations into less available forms (Barrow, 1974), with rates varying with soil types and properties and several environmental conditions, e.g., temperature and moisture. Larsen et al. (1965) investigated the fate of residual P in soils by following the changes in the levels of labile soil P under field conditions after addition of a heavy superphosphate dressing; *L*-values decreased exponentially with time at a rate that was expressed as the time in months for half of the applied P to become nonlabile or what is designated as the half-life of applied P. Expressing the change in soil test levels with time provides a reliable and rapid method for assessing the residual value of a nutrient (Mattingly, 1971).

The half-life ($t_{1/2}$) of available P in some typical calcareous Mediterranean soils has been measured under natural conditions (ICARDA, 1985); values ranged from 5.5 months for Vertisols to 11.8 months for Typic Calciorthids. A descriptive model based on chemically extractable soil P, similar to that of Lins et al. (1985), was developed for Syria (Matar, 1990). This model requires information on soil P buffer capacity, rate of P immobilization, and an estimate of crop P uptake the previous season.

$$Pt = Peq + [(A + BF) - Pdq] \exp(-k_1 t) - TPU \qquad (4)$$

where Pt is the available P (Olsen-P) at any time (t), ppm; Peq is the extractable P at equilibrium state, ppm; F is the P applied, kg P ha^{-1}; k_1

Table 7. Observed and predicted soil $NaHCO_3$-P values 1 year after application to wheat at two sites

Applied P ($kg\,ha^{-1}$)	Breda Uptake ($kg\,ha^{-1}$)	Breda Observed (ppm)	Breda Predicted (ppm)	Tel Hadya Uptake ($kg\,ha^{-1}$)	Tel Hadya Observed (ppm)	Tel Hadya Predicted (ppm)
0	2.44	4.8	4.8	6.26	3.4	3.4
17.5	2.87	8.0	6.7	6.81	6.6	5.9
35.0	3.18	10.0	11.8	6.85	11.2	11.7
52.5	3.74	15.7	16.7	7.47	15.3	17.2

is the rate of P immobilization, $month^{-1}$; TPU is the total crop P uptake at harvest, $kg\,P\,ha^{-1}$; and AB are the constants. The model was tested at some sites and proved quite promising (Table 7). Further work is still needed for its generalization (Matar, 1990).

Because of the importance of residual P in soils, calculations of economics of fertilizer application should not be restricted to the year of application. This is a pertinent consideration where rotations are involved and where only one phase would normally be fertilized, e.g., by farming where a fertilized cereal is alternated with a medic pasture. The evidence presented here indicates that P carryover is sufficient for the succeeding crop. While farmers in general are convinced of the need for yearly N application, that P can be applied less frequently is less appreciated.

E. Phosphate Application Methods

Broadcasting P fertilizer and incorporating it at sowing is the common practice used by most farmers in the WANA region. However, evidence has accumulated that P banded, or drilled with the seeds, is more efficient than broadcasting. Soltanpour et al. (1988), working with wheat grown on shallow Petrocalcic Palexeroll with 4.5 ppm Olsen-P, found that grain response to broadcasting, followed by disking, reached a plateau at $9\,kg\,P\,ha^{-1}$, but response to banded P continued further. They postulated that banding stimulates growth and early P uptake, leading to increased crop yield potential. In a dry site in Syria (280 mm), banded P produced relatively higher grain yield increases of durum wheat equivalent to 34%, 63%, and 23% in 3 successive years compared with being broadcasted (Matar and Brown, 1989a). It is important to emphasize that the advantage of banding over broadcasting depends on the extent of P deficiency. Where soil P test values are beyond the critical level, no yield differences between the two methods can be expected. In a recent comparison in Morocco with wheat, Abdel Monem et al. (1990c), found that banding increased P uptake but not yield.

Two mechanisms are postulated to explain the advantageous effect of banding on P fertilizer efficiency under field conditions, i.e., (i) banding

Table 8. Trends in phosphorus fertilizer use for some countries of the Mediterranean region. From *FAO Yearbooks* (1979, 1989)

	Morocco		Libya		Syria		Cyprus	
	1979	1989	1979	1989	1979	1989	1979	1989
Fertilizer	P-tons × 10^3							
Single Superphosphate	—	5.6	—	—	—	—	—	—
Triple Superphosphate	—	6.8	—	0.3	15.6	52.5	—	0.3
Ammonium phosphate	—	19.8	—	22.3	—	—	—	—
Complex	—	24.0	—	1.0	—	—	—	3.3
	—							
Total	31.0	54.2	9.4	23.6	15.6	52.5	3.1	3.6

reduces soil-fertilizer contact, which in turn decreases P immobilization and (ii) increases the root-P fertilizer contact and, thus, P concentration, resulting in greater P uptake. Other observations indicate that (i) reduction in P availability with time is independent of the method of P placement, whether banded or broadcast; and (ii) the root growth increases in the soil zone of P enrichment particularly where P is banded (Matar and Brown, 1989b).

F. Types of Phosphorus Fertilizers

Fertilizer use varies from country to country in terms of amount and type. Selected data extracted from *FAO Yearbooks* (1979, 1989) illustrate trends (Table 8.). Over a 10-year period, total P fertilizer use more than doubled in Libya and Syria and substantially increased in Morocco—although the initial use base was low. Cyprus, with a higher P use intensity, increased by about 10% in the same period. Corresponding figures for the Near East (Syria, Jordan—Iraq, Arabian Peninsula zone) increased from 0.98 to 1.69 M tons, while global P use only increased from 30 to 38 M tons.

Though somewhat sketchy, the data also illustrate fertilizer type usage. It showed that consistent with global trends, single superphosphate (8%P) is of minimal importance. The more concentrated (20%) or triple superphosphate dominates the Syrian market and is the major single P carrier in Morocco. However, the major forms of P in Morocco are ammonium phosphate and compounds, usually N-P-K. Ammonium phosphate is the dominant P source in Libya, but is not used in Syria. Use of more concentrated P fertilizer is based on cheaper per unit nutrient costs, while compounds are favored by simultaneous need for N. While K is often included in compounds, it is rarely needed for dryland cereals.

Triple or concentrated superphosphate has been the standard P source for field trials (Soltanpour et al., 1988; Matar and Brown, 1989a; Ryan and Matar, 1990). While compound N-P fertilizers have been used in such trials, e.g., Nitriphos (Rashid et al., 1990), additional N must be added to standardize its application, and different N source may reduce the validity of the trial. Little research has been conducted in the region on P fertilizers per se. This was limited to incubation studies of pyrophosphate reactions in calcareous soils (Al-Khateeb et al., 1986). Pyrophosphate rapidly hydrolizes in the soil to orthophosphate—the form in common P fertilizers. Another experimental material, urea phosphate, was examined in a series of experiments in Lebanon (e.g., Ryan and Tabarra, 1989). Limited work was also done on the influence of potentially acidifying sulfur compounds on native soil P availability (Dawood and Murtathee (1986). However, neither approach has a significant implication for dryland agriculture.

VII. Regional Research Infrastructure

Knowledge acquisition does not occur in a vacuum. It occurs, by and large, within the public domain—especially in developing countries. The extent of research activity in any one country reflects its political and socioeconomic development. Research on P—indeed on agriculture in general—also reflects the importance of agriculture in the national economy and the extent of governmental support for agriculture. In that respect, countries of the WANA region represent the full spectrum of support, from a strong commitment of resources to little more than lip-service. What has been achieved in P research and what is likely to be in the future is related to such support. It is therefore pertinent to briefly examine the varying institutions involved with P research.

A. National Organizations

All countries of the WANA region have some sort of research infrastructure, usually the Ministry of Agriculture. Frequently, research and extension activities are under separate and independent branches of the ministry. One or two examples will suffice to illustrate this relationship. In Morocco, research falls within the mandate of the Institut National de la Récherche Agronomique (INRA), while the extension function—designed to implement the findings of applied research—lies with the separate Department de Vulgarisation (DPV). Since its inception, INRA has contributed to knowledge of P in Morocco's dryland cropping—largely through a network of research stations. Most countries of the region have separate divisions dealing with research. In Syria, the Soils Directorate fulfills this mission.

Though small in size, and with a significant proportion of its land under rainfed conditions, Lebanon represents the other end of the institutional support spectrum. Due to tragic internal circumstances research in the public domain has all but ceased. It is clear from an examination of the region's public agricultural institutions that there is a greater need for increased investment in research facilities and scientist training. While separation of research generation from its application is not always inimical to effective technology transfer, more often than not it is. Fortunately, current deficiencies are mitigated to some extent by national universities. Again, a few examples will suffice.

In Syria, a modest dryland research program has been undertaken by the University of Aleppo and the University of Damascus. Studies on P fertilization of native dryland medics were already under way at the University of Mosul in northern Iraq in the 1970s (El Fakhry, 1980). Though predominantly oriented towards irrigated agriculture, the Institut d'Agricutlure et Veterinaire (IAV) Hasan II in Morocco, has a program on P in medics. Similarly, several studies of soil P have emanated from the University of Jordan in Amman, and the Al-Fatah University in Tripoli has added to knowledge of P in Libyan soils. Significant strides have been made in Lebanon at the American University of Beirut in P research (Ryan, 1983, 1990), though efforts have, of necessity, been curtailed in recent years.

Of the countries of the region, the only one without a university to supplement ministry research is Cyprus. Despite similarities in environments and common farming systems—and problems with P fertilization or the lack of it—it was apparent that little coordination of effort existed between the disparate national organizations. The awareness that coordination and information-sharing would greatly enhance research efficiently lead to the birth of a regional research network.

B. Regional Soil-Test Calibration Network

As the mandate of ICARDA is to promote, through collaboration with national scientists, agricultural research on dryland crops and its application at farm level in countries of the WANA region, a network system was developed. The most prominent aspect of this was that dealing with soil-test calibration. Together with N, P was a primary focus of attention in this effort. The broad objective was to promote soil testing as a basis for efficient fertilizer use.

The initial meeting in Aleppo (Soltanpour, 1987), Syria, in June 1986, was jointly sponsored by ICARDA, Mid-America International Agricultural Consortium (MIAC) in Morocco, and the International Development Research Center (IDRC). Along with some general background papers, reports dealt with P tests and responses for cereals and legumes in Morocco, Cyprus, Syria, Jordan, Turkey, and Pakistan. Most reports illustrated consistent crop responses with P fertilizer. Protocols for conducting fertilizer trials for participating scientists were established, implementing factorial N-P design with the main focus on wheat.

A second meeting in 1987 in Ankara, Turkey (Matar et al., 1988d) brought an expanded range of participants. Papers dealing with P considered issues such as residual P for barley (Cyprus), plant P diagnostic norms (Tunisia), calibration of different P tests for cereals, legumes, and oilseeds (Pakistan), P fertilization of legumes (Syria), and wheat responses in Morocco, Jordan, and Syria. The protocols for P trials were expanded; standardized procedures were developed for site selection, soil sampling and analysis, fertilizer treatments, and experimental design. A procedure was developed for quality control with the region's soil-testing laboratories.

The third annual meeting of the Network took place in Amman, Jordan, in 1988 (Ryan and Matar, 1990). Additional national delegations attended, including those from Iraq and Yemen for the first time. In addition to P calibration studies with cereals and soil testing procedures, topics included modeling of residual P responses, fertilizer P placement, i.e., banding vs broadcasting, P in relation to rotations, and the use of isotherms to evaluate crop P requirements. The protocols were modified to have separate P trials with basal N, rather than more elaborate and expensive factorial trials, since many sites are not consistently responsive to both elements. A fourth meeting of the network was held in Agadir, Morocco, in May, 1991. Increasing emphasis is being placed on implementation at the farmer level of what has already been established. To this end, crop P response and fertilizer use data are being interpreted in economic terms.

The consensus of this network, arrived at after the first three regional meetings, was to adopt the $NaHCO_3$ or Olsen procedure as the official soil P test for the region and to consider between 5 to 7 ppm the critical value below which a response is likely. The convention of taking soil samples to a depth of 20 cm was adopted. Fertilizer P rates of 10 to 20 $kg\,ha^{-1}$ are recommended, depending on the extent of deficiency. In most countries of the region few farmers test soils. Indeed, facilities for testing are poorly developed and often beyond the reach of the small farmer because of costs. In demonstrating the benefits of soil testing, the Soil Test Calibration Network can serve as a catalyst for governmental institutions and the private sector to provide such services. Future efforts of the network will be as much in the educational domain as in the technical. It will continue to be a forum for dryland soil scientists and agronomists of the region to share information and a means of fostering their professional development.

C. Other International Research Institutions

While many institutions of an international or regional character are involved in P research, the following are worthy of mention.

1. Arab Center for the Studies of Arid Zones and Dry Lands (ASCAD). This international institution, based in Damascus, Syria, though largely concentrating on Syria, has collaborative fertilization programs with countries of the WANA region. Its primary concern is, however, with soil mapping and resource inventory. It currently has an on-going program of cereal fertilization with dryland scientists from INRA in Morocco.
2. Mid-America International Agricultural Consortium (MIAC). This project which established the Aridoculture Center in Settat, Morocco, as an example of an international assistance program (USAID) for a limited period (1980 to 1994). Research on soil P and crop responses to P fertilizers formed the core of the research program (Soltanpour et al., 1989; Abdel Monem et al., 1990a; Ryan et al., 1991b). Research at this center is integrated with that of the INRA national network.
3. Food and Agricultural Organization of the United Nations (FAO). Operating from its Rome headquarters and regional offices, FAO provides a clearing-house of information on P research in soils and crops. It also promotes and coordinates field work in most countries of the region (FAO, 1980).

VIII. Future Research and Extension

While research on all phases of P in soils will continue, the bulk of it will be towards application and technology transfer. Basic research is essential to answer the many gaps in our knowledge of this tantalizing element. However, for the most part, research institutions in the Mediterranean area lack the necessary resources to do such research, which often appears an academic luxury given the backlog of information which has yet to reach the farmer. The following list of concerns could be addressed in future by soil scientists concerned with P use in Mediterranean-region soils.

A. Applied Soil Phosphorus Research

While the list of potentially researchable topics in the area of soil and fertilizer P is extensive, the concerns identified here have more immediate urgency.

1. Soil-Test Calibration

While strides have been made in identifying the Olsen test and establishing critical crop-P response levels for rainfed crops, the process of calibration is a continuous one; the soil is not static nor are crop cultivars or yields. With higher yields from improved management and plant breeding, critical soil P levels in relation to crop response will also change.

2. Residual Phosphorus

While it is easy to identify build-up of residual soil P, the question "how long does it persist in the soil" is more difficult to answer. Though several studies addressed residual P and changes with time (Ryan et al., 1985b; Matar, 1990), they usually involved pots in the greenhouse or pot-like conditions in the field. Most were of limited duration with extrapolation to longer periods using predictive models. What is clearly needed is long-term trials in which soils with above adequate P levels are cropped without fertilizer for several years. This information is all the more crucial in areas where some farming with minimum tillage is practical.

3. Spatial and Temporal Variability

The soil sampling phase is crucial to the effectiveness of any soil-testing program. Only recently has emphasis been given to the fact that soil-test values change throughout the year in a definite cyclical pattern and that considrable variation exists in sample values even with a single field or part thereof (Sabbe and Marx, 1987). While temporal changes are more easily predicted, wide spatial variability presents more of a problem. If the sample taken does not accurately reflect the fertility status of a field, the consequent fertilizer recommendation will be inaccurate, i.e., either too little or too much. Lack of site representativeness may even explain inconsistent field crop P fertilizer responeses.

Most field trials in the Mediterranean area at most only involve P analysis at planting time. Measurements over time are rare. The issue of field-test variability has been recently addressed in Morocco (Ryan and Abdel Monem, 1991). On a shallow soil, visually uniform, 200 × 200 m area, comprising seven fields or parts of fields, each with a different cropping history, test values for P were highly variable and exhibited a consistent pattern related to cropping history. A similar adjacent site comprising only one field with presumed uniform management was more variable. A number of factors were attributed to the random variability: inherent soil variability, irregular crop and weed stands, animal droppings, and nonuniform hand-spreading of fertilizer, when used. While detailed site variability studies are not necessary for each trial site, it is important that P fertility be uniform. Nevertheless, field variability has to be considered when recommendations are being made for sampling schemes. Economics and practicality—small fragmented farms, minimum financial resources, possible crop failures—must dominate such decisions.

4. Organic Phosphorus Mineralization

While most soils of the Mediterranean zone are low in organic matter, some extensive areas of Mollisols and mollic-type soils do occur. However, little is known about organic P forms or their occurrence in relation to total soil P. Most organic P compounds have yet to be identified (Dalal, 1977). Though sophisticated research is needed to study soil organic P biochemistry, gross effects on crop growth are easily detected. As temperatures and rainfall conditions in most countries of the region are conducive to microbial activity throughout the growing season, the pool of available P is likely to vary as a result of mineralization and mobilization. Characterization of this fraction may help to elucidate field-crop response to P fertilizers. Recent observations in Morocco (Azzaoui et al., 1990) that available P in an unfertilized organic-matter-rich soil (Petrocalcic Palexeroll) increased throughout the growing season may reflect that phenomenon.

5. Sampling and Conservation Tillage

While most soil sampling studies from the Mediterranean region assume uniform broadcast application of P fertilizers, sampling schemes have to be modified where fertilizer is drilled in rows and where the band remains undisturbed until the following season. This is particularly true where direct drilling with minimum or no tillage is used. While such conservation approaches are in their infancy, they will no doubt be more important in the future. Sampling strategies should be modified accordingly.

6. Crop Modeling

While crop models are currently in vogue in agricultural research, their use with respect to P in the Middle East region has been limited (Matar, 1990). However, with increasing climatic data bases integrated with information from soil-crop-P fertilizer trials, the potential for effective use of models is greatly enhanced.

B. Plant Analysis

Determination of P in plant material is a simple process. Knowledge of plant P concentrations, while primarily of concern for fertilizer response, can also yield valuable information on crop quality and the diagnosis of nutritional problems.

1. Tissue-Concentration Levels

While the percentage of P in plant tissue is influenced by numerous soil, plant, and environmental factors, it is clear that for crops of the Mediterranean region, i.e., cereal and food and forage legumes, the minimum P concentration tolerable for adequate growth, given a favorable supply of the other factors, needs to be identified. It further needs to be clarified if such levels are similar for all species or cultivars within any crop grouping. Furthermore, the relationship between P tissue concentration and critical soil-test levels is poorly understood. By comparison with soil P analysis, use of plant P analyisis in the interpretation of crop-fertilizer responses is not well developed: Theoretically, it should complement the soil approach. Given the paucity of references to plant P analysis in studies from the region that awareness is now only dawning.

2. Phophorus Imbalances

While little or no reference has been made to excess levels of P in relation to other elements in the research from the Mediterranean area, the possibility of such adverse interactions cannot be ruled out, especially in view of the fact that some soils are high in available P, either naturally or from residual P accumulation from fertilization (Abdel Monem et al., 1990b). While the balance between P and calcium is important for the grazing animal, the most significant P interaction is with zinc. This might have accounted for the decreased growth of medic cultivars at high levels of applied P in Derkaoui et al.'s study (1990). However, Ryan et al. (1991b) found that though P induced Zn deficiency in greenhouse-grown corn cultivars, the effect was due to dilution from increased yield rather than an adverse effect of P on Zn translocation or uptake. Examples of P-Zn interaction for dryland field crops have not been reported in the region.

3. Nutritional Aspects of Fertilization

As an essential element for annual growth and health as well as crop growth, it is important that fertilization consider both aspects. As most farm animals in Middle Eastern livestock-grazing systems rarely receive enriched supplementary feed and are primarily dependent on P intake from grazing of animal forage, weedy fallows, or cereal stubble, the P concentration in that supply is frequently less than desireable. Concentrations of 0.17% to 0.25% P are necessary in the feed of grazing animals (Ozanne, 1980). It is clear that such mineral deficiencies can affect weight gain and lambing percentage and other health-related factors. The problem is compounded by the fact that plant P concentration decreases steadily with age or as the season progresses. Thus, while P fertilization may not always increase crop yields, it nevertheless increases plant P concentration and therefore nutritional quality. For instance, Derkaoui et al. (1990) showed that P, at 200 ppm, not only increased medic yields several-fold, but also increased P from 0.29% to 0.68%. While little attention has been devoted to fertilization and animal performance in the region, collaborative efforts are needed between animal and crop-soil scientists to address this issue.

C. Resource Development

As P is vital to the economies of the Mediterranean region, we must adopt a broader perspective in considering this resource. For most farmers, commercial fertilizers will continue to be expensive, which will often deter their use. This brief section pinpoints some means to improve our use of soil P and identify other sources.

1. Soil Inventories

In the past, agronomists and soil scientists who are concerned with fertilizer-crop responses have tended to ignore the potential ability of soil classification systems for extension of their findings. The key to effective international agrotechnology transfer is a common soil classification system (Beinroth et al., 1980). Despite the various systems in use throughout the world, soil taxonomy has emerged as the de facto international system. It enables combining both soil and climatic data and permits generalizations regarding crop behavior to a larger population of similar environments.

As soil types can differ widely with respect to available P, i.e., inherently rich in P, high P-fixation capacity, potentially mineralizable P, etc., such differences should be recognized in order to formulate general recommendations for farmer use of P fertilizers. Theoretically, soil maps can be useful in this regard. For the Mediterranean area of Italy, Paris (1986) cited the lack of suitably detailed maps as a basis for advisory purposes. Indeed, this generalization could be applied to all countries of the region. Though Abdel Monem et al. (1990b) identified categories of P availability acording to soil type in Morocco's dryland zone, only a small percentage of that area is mapped at any level. While mapping is a slow and expensive process, it is important that, if available, as much practical information as possible regarding P be extracted and used from soil maps.

Other more localized soil fertility inventories can be vital to the fertilizer-crop relationship and recommendation. For instance, the experimental station surveys of Ryan et al. (1990) in Morocco and previously in Lebanon (1980) highlight situations where conditions are inappropriate for testing P fertilizer responses or differences between methods of P application. Fertility build-up at such stations can lead to invalid fertilizer recommendations as they bear no meaningful relationship with on-farm low fertility conditions. Such surveys are simple and inexpensive to conduct and should be periodically updated since P in soil does not remain constant over time.

2. Phosphorus Sources and Organic Wastes

While organic manures—from sheep, goats, and cattle—have been used from time immemorial in the Middle East, such materials have not featured in field trials. Most organic fertilizers are primarily considered suppliers of N, with little attention to P (Ryan et al., 1985c). Though such materials are not competitive with commercial sources, they are, nevertheless, important in the traditional agriculture of the region and should be evaluated in longer-term trials.

With increasing urbanization, municipal waste will also increase. With adequate processing, such materials are potentially useful in agriculture (Khuri et al., 1987)—as sources of nutrients, including P and for increasing soil organic matter. Though socio-cultural values have militiated against using such materials in agriculture, they can enhance fertility and improve the region's soil resources.

3. New Cultivar Development

The efficiency with which plants use soil P is influenced by root growth and root hair characteristics (Barber, 1980). As these characteristics vary between species and even cultivars of a species, the possibility is raised that new cultivars could be developed that are more efficient at exploiting the limited supply of P normally found in soils or can produce optimum growth at lower critical plant P concentrations. While research aimed at answering the many basic root-P questions is still unclear and may not be appropriate to many countries of the region, it is possible, however, to test new crop cultivars in terms of response to given levels of soil P. In this respect, a joint effort of breeders and soil fertility scientists is called for.

D. Extension and Education

Despite the considerable P research in the Mediterranean area, a few sad facts remain: fertilizer use is still a haphazzard affair, and soil testing has yet to impinge on the region's agriculture. Greater and more intensified efforts are needed in order to implement what researchers already know—and have known for a long time. Closer integration of the region's research and extension services is sorely needed along with improved extension personnel training. Modern communications technology offers possibilities to redress past institutional weaknesses.

References

Abdel Monem, M., A. Azzaoui, M. El Gharous, J. Ryan, and P. Soltanpour. 1990a. Residual nitrogen and phosphorus for dryland wheat in central Morocco, pp. 163–175. In: Ryan, J., Matar, A. (eds.) *Proc. Third Regional Soil Test Calibration Workshop*. ICARDA, Aleppo, Syria.

Abdel Monem, M., J. Ryan, and M. El. Gharous, 1990b. Preliminary assessment of the soil fertility status of the mapped area of Chaouia. *Al Awamia* 72: 85–107.

Abdel Monem, M., A. Azzaoui, M.E. Gharous, J. Ryan, and P. Soltanpour. 1990c. Fertilizer placement for dryland wheat in Central Morocco, pp. 149–162. In: Ryan, J., Matar, A. (eds.) *Proc. Third Regional Soil Test Calibration Workshop*. ICARDA, Aleppo, Syria.

Abedi, M.J., and O. Talibudeen. 1974. The calcareous soils of Azerbaijan. I. Catena development related to the distribution and properties of soil carbonate. J. *Soil Sci.* 25: 357–372.

Abreu, M.M., and M. Robert. 1985. Characterization of maghemite in B horizons of three soils from southern Portugal. *Geoderma* 36: 97–108.

Abu Rub, N., and T.I. Ashab. 1987. Fertilization of wheat and barley under rainfed agriculture of Jordan, pp. 151–153. In: Soltanpour, P.N. (ed.) *Proc. First Regional Soil Test Calibration Workshop*. ICARDA, Aleppo, Syria.

Adepoju, A.Y., P.F. Pratt, and S.V. Mattigod. 1986. Relationships between probable dominant phosphate compound in soil and phosphorus availability to plants. *Plant Soil* 92: 47–54.

Aggarwal, R.K., M.K. Sharma, and P. Raina. 1987. Phosphate adsorption characteristics of some Aridisols as affected by soil physicochemical properties. *Indian J. Agric. Res.* 21: 164–170.

Ahmad, N., J.G. Davide, and T. Saleem. 1988. Fertility status of soils in dryland areas of Pakistan, pp. 22–49. Proc. Dryland Agric. Int. Seminar, Lahore, Pakistan, Nov. 6–8. FFC Fauji Fert.

Agbani, M., and K. El Mejahed. 1983. Comparaison de methodes d'analyse du phosphore dans quelques sols marocains, pp. 249–261. Proc. Third Int. Cong. P Cpds., Inst. Mond. Phos. Casablanca, Morocco.

Al-Khateeb, I.K., M.J. Raihan, and S.R. Asker. 1986. Phase equilibria and kinetics of orthophosphate in some Iraqi soils. *Soil Sci.* 141: 31–37.

Amar, B., and A. Ait Houssa. 1990. Comparison of different soil testing methods in various Mediterranean soils, pp. 61–52. In: Ryan, J., Matar, A. (eds.) *Proc. Third Regional Soil Test Calibration Workshop*. ICARDA, Aleppo, Syria.

Amer, F., D.R. Bouldin, C.A. Black, and F.R. Duke. 1955. Characterization of soil phosphorus by anion exchangeable resin adsorption and ^{32}P equilibration. *Plant Soil* 6: 391–408.

Amer, F.A., A.A. Mahmoud, and V. Sabet. 1985. Zeta potential and surface area of calcium carbonate as related to phosphorus sorption. *Soil Sci. Soc. Am. J.* 49: 1137–1142.

Arif, A., M. Abdel Monem, and J. Ryan. 1989. Impact of N and P fertilization of perennial grasses in Morocco. *Agron. Abst.*, p. 298.

Aslyng, H.C. 1954. The lime and phosphate potentials of soils; the solubility and availability of phosphates. pp. 1–50. *Roy. Vet. Agric. Coll. Yearbook*, Copenhagen, Denmark.

Azzaoui, A., R.G. Hanson, and P.N. Soltanpour. 1989. Wheat P requirements on calcareous Moroccan soils. 1. A comparison of Olsen, Soltanpour, and $CaCl_2$ soil tests. *Commun. Soil Sci. Plant Anal.* 20: 869–891.

Azzaoui, A., M. Abdel Monem, and J. Ryan. 1990. Phosphorus response of wheat, barley, and triticale in semi-arid conditions in Morocco. *Agron. Abst.*, p. 97.

Bache, B.W., and C. Ireland. 1980. Desorption of phosphate from soils using anion exchange resins. *J. Soil Sci.* 31: 297–306.

Badawy, F.M. 1976. Effect of phosphate fertilization and seed incubation with Okadun at high rate on yield of broad bean and lentil. *Zentrablatt F. Bakteriologies, Parasiten kinde, Infektions Krankenkeit und Hygiene Zweite-Naturwissen Schaftliche Abteilung*, 131: 655–670.

Barber, S.A. 1980. Soil-plant interactions in the phosphorus nutrition of plants, pp. 591–615. In: Khasawneh, F.E., Sample, E.C., Kamprath, E.J. (eds.) *The Role of Phosphorus in Agriculture*. Soil Sci. Soc. Am., Madison, Wis.

Barrón, V., M. Herruzo, and J. Torrent. 1988. Phosphate adsorption by aluminous hematites of different shapes. *Soil Sci. Soc. Am. J.* 52: 647–651.

Barrow, N.J. 1974. The slow reaction between soil and anions. I. Effect of time, temperature and water content on the decrease in effectiveness of phosphates for plant growth. *Soil Sci.* 118: 380–386.

Barrow, N.J. 1978. The description of phosphate adsorption curves. *J. Soil Sci.* 29: 447–462.

Barrow, N.J. 1980. Evaluation and utilization of residual phosphorus in soils, pp. 333–359. In: Stelly, M. (ed.) *The Role of Phosphorus in Agriculture*. Soil Sci. Soc. Am., Madison, Wis.

Barrow, N.J. 1983. On the reversibility of phosphate sorption by soils. *J. Soil Sci.* 34: 751–758.

Barrow, N.J. 1987. Reactions with variable charge soils. *Developments in Plant and Soil Sciences*. Martinus Nijhoff, Dordrecht.

Barshad, I. 1966. The effect of variation in precipitation on the nature of clay mineral formation in soils from acid and basic igneous rocks, Vol. 1, pp. 167–173. *Proc. Int. Clay Conf. (Jerusalem)*.

Bar-Yosef, B., U. Kafkafi, R. Rosenberg, and G. Sposito. 1988. Phosphorus adsorption by kaolinite and montmorillonite: I. Effect of time, ionic strength, and pH. *Soil Sci. Soc. Am. J.* 52: 1580–1585.

Bech, J., N. Fedoroff, P. Quantin, and P. Segalen. 1982. Etude des sols fersiallitiques lessivés formés sur des arènes granitiques de la Selva (Catalogne, Espagne). *Cah. ORSTOM, sér. Pédol.* 19: 233–256.

Beek, J., and W.H. Van Riemsdijk. 1979. Interaction of orthophosphate ions with soil, pp. 259–284. Soil Chemistry. B. Physico-Chemical models. *Developments in Soil Science*. Elsevier Science, New York.

Beinroth, F.H., G. Uehara, J.A. Silva, R.W. Arnold, and F.B. Cady. 1980. Agrotechnology transfer in the tropics based on soil taxonomy. *Adv. Agron.* 33: 304–339.

Berigari, M.S., M.J. Raihan, and M.A. Umran. 1985. Phosphate adsorption isotherm for assessing phosphorus requirements of soils in Iraq. *J. Agric. Water Resources Res.* 4: 97–117.

Boero, V., and U. Schwertmann. 1989. Iron oxide mineralogy of terra rossa and its genetic implications. *Geoderma* 44: 319–327.

Bohn, H.L., B.L. McNeal, and G.A. O'Connor. 1979. *Soil Chemistry*. Wiley, New York.

Borggaard, O.K. 1983a. The influence of iron oxides on the surface area of soils. *Soil Sci.* 32: 427–432.

Borggaard, O.K. 1983b. Effect of surface area and mineralogy of iron oxides on their surface charge and anion-adsorption properties. *Clay Clay Miner.* 31: 230–232.

Borrero, C., F. Peña, and J. Torrent. 1988. Phosphate sorption by calcium carbonate in some soils of the Mediterranean part of Spain. *Geoderma* 42: 261–269.
Bousselham, L. 1986. Niveau de carence en phosphore et le phosphore assimilable des sols de la region de Meknes. *Bull. de l Ecole Nat. d'Agric. de Meknes. Morocco.* 2: 23–29.
Bowman, R.A., and S.R. Olsen. 1985. Assessment of phosphate buffering capacity: 2. Greenhouse methods. *Soil Sci.* 140: 387–392.
Bray, R.H. 1958. The correlation of a phosphorus soil test with the response of wheat through a modified Mitscherlich equation. *Soil Sci. Soc. Am. Proc.* 22: 314–337.
Brown, S.C., J.D.H. Keatinge, P.J. Gregory, and P.J.M. Cooper. 1987. Effects of fertilizer, variety and location on barley production under rainfed conditions in northern Syria. I. Root and shoot growth. *Field Crops Res.* 16: 53–66.
Chang, S.C., and M.L. Jackson. 1957. Fractionation of soil phosphorus. *Soil Sci.* 84: 133–144.
Chen, Y.S.R., J.N. Butler, and W. Stumm. 1973. Adsorption of phosphate on alumina and kaolinite from dilute aqueous solutions. *J. Colloid Interface Sci.* 43: 421–436.
Clawson, M., H.H. Landsberg, and L.S. Alexander. 1971. The agricultural potential of the Middle East. Elsevier Science, New York.
Cole, C.V., S.R. Olsen, and C.O. Scott. 1953. The nature of phosphate sorption by calcium carbonate. *Soil Sci. Soc. Am. Proc.* 17: 352–356.
Cooper, P., M. Jones, H. Harris, and A. Matar. 1988. Agroecological constraints to crop production in West Asia and North Africa and their impact on fertilizer use. Workshop on Fertilizer Sector Development and Agricultural Production in Selected Countries of the Mediterranean, Middle East, and North Africa, May 1–14, Muscle Shoals, Al., USA.
Cooper, P.J.M. 1983. Crop management in rainfed agriculture with special response to water use efficiency, pp. 63–81. Nutrient balances and the need for fertilizer in semi-arid and arid regions. Proc. 17th Colloq. Intl. Potash Inst., Rabat and Marrakech, Morocco.
Cooper, P.J.M., P.J. Gregory, J.D.H. Keatinge, and S.C. Brown. 1987. Effects of fertilizer, variety, and location on barley production under rainfed conditions in northern Syria. 2. Soil water dynamics and crop water use. *Field Crops Res.* 16: 67–84.
Cyprus Agric. Res. Inst. 1963–1978. *Annual Reports*. Nicosia, Cyprus.
Dalal, R.C. 1977. Soil organic phosphorus. *Adv. Agron.* 29: 83–117.
Dawood, F.A., and M.S. Murtathee. 1986. Effect of sulfur on the availability of phosphorus in calcareous soils, vol. 1., pp. 256–258. Proc. Sci. Conf., Sci. Res. Council, Bagdad, Iraq.
Digdigoglu, A. 1980. A study on the calibration for some soil tests with barley response under Central Anatolia soil conditions. Ph.D. thesis, University of Ankara, Turkey, (in Turkish with English summary).
Derkaoui, M., J. Ryan, and M. Abdel Monem. 1990. Field and greenhouse response of Moroccan medics (*Medicago* spp) to phosphorus. *Agron. Abst.*, p. 111.
Drees, L.R., and L.P. Wilding. 1987. Micromorphic record and interpretation of carbonate forms in the rolling plains of Texas. *Geoderma* 40: 157–175.

Dregne, H.E. 1976. *Soils of Arid Regions*. Elsevier Science, New York.
Drouineau, G. 1942. Dossage rapide du calcaire actif du sol; nouvelles données sur la séparation et la nature des fractions calcaires. *Ann. Agron.* 12: 441–450.
Eleizalde, B. 1976. Contribución al conocimiento del fósforo en algunos suelos de la Provincia de Zaragoza (España). *An. Aula Dei* 13: 451–480.
Eleizalde, B. 1977. Distribución de fósforo total, orgánico e inorgánico en las fracciones granulométricas de los horizontes de cuatro grupos de suelos. *ITEA* 26: 2–14.
Eleizalde, B. 1978. Adsorción de fósforo en los suelos aluviales de la Provincia de Zaragoza. *ITEA* 30: 25–34.
Eleizalde, B. 1983. Influencia de las características químicas del suelc sobre el status del P en el Valle del Ebro. *Agrochim.* 27: 487–497.
Eleizalde, B., and M. Fernández. 1982. Q/I ratio in saline soils belonging to Ebro Valley (Spain). *Anal. Edaf. Agrob.* 41: 271–281.
El-Fakhry, A.K. 1980. Studies on dryland farming at the College of Agriculture and Forestry, University of Mosul, pp. 66–70. *Proc. FAO Regional Seminar on Rainfed Agriculture in the Near East and North Africa.* Food Agric. Organization, Rome, Italy.
Elrashidi, M.A., A. Van Diest, and A.M. El Damaty. 1975. Phosphorus determination in highly calcareous soils by the use of anion exchange resin. *Plant Soil* 42: 273–286.
El-Zahaby, E.M., and S.H. Chien. 1982. Effect of small amounts of pyrophosphate on orthophosphate sorption by calcium carbonate as related to phosphate sorption. *Soil Sci. Soc. Am. J.* 46: 38–46.
FAO. 1970. Soils and fertilizers. Report Damascus Agric. Station Project ESR/SF/SYR 14 No. 1. FAO, Rome, Italy.
FAO. 1979. *Fertilizer Yearbook*. FAO, Rome, Italy.
FAO. 1980. *Proc. Regional Seminar on Rainfed Agriculture in the Near East and North Africa*. FAO, Rome, Italy.
FAO. 1984. *Food Balance Sheets, 1979–81 Average*. FAO, Rome, Italy.
FAO. 1989. *Fertilizer Yearbook*. FAO, Rome, Italy.
Fixen, P.E., A.E. Ludwick, and S.R. Olsen. 1983. Phosphorus and potassium fertilization of irrigated alfalfa in calcareous soils. II. Soil phosphorus solubility relationships. *Soil Sci. Soc. Am. J.* 47: 112–117.
Freeman, J.S., and D.L. Rowell. 1981. The adsorption and precipitation of phosphate onto calcite. *J. Soil Sci.* 32: 75–84.
Gachon, L. 1966. Phosphore isotopiquement diluable et pouvoir fixateur des sols en relation avec la croissance des plantes. *C.R. Acad. Agric. France* 52: 1108–116.
Gal, M., A.J. Amiel, and S. Ravikovitch. 1974. Clay mineral distribution and origin in the soil types of Israel. *J. Soil Sci.* 25: 79–89.
Gharbi, A., L. Ettounsi, and A. Haddad. 1990. Soil test calibration for wheat cropped under Tunisian rainfed conditions, pp. 83–90. In: Ryan, J., Matar, A. (eds.) *Proc. Third Regional Soil Test Calibration Workshop*. ICARDA, Aleppo, Syria.
Goldberg, S., and G. Sposito. 1984. A chemical model of phosphate adsorption by soils. II. Non-calcareous soils. *Soil Sci. Soc. Am. J.* 48: 779–783.
Gradusov, B.P. 1974. A tentative study of clay mineral distribution in soils of the world. *Geoderma* 12: 49–55.

Gregory, P.J., K.D. Shepherd, and P.J. Cooper. 1986. Effects of fertilizer on root growth and water use of barley in northern Syria. *J. Agric. Sci.* 103: 429–438.

Griffin, R.A., and J.J. Jurinak. 1973. The interaction of phosphate with calcite. *Soil Sci. Soc. Am. Proc.* 37: 847–850.

Haddad, N. 1986a. Recommendations for growing lentil in Jordan. *Agric. Ext. Bull. No 3.* Min. Agric., Amman, Jordan. (in Arabic).

Haddad, N. 1986b. Recommendations for growing chickpea in Jordan. *Agric. Ext. Bull. No 4.* Min. Agric., Amman, Jordan. (in Arabic).

Hagin, J., and B. Tucker. 1982. *Fertilization of Dryland and Irrigated Soils.* Springer-Verlag, New York.

Harmsen, K., K.D. Shepherd, and A.Y. Allan. 1983. Crop reponse to nitrogen and phosphorus in rainfed agriculture, pp. 223–249. Nutrient balances and the need for fertilizers in semi-arid and arid-regions. Proc. 17th Colloq. Intl. Potash Inst., Rabat and Marrakech, Morocco.

Hassan, N., F. Aziz, T. Al-Tamimi, S. Asker, and E. Rabban. 1974. Limits of phosphorus availability in representative Iraqi soils as measured by crop response and soil test values. *Inst. Appl. Res. Nat. Res. Tech. Bull. No. 74.* Sci. Res. Found., Baghdad, Iraq.

Hasan, H. 1980. Absorption and transformation of phosphorus in calcareous Lebanese soils. M.S. thesis, American University, Beirut, Lebanon.

Havlin, J.L., and D.G. Westfall. 1984. Soil test phosphorus and solubility relationships in calcareous soils. *Soil Sci. Soc. Am. J.* 48: 327–330.

Hernando, V., V. Lombardía, and R. de Clerk. 1968. Relación entre el pH, la materia orgánica y las formas de fósforo en tres catenas de suelos. *Anal. Edaf. Agrob.* 27: 779–816.

Holford, I.C.R. 1988. Buffering of phosphate in the soil solution during growth of ryegrass compared with buffering in soil solution. *Plant Soil* 111: 3–9.

Holford, I.C.R., and G.E.G. Mattingly. 1975a. Surface areas of calcium carbonate in soils. *Geoderma* 13: 247–255.

Holford, I.C.R., and G.E.G. Mattingly. 1975b. The high-and low-energy phosphate adsorption surfaces in calcareous soils. *J. Soil Sci.* 26: 407–417.

Hooker, M.L., G.A. Peterson, D.H. Sander, and L.A. Daigger. 1980. Phosphate fractions in calcareous soils as altered by time and amounts of added phosphate. *Soil Sci. Soc. Am. J.* 44: 269–277.

Hundal, H.S. 1988. A mechanism of phosphate adsorption on Narrabi medium clay loam soil. *J. Agric. Sci.* 111: 155–158.

ICARDA. *Annual Reports, 1983–1988.* ICARDA, Aleppo, Syria.

ICARDA, Soils Directorate, and Syrian Min. Agric. 1989. *Collaborative Research Project Report on Fertilizer Use on Wheat in Northern Syria 1986–88.* ICARDA, Aleppo, Syria.

Itoh, S. 1987. Characteristics of phosphorus uptake of chickpea in comparison with pigeon pea, soybean, and maize. *Soil Sci. Plant Nutr.* 33: 417–422.

Jackson, T.L., A.D. Halvorson, and B.B. Tucker. 1983. Soil fertility in dryland agriculture, pp. 299–332. In: Dregne, H.E., Willis, W.O. (eds.) *Dryland Agriculture.* Agron 23. Amer. Soc. Agron., Madison, Wis.

Jones, M., and A. Matar. 1990. A note on the effect of regular fertilizer use within different two-year cropping rotations on soil phosphate and organic matter status, pp. 211–220. In: Ryan, J., Matar, A. (eds.) *Proc. Third Regional Soil Test Calibration Workshop.* ICARDA, Aleppo, Syria.

Joret, G., and J. Hebert. 1955. Contribution a la determination du besoin des sols en acide phosphorique. *Ann. Agron.*, pp. 233–299.
Kacar, B. 1967. A study of phosphorus fixation in some Turkish soils, and the factors affecting fixation. *Ankara Univ. Zir. Fak. Yill* 17: 215–234.
Kacar, B. 1969. Phosphorus fractionation and correlation of phosphorus fractions with phosphorus availability by various tests in Cukurova soils, pp. 85–88. Univ. Ankara Yearbook, Fac. Agric., Ankara, Turkey.
Kacar, B. 1972. A study of the residual effects of phosphorus fertilizer, pp. 207–232. Univ. Ankara Yearbook, Fac. Agric., Ankara, Turkey.
Kacar, B., F. Didehvar, and E. Shokravi. 1967. Evaluation of various methods for the estimation of plant available phosphorus in the soils of Caspian Sea, pp. 140–150. Univ. Ankara Yearbook, Fac. Agric., Ankara, Turkey.
Kafkafi, U., A.M. Posner, and J.P. Quirk. 1967. The desorption of phosphate from kaolinite. *Soil Sci. Soc. Am. Proc.* 31: 348-353.
Karim, M.I., M.S. Berigari, F.M.S. al-Any, and L.H. Ibrahini. 1989. Effect of citric, tartaric and oxalic acids on phosphate sorption by some calcareous soils of Iraq. *J. Agric. Resources Res.* 8: 51–67.
Kassam, A.H. 1981. Climate, soil and land resources in North Africa and West Asia. Plant Soil 58: 1–28.
Kaushansky, P., S. Levin, and A.J. Amiel. 1984. The precipitation of calcium oxalate on carbonate mineral surfaces. *Soil Sci.* 138: 397–402.
Keatinge, J.D.H., M.D. Dennett, and J. Rodgers. 1986. The influence of precipitation regime on the crop management of dry areas in northern Syria. *Field Crops Res.* 13: 239–249.
Khader, S., and T. Abu Sharar. 1979. Phosphorus adsorption isotherm by a Jordanian soil. *Dirasat.* 6: 139–151.
Kishk, M.A., and A.Y. Lashin. 1978. Phosphate retention by soils and its relation to soil properties. *Beiträge zur Tropischen Landwirtschaft und Veterinärmedizin* 16: 145–153.
Khuri, N., A.T. Shammas, and J. Ryan. 1987. Greenhouse evaluation of Beirut municipal compost. *Leb. Sci. Bull.* 3: 53–63.
Krentos, V.D., and P.I. Orphanos. 1979. Nitrogen and phosphorus fertilizers for wheat and barley in a semi-arid region. *J. Agric. Sci.* 93: 711–717.
Lajtha, K., and S.H. Bloomer. 1988. Factors affecting phosphate sorption and phosphate retention in a desert ecosystem. *Soil Sci.* 146: 160–167.
Larsen, S., D. Gunary, and C.D. Sutton. 1965. The rate of immobilization of applied phosphate in relation to soil properties. *Soil Sci.* 16: 141–148.
Larsen, S. 1967. Soil phosphorus. *Adv. Agron.* 19: 151–210.
Lindsay, W.L. 1979. Chemical Equilibria in Soils. Wiley, New York.
Lindsay, W.L., and P.L.G. Vlek. 1979. Phosphate Minerals. pp. 639–672. In: Dixon, J.B., Weed S.B. (ed.) *Minerals in Soil Environments*. Soil Sci. Soc. Am., Madison, Wis.
Lins, I.D.G., F.R. Cox, and J.J. Nicholaides III. 1985. Optimizing fertilization rates for soybean grown on Oxisols and associated Entisols. *Soil Sci. Soc. Am. J.* 49: 1457–1460.
Loizides, P. 1970. Experiments with dryland rotation in the Syrian Arab Republic. Soils Fert. Report Damascus Agric. Station Project ESR/SF/SYR, 14 (1). FAO, Rome, Italy.
Loizides, P. 1980. Crop rotations under rainfed conditions in a Mediterranean climate in relation to soil moisture and fertilizer requirements, pp. 23–38. *Proc.*

FAO regional seminar on Rainfed Agriculture in the Near East and North Africa. FAO, Rome, Italy.
Luque, T., and P. de Arambarri. 1983. Dinamica del fośforo en los suelos de las marismas del Guadalquivir. *Anal. Edaf. Agrob.* 42: 1723–1735.
Macías Vázquez, F. 1981. Formation of gibbsite in soils and saprolites of temperate-humid zones. *Clay Miner.* 16: 43–52.
Madrid, L., and P. de Arambarri. 1985. Adsorption of phosphate by two iron oxides in relation to their porosity. *J. Soil Sci.* 36: 523–530.
Majeed, Y.A. 1986. Anti-desertification technology and management. United Nations Environ. Prog., Nairobi, Kenya.
Martin, R.R., R. St. C. Smart, and K. Tazaki. 1988. Direct observation of phosphate precipitation in the goethite/phosphate system. *Soil Sci. Soc. Am. J.* 52: 1492–1500.
Matar, A. 1976a. Direct and cumulative effects of phosphates in calcareous soils under dry farming agriculture of southern Syria. ACSAD, Damascus, Syria.
Matar, A. 1976b. Correlation between $NaHCO_3$-extractable P in soil and yield of wheat and lentil grown under dry farming conditions. ACSAD, Damascus, Syria.
Matar, A.E. 1977. Yields and response of cereal crops to phosphorus fertilization under changing rainfall conditions. *Agron. J.* 69: 879–881.
Matar, A.E. 1990. A descriptive model for prediction of residual phosphorus in soil after phosphate fertilization, pp. 29–60. In: Ryan, J., Matar, A. (eds.) *Proc. Third Regional Soil Test Calibration Workshop*. ICARDA, Aleppo, Syria.
Matar, A.E., and S.C. Brown. 1989a. Effect of rate and method of phosphate placement on productivity of durum wheat in Mediterranean environments. I. Crop yields and P uptake. *Fert. Res.* 20: 75–82.
Matar, A.E., and S.C. Brown. 1989b. Effect of rate and method of phosphate placement on productivity of durum wheat in a Mediterranean climate. II. Root distribution and P dynamics. *Fert. Res.* 20: 83–88.
Matar, A.E., M. Saxena, and S.N. Silim. 1988a. Soil testing as a guide to phosphate fertilization of five legumes in Syria, pp. 94–103. In: Matar, A.E., Soltanpour, P.N., Chouinard, A. (eds.) *Proc. Second Regional Soil Test Calibration Workshop*. ICARDA, Aleppo, Syria.
Matar, A.E., S. Garabet, S. Riahi, and A. Mazid. 1988b. A comparison of four soil test procedures for determination of available phosphorus in calcareous soils of the Mediterranean region. *Commun. Soil Sci. Plant Anal.* 19: 127–140.
Matar, A.E., J. Abdel Karim, and K. El Hajj. 1987. Studies on response of cereals and food legumes to phosphate fertilization in Syria as related to available P in soils, pp. 133–151. In: Soltanpour, P.N. (ed.) *Proc. First Regional. Soil Test Calibration Workshop*. ICARDA, Aleppo, Syria.
Matar, A.E., E. Jabbour, and K. El Hajj. 1988c. Prediction of barley response to fertilizers by means of soil nitrogen and phosphorus tests, pp. 12–23. In: Matar, A.E., Soltanpour, P.N., Chouinard, A. (eds.) *Proc. Second Regional Soil Test Calibration Workshop*. ICARDA, Aleppo, Syria.
Matar, A.E., P.N. Soltanpour, and A. Chouinard. 1988d. *Proc. Second Regional Soil Test Calibration Workshop*. ICARDA, Aleppo, Syria.
Mattingly, C.E.G. 1971. Residual value of phosphorus fertilizer on neutral and calcareous soils, pp. 1–15. Residual value of applied nutrients. *Tech Bull. 20.*, Min. Agric., Fish, Food, London.

Mehlich, A. 1984. Mehlich 3 soil test extractant: a modification of Mehlich 2 extractant. *Commun. Soil Sci. Plant Anal.* 15: 1409–1416.

Mehra, O.P., and M.L. Jackson. 1960. Iron oxide removal from soils and clays by a dithionite-citrate system buffered with sodium bicarbonate, pp. 317–327. In: Swineford, A. (ed.) *Clays and Clay Minerals, Proc. 7th Natl. Conf.* Pergamon Press, New York.

Meixner, R.E., and M.J. Singer. 1985. Phosphorus fractions from a chronosequence of alluvial soils, San Joaquin Valley, California. *Soil Sci.* 139: 37–46.

Mergoum, M., J. Ryan, and M. Abdel Monem. 1990. Response of high-yielding triticale to N and P in a rainfed Mollisol and Vertisol in Morocco. *Agron. Abst.*, p. 80.

Michel, C., A. Oudghiri, and A. Dardari. 1967. Diagnosis of mineral deficiencies of Moroccan soil in pot culture. *Al Awamia* 23: 1–58.

Michel, C., and A. Bouzoubaa. 1980. Premiers resultats des essais de fumure de redressement phosphate au Maroc. *Al Awamia* 60: 45–64.

Millot, G. 1970. *Geology of Clays*. Springer-Verlag, New York.

Millot, G., H. Paquet, and A. Ruellan. 1969. Néoformation de l'attapulgite dans les sols à caparaces calcaires de la Basse Moulouja (Maroc Oriental). *C.R. Acad. Sci. Paris* 268: 2271–2274.

Moore, D. 1974. Soil fertility research on wheat: Jordan wheat research and production. Final Report. Min. Agric., Amman, Jordan.

Moore, T.J., R.C. Hartwig, and R.H. Loeppert. 1990. Steady-state procedure for determining the effective particle-size distribution of soil carbonates. *Soil Sci. Soc. Am. J.* 54: 55–59.

Moreno, E.C., W.L. Lindsay, and G. Osborn. 1960. Reactions of dicalcium phosphate dihydrate in soils. *Soil Sci.* 90: 58–68.

Muljadi, D., A.M. Posner, and J.P. Quirk. 1966. The mechanism of phosphate adsorption by kaolinite, gibbsite and pseudoboehmite. Part I. The isotherms and the effect of pH on adsorption. *J. Soil Sci.* 17: 212–229.

Nat. Cent. Agric. Res. Tech. Trans. (NCARTT). 1975–1984. *Annual Reports*. Min. Agric., Amman, Jordan.

Nakos, G. 1987. Phosphorus adsorption by forest soils. *Commun. Soil Sci. Plant Anal.* 18: 279–286.

Nettleton, W.D., K.W. Flach, and R.E. Nelson. 1970. Pedogenic weathering of tonalite in southern California. *Geoderma* 4: 387–401.

Novais, R., and E.J. Kamprath. 1978. Phosphorus supplying capacities of previously heavily fertilized soils. Soil Sci. Soc. Am. J. 42: 931–934.

Nychas, E., and C.S. Kosmas. 1984. Phosphate adsorption by dark alkaline Vertisols in Greece. *Geoderma* 32: 319–327.

Olsen, S.R., C.V. Cole, F.S. Watanabe, and L.A. Dean. 1954. Estimation of available phosphorus in soils by extraction with sodium bicarbonate. *USDA Circ. No 939*, Washington, D.C.

Olsen, S.R., F.S. Watanabe, and R.E. Danielson. 1961. Phosphorus absorption by corn roots as affected by moisture and phosphorus concentration. *Soil Sci. Soc. Am. Proc.* 25: 282–294.

Olsen, S.R., and A.D. Flowerday. 1971. Fertilizer phosphorus interactions in alkaline soils, pp. 153–185. In: Olsen, R.A., Army, T.J., Hanway, J.J., Kilmer, V.J. (eds.) *Fertilizer Technology and Use*. Soil Sci. Soc. Am., Madison, Wis.

Oskay, K.S. 1986. The influence of time on phosphate retention and the description of P adsorption by means of the Langmuir isotherms in calcareous soils. *Doga, Türk Tarimve Ormancilic Dergisi* 10: 252–262.

Orphanos, P.I. 1987. Response to fertilizer phosphorus by barley under varying rainfall and available soil phosphorus, pp. 115–131. In: Soltanpour, P.N. (ed.) *Proc. First Regional Soil Test Calibration Workshop*. ICARDA, Aleppo, Syria.

Orphanos, P.I. 1988. Direct and residual effect of fertilizer phosphorus on barley in Cyprus, p. 44–49. In: Matar, A.E., Soltanpour, P.N., Chouniard, A. (eds.) *Proc. Second Regional. Soil Test Calibration Workshop*. ICARDA, Aleppo, Syria.

Orphanos, P.I., and Krentos. 1980. Concentration of N, P and K in leaves; straw and grain of wheat and barley as influenced by N and P fertilizers under semi-arid conditions. *J. Agric Sci.* 94: 551–556.

Osman, A., F. Russi, M. Pagnota, and P. Cocks. 1991. Response to phosphate application by Mediterranean grasslands grazed at two stocking rates, biomass production, and change in botanical composition. *J. Appl. Ecol.* (in press).

Ozanne, P.G. 1980. Phosphate mutrition of plants—a general treatise, pp. 559–589. In: Khasawneh, F.E., Sample, E.C., Kamprath, E.J. (eds.) *The Role of Phosphorus in Agriculture*. Soil Sci. Soc. Am., Madison, Wis.

Parfitt, R.L. 1978. Anion adsorption by soil and soil materials. *Adv. Agron.* 30: 1–50.

Parfitt, R.L. 1989. Phosphate reactions with natural allophane, ferrihydrite and goethite. *J. Soil Sci.* 40: 359–369.

Paris, P. 1986. Fertilizers and yield improvement in Mediterranean Italy. *Fert. Agric.* 92: 13–37.

Parra, M.A., J. Torrent, J. Barrios, and L. Montealegre. 1983. Balances mineralógicos y texturales en la formación de suelos de toposecuencias típicas de la parte central del Valle de los Pedroches (Córdoba). *Anal. Edaf. Agrob.* 42: 945–954.

Pavel, L., and G. Toma. 1972. Adsorption of phosphate anions in Iraqi soils. *Sbornik Visoké Skoly Zemedelské v Praze. Fakulta Agronomická* A(2): 53–64.

Peña, F. 1990. Influencia de diversos factores mineralógicos en la adsorción de fosfato en suelos de áreas mediterráneas. Ph.D. thesis, Univ. Córdoba, Spain.

Peña, F., and J. Torrent. 1984. Relationship between phosphate sorption and iron oxides in alfisols from a river terrace sequence of Med, terranean Spain. *Geoderma* 33: 283–296.

Pissarides, A., J.W.B. Stewart, and D.A. Rennie. 1968. Influence of cation saturation on phosphorus adsorption by selected clay minerals. *Can. J. Soil Sci.* 48: 151–157.

Rashid, N., N. Bughio, and M. Salim. 1988. Calibration of three soil tests for determining phosphorus fertility of soils to support cereals, legumes and oilseeds, pp. 86–94. In: Matar, A.E., Soltanpour, P.N., Chouniard, A. (eds.) *Proc. Second Regional Soil Test Calibration Workshop*. ICARDA, Aleppo, Syria.

Rashid, A., F. Hussain, G. Ahmad, and M. Salim. 1990. Use of a universal soil test and plant analysis for diagnosing P deficiency in rainfed wheat, pp. 98–110. In: Ryan, J., Matar, A. (eds.) *Proc. Third Regional Soil Test Calibration Workshop*. ICARDA, Aleppo Syria.

Rice, W.A., M.E. Akhtar, Y. Rohul Amin, and J.A. Campbell. 1990. Wheat responses to nitrogen and phosphorus in rainfed areas of Pakistan. pp. 66–75. In: Ryan, J., Matar, A. (eds.) *Proc. Third Regional Soil Test Calibration Workshop*. ICARDA, Aleppo, Syria.

Rich, C.I. 1968. Hydroxy interlayers in expansible layer silicates. *Clays Clay Miner.* 16: 15–30.

Roca, J., and F. Pomares. 1987. Movilidad y transformaciones en el suelo del fertilizante fosforado aplicado en un ensayo de campo en cítricos sometido a no laboreo. *Invest. Agr.: Prod. Prot. Veg.* 2: 237–242.

Russell, J.S., E.J. Kamprath, and C.S. Andrew. 1988. Phosphorus sorption of subtropical acid soils as influenced by the nature of the cation suite. *Soil Sci. Soc. Am. J.* 52: 1407–1410.

Ryan J. 1983. Phosphorus in soils of arid regions. *Geoderma* 19: 341–356.

Ryan, J. 1990. Soil and fertilizer studies in Lebanon, pp. 6–28. In: Ryan, J., Matar, A. (eds.) *Proc. Third Regional Soil Test Calibration Workshop*. ICARDA, Aleppo, Syria.

Ryan, J., and M. Alem Zghard. 1980. Phosphorus transformations with age in a calcareous soil chronosequence. *Soil Sci. Soc. Am. J.* 44: 168–169.

Ryan, J., G. Musharrafieh, and A. Barsumian. 1980. Soil fertility characterization of the Agricultural Education and Research Center of the American University of Beirut. *FAFS Bull. No. 64*. Amer. Univ. Beirut, Lebanon.

Ryan, J., and A.G. Ayubi. 1981. Phosphorus availability indices in calcareous Lebanese soils. *Plant Soil* 62: 141–145.

Ryan, J., D. Curtin, and M.A. Cheema. 1985a. Significance of iron oxides and calcium carbonate particle size in phosphate sorption by calcareous soils. *Soil Sci. Soc. Am. J.* 49: 74–76.

Ryan, J., H.M. Hasan, M. Baasiri, and H.S. Tabbara. 1985b. Availability and transformation of applied phosphorus in calcareous Lebanese soils. *Soil Sci. Soc. Am. J.* 49: 1215–1220.

Ryan, J., R. Shwayri, and S.N. Hariq. 1985c. Short-term evaluation of non-conventional organic wastes. *Agric. Wastes* 12: 241–249.

Ryan, J., and H. Tabbara. 1989. Influence of urea phosphate on infiltration and sodium parameters of a calcareous sodic soil. *Soil Sci. Soc. Am. J.* 53: 1531–1536.

Ryan, J., and A. Matar. (eds.) 1990. *Proc. Third Regional Soil Test Calibration Workshop*. ICARDA, Aleppo, Syria.

Ryan, J., M. Abdel Monem, and M. El Gharous. 1990. Soil fertility assessment at agricultural experiment stations in Chaouia, Abda, and Doukkala. *Al Awamia* 72: 1–47.

Ryan, J., and M. Abdel Monem. 1991. Implications of spatial variability for soil sampling and fertilizer use. Proc. Fourth Regional Soil Test Calibration Workshop, May 5–11, Agadir, Morocco. ICARDA, Aleppo, Syria.

Ryan, J., M. Abdel Monem, and M. Mergoum. 1991a. Responses of triticale varieties to N and P in semi-arid Morocco. Al Awamia (in press).

Ryan, J., M. Abdel Monem, M. Dafir, M. Mergoum, and S. Belaid. 1991b. Response of local and improved corn varieties in Morocco to phosphorus and zinc. Al Awamia (in press).

Ryan, J., M. Abdel Monem, M. Mergoum and D. Haderbach. 1991c. Impact of phosphorus fertilization on barley, wheat, and triticale in Morocco's dryland zone. Al Awamia (in press).

Ryan, J., M. Abdel Monem, and J.P. Shroyer. 1992. Visual assessment of nitrogen deficiency in dryland cereals: a basis for action in Morocco. *J. Agron. Educ.* (in press).
Saad, H.S., A. Bamatraf, and A. Haidra. 1990. Wheat response to soil P content in the Central Highlands of Yemen. pp. 91–97. In: Ryan, J., Matar, A. (eds.). *Proc. Third Regional Soil Test Calibration Workshop*. ICARDA, Aleppo, Syria.
Sabbe, W., and D.B. Marx. 1987. Soil sampling: spatial and temporal variability, pp. 1–14. In: Brown, J.R. (ed.) *Soil Testing: Sampling, Correlation, Calibration, and Interpretation*. Spec. Publ. 21. Soil Soc. Am., Madison, Wis.
Sample, E.C., R.J. Soper, and G.C. Racz. 1980. Reaction of phosphate fertilizers in soils, pp. 263–310. In: Stelly, M. (ed.) *The Role of Phosphorus in Agriculture*. Soil Sci. Soc. Am., Madison, Wis.
Santa Cruz, F., M.C. Bolarín, M. Caro, and M. Romero. 1981. Adsorción de fosfato en suelos calizos. II. Correlaciones entre parámetros de adsorción y algunas características del suelo. *Anal. Edaf. Agrob.* 40: 601–608.
Santa Cruz, F., M.C. Bolarín, M. Romero, and F.G. Fernández. 1982. Adsorción de fosfato por minerales de la arcilla. *Anal Edaf. Agrob.* 41: 2235–2243.
Saxena, N.P. 1984. Chickpea, pp. 491–452. In: Goldsworthy, P., Fisher, T. (eds.) *The Physiology of Tropical Field Crops*. Wiley, Chichester, U.K.
Sayegh, A.H., and A.A. Majid. 1969. Phosphorus fractionation and retention in alkaline Lebanese soils. *Agrochim.* 13: 265–276.
Schulze, D.G. 1981. Identification of soil iron oxide minerals by differential X-ray diffraction. *Soil Sci. Soc. Am. J.* 45: 437–440.
Schulze, D.G. 1988. Separation and concentration of iron-containing phases, pp. 63–81. In: Stucki, J.W., Goodman, B.A., Schwertmann, U. (eds.) *Iron in Soils and Clay Minerals*. Reidel, Dordrecht.
Schwertmann, U. 1964. Differenzierung der Eisenoxide des Bodens durch Extraktion mit Ammoniumoxalat Lösung. *Z. Pflanzenernähr. Düng. Bodenk.* 105: 194–202.
Schwertmann, U. 1985. The effect of pedogenic environments on iron oxide minerals. *Adv. Soil Sci.* 1: 171–200.
Seklani, H. 1983. Essais de fertilisation phosphatée en vue d'une amelioration pastorale sur des terres marginales des étages humide et semi-aride de la Tunisie, pp. 375–387. *Proc. Third Intl. Cong. P Cpds.* Inst. Mond. Phos. Casablanca, Morocco.
Sharar, M.S., M.A. Gill, and A.A. Shafqat. 1976. Lentil yield and quality as influenced by irrigation and fertilizer levels. *Pak. J. Agric. Sci.* 13: 231–234.
Shaviv, A., N. Shachar, and J. Hagin. 1989. Kinetics of phosphorus reactions in calcareous soils. *Commun. Soil. Sci. Plant Anal.* 20: 465–482.
Shepherd, K., P.J.M. Cooper, A. Allan, D. Drennan, and J.D.H. Keatinge. 1987. Growth, use and yield of barley in Mediterranean-type environments. *J. Agric. Sci.* 108: 365–378.
Shroyer, J.P., J. Ryan, M. Abdel Monem, and M. El Mourid. 1990. Production of fall-planted wheat in Morocco and technology of its improvement. *J. Agron. Educ.* 19: 32–60.
Sibbensen, E. 1978. An evaluation of the anion-exchange resin method for soil phosphate extraction. *Plant Soil* 50: 305–321.

Singer, A. 1978. Phosphorus retention in some basalt and tuff-derived Mediterranean soils. *Agrochim.* 22: 75–82.

Solís, P. 1988. Dinámica del fósforo en suelos de campiñas andaluzas. Ph.D. thesis, Univ. Córdoba, Spain.

Solís, P., and J. Torrent. 1989a. Phosphate fractions in calcareous Vertisols and Inceptisols of Spain. *Soil Sci. Soc. Am. J.* 53: 462–466.

Solís, P., and J. Torrent. 1989b. Phosphate sorption by calcareous Vertisols and Inceptisols of Spain. *Soil Sci. Soc. Am. J.* 53: 456–459.

Soltanpour, P.N. 1985. Use of ammonium bicarbonate—DTPA soil test to evaluate elemental use availability and toxicity. *Commun. Soil Sci. Plant Anal.* 16: 323–338.

Soltanpour, P.N. (ed.). 1987. *Proc. First Regional Soil Test Calibration Workshop*. ICARDA, Aleppo, Syria, and Mid Amer. Agric. Cons., Settat, Morocco.

Soltanpour, P.N., M. El Gharous, A. Azzaoui, and M. Abdel Monem. 1988. Nitrogen and phosphorus soil test calibration studies in the Chaouia region of Morocco, pp. 67–81. In: Matar, A.E., Soltanpour, P.N., Chouinard, A. (eds.) *Proc. Second Regional Soil Test Calibration Workshop*. ICARDA, Aleppo, Syria.

Soltanpour, P.N., M. El Gharous, A. Azzaoui, and M. Abdel Monem. 1989. Response of dryland wheat to P rates and placement methods. *Commun. Soil Sci. Plant Anal.* 20: 597–605.

Steiner, J.L., J.C. Day, R.I. Papendick, R.E. Meyer, and A.R. Bertrand. 1988. Improving and sustaining productivity in dryland regions of developing countries. *Adv. Soil Sci.* 8: 79–122.

Stumm, W., and J.O. Leckie. 1971. Phosphate exchange with sediments: its role in the productivity of surface waters. Proc. 5th Int. Water Pollution Res. Conf., San Francisco.

Syers, J.K., G.W. Smillie, and J.D.H. Williams. 1972. Calcium fluoride formation during extraction of calcareous soils with fluoride: I. Implications to inorganic phosphorus fractionation schemes. *Soil Sci. Soc. Am. Proc.* 36: 20–24.

Syrian Min. Agric./ICARDA. 1985–1988. Collaborative Research Project. *Annual Reports on Fertilizer Use on Barley in North Syria.* Syrian Arab Republic, Min. Agric. Agrarian Reform, Soils Directorate, and ICARDA, Aleppo, Syria.

Taimeh, A.Y., and B. Hattar. 1988. Phosphorus fixation in some calcareous Vertisols in Jordan. *Dirasat.* 15: 7–29.

Talibudeen, O. 1981. Precipitation, pp. 81–116. In: Greenland, D.J., Hayes, M.H.B. (eds.) *The Chemistry of Soil Processes.* Wiley, Chichester, U.K.

Talibudeen, O., and P. de Arambarri. 1964. The influence of the amount and the origin of calcium carbonate on the isotopically exchangeable phosphate in calcareous soils. *J. Agric. Sci.* 62: 93–97.

Tnani, T., and J. Kanenberg. 1971. Examen des differentes méthodes de determination du P et K assimilables dans les sols riches en carbonates de la Tunisie: et interpretation des resultats d'analyses. *Ann. de l'Inst. Nat. Rech. Agron. de Tunisie* vol. 44 (5): 32.

Torrent, J. 1976. Génesis de un suelo desarrollado en una arcosa al sur de la Provincia de Madrid. *Anal. Edaf. Agrob.* 35: 667–686.

Torrent, J. 1987. Rapid and slow phosphate sorption by Mediterranean soils: effect of iron oxides. *Soil Sci. Soc. Am. J.* 51: 78–82.

Torrent, J., V. Barrón, and U. Schwertmann. 1990. Phosphate adsorption and desorption by goethites differing in crystal morphology. *Soil Sci. Soc. Am. J.* 54: 1007–1012.

Torrent, J., and J. Benayas. 1977. Origin of gibbsite in a weathering profile from granite in West-Central Spain. *Geoderma* 19: 37–49.

Torrent, J., U. Schwertmann, and D.G. Schulze. 1980. Iron oxide mineralogy of some soils of two river terrace sequences in Spain. *Geoderma* 23: 191–208.

Truog, E. 1930. The determination of the readily available phosphorus of soil. *J. Am. Soc. Agron.*, pp. 874–882.

Turan, C., B. Kacar, and M. Sagatay. 1976. Phosphorus fixation in soils of the Antalya coastal region. *Yayinlari Ziraat Fakültesi, Ankara Üniversitesi* 588: 1–53.

Wada, K. 1977. Allophane and imogolite, pp. 603–638. In: Dixon, J.B., Weed, S.W. (eds.). *Minerals of Soil Environments*. Soil Sci. Soc. Am. J., Madison, Wisc.

Walker, T.W., and J.K. Syers. 1976. The fate of phosphorus during pedogenesis. *Geoderma* 15: 1–19.

Walsh, L.M., and J.D. Beaton. 1973. Soil Testing and Plant Analysis. Soil Sci. Soc. Am., Madison, Wis.

White, R.E. 1981. Retention and release of phosphate by soil and soil constituents. *Soils Agric. Crit. Rep. Appl. Chem.* 2: 71–114.

Yadav, B.R., K.V. Paliwal, and N.M. Nimgade. 1984. Effect of magnesium-rich waters on phosphate adsorption by calcite. *Soil Sci.* 138: 153–157.

Yurtsever, N. 1965. A tentative correlation for the Olsen bicarbonate phosphorus soil test with wheat responses under Turkish soil conditions. *Soil Sci.* 100: 163–167.

Yurtsever, N. 1987. A study on calibration for the Olsen phosphorus soil test with wheat responses under Turkish soil conditions, pp. 97–115. In: Soltanpour, P.N. (ed.) *Proc. First Regional Soil Test Calibration Workshop*. ICARDA. Aleppo, Syria.

Yurtsever, N. 1988. Two soil tests for phosphorus calibrated with barley responses in rainfed conditions of Turkey, pp. 36–44. In: Matar, A.E., Soltanpour, P.N., Chouinard, A. (eds.) *Proc. Second Regional Soil Test Calibration Workshop*. ICARDA, Aleppo, Syria.

Distribution and Management of Drylands in the People's Republic of China

Li Shengxiu and Xiao Ling

Advances in Soil Science, Volume 18

I. Introduction

The drylands, referred to in this review as arid and semiarid lands, as well as subhumid lands prone to drought, are widely distributed in China. The arid and semiarid areas cover 52.2% of its total (arid area 30.8% and semiarid area 21.4%), based on either precipitation of 400 mm as the rainfall line or 1.5 aridity (defined as evaporation/precipitation). Although the arid and semiarid areas occupy a wide range of the total lands, the arable portion of these areas is much less, only 30 million ha, about 30% of the total arable land in China. Together with the subhumid area, 74.24% of the total area of the country falls into this category of drylands (Working Committee of Natural Regionalization, Academia Sinica, 1959); these are mainly distributed in north China, including 15 provinces, municipalities, and autonomous regions with a population of 140 million, representing 17% of the total population of the country. As the lands in most of these areas cannot be irrigated, rainfed agriculture remains the main system of farming. The development of agriculture in these vast areas is very important for the national economy.

In ancient times, most of this area was covered by thick and vigorous forests and grasses. Cattle and sheep could be found everywhere, and the lands were extremely fertile. Some 6000 years ago, in their search for food, the ancestors of the Chinese people began cropping the fertile lands along the valleys of the Fenhe and Weihe rivers, tributaries of the Yellow River. During the Western Zhou Dynasty (ca. 11th century to 770 B.C.) and the Spring and Autumn Period (770–476 B.C.), the Yellow River Valley was already established as the country's main farming region, which later spread to the Yangtze River Valley, the southeastern coast, the regions north of the Great Wall, and the southwestern highlands. Since 121 B.C., when more than 100,000 Chinese soldiers under General Huo Qubing occupied the Hexi Corridor and established a civil administration, irrigated agriculture has flourished along the middle and lower reaches of all perennial streams. The middle reaches of the Black and Shiyang rivers have been particularly well developed and have become known as the "Golden Zhangye" and the "Silvery Wuwei" (Zhao Songqiao, 1985). These areas were not only the birthplace of Chinese agriculture but also the cradle of Chinese civilization.

The famous "Silk Road" which connected China with the West passed through these dryland areas. The "Dunhuang Grotto," symbolizing the second peak of ancient Chinese culture, is also located in this region and has become of historical importance. Over the centuries, in their struggle against nature, the people accumulated valuable experience in the management of drylands, especially land reclamation, water and soil conservation, and the construction of water conservancy projects. In addition, they invented farm tools, seasonally planned farm work, and coped with plant diseases and pests. The Karez underground irrigation systems in

Xinjiang were created long ago, and still benefit a vast area today. The water resources have been exploited profitably for several thousand years, and a number of fertile lands have been created, such as the fertile oases in desert areas. The Hexi Corridor is an example of such a large-scale oasis created in ancient times; the Manas Valley in the Hunggar Basin is a famous example of a recently reclaimed area; and the middle-low reaches of the Tarim River support a string of ancient oases.

Management of drylands has created productive oases and improved the natural environment as a whole. For example, in the Dunhuang oasis, the irrigated and fertilized agricultural soils have good tilth more than 2 m deep following 2000 years of continuous farming. In the Loess Plateau, through thousands of years of applying organic fertilizers, the top layers of the soils, which are as deep as 20 to 30 cm, even 50 cm in some places, are man-made. Human activities, however, have also had a significantly negative impact on the drylands; over a long period of cultivation, the lands have been poorly managed and often abused. As a result, the vegetation in many areas has been destroyed and the once-productive ecosystem has deteriorated. Persistent hazards, such as drought, salinization, and erosion by wind and water have become widespread. Also, the natural and social changes have intensified the problems that accompany human activities. Consequently, many of the dryland areas are at present poorly developed.

Since the establishment of "New China," a series of important policies and measures have been adopted in developing and realigning these areas. Through about 40 years of hard work, major achievements have been attained in industry, agriculture, transportation, communication, and culture. Some important bases of agriculture, forestry, and animal husbandry have been gradually formed, and coal, petroleum, and minor minerals have been exploited. Also, some important modern manufacturing industries have been established. Obviously, these lands have the potential to become the next frontier for agricultural and other industries. In fact, some regions have considerable promise for development.

II. Main Characteristics of the Drylands

The drylands are located in China in the latitude between 30° and 50° N, spreading from the warm temperate belt in the south to the temperate zone in the north. These lands are characterized by the following features.

A. Climate

The climate is rather different among arid, semiarid, and subhumid areas. As most of the drylands are located in the central Eurasian continent with surrounding high mountains and high plateaus, the moisture-laden summer monsoon from the Pacific Ocean cannot penetrate deep into the northwest hinterland; nor can the wet summer monsoon from the Indian Ocean, because the Himalayas (Ximalayas) and the Tibet Plateau stand in their way. As a result, precipitation is much less than in other areas. The lowest annual rainfall in the country is recorded here, and in some parts of the deserts there is no rain all year.

Rainfall varies greatly from arid to semiarid and subhumid areas, increasing gradually from the northwest towards the east, south, and southeast with 300 to 500 mm as an average annual precipitation for the most part. Rainfall distribution is highly variable within a season, and it is generally concentrated during June to September, accounting for 70% to 80% of the total rainfall.

Temperatures in the northwestern part of China, where most of the dryland areas are located, are lower in winter than are those in the southern part of the country. The northwestern part of these areas is ice-bound and covered by snow. The January mean temperature in the southern part is below 0° C; while in the northernmost part it drops as low as −30° C. In summer, thanks to southerly winds and more hours of sunshine, the north-south difference in temperature is less marked, and temperatures for the area as a whole are relatively high. For example, in the south the July mean temperature is about 30° C, while in the north it also exceeds 20° C, being even higher in some places in the north than in the south. Yet the summer season is short and the severe winter is long. Altitude and terrain both exert a strong influence on temperatures. Generally, with 1000-m rise in altitude, the temperature drops by 5° to 6° C. Most of the Qinghai-Tibet Plateau stands at 4000 to 5000 m above sea level. In the northern Tibet Plateau, the average temperature even in July, the warmest part of summer, stays below 10° C; sometimes it even snows. Winter turns the plateau into a world of gleaming snow. Both the annual and diurnal temperature ranges are great.

The highest maximum temperature (48.9° C), recorded in China was documented in the famous "Fire Prefecture" of Turfan, while nearby Fuyun has one of the lowest minimum temperature records (−51.5° C). In the Tarim Basin, as a typical example, the temperature difference between night and day may be as much as 15° to 20° C. In late spring and early summer, or late autumn and early winter, while the mornings and evenings are cold enough to require thick clothing, midday is so hot that even the lightest summer clothes seem too heavy. A local saying goes, "Fur coats in the morning, gossamer at noon; a place where you enjoy watermelon by the fireside." As a whole, the frost-free period is short and the average temperature from June to September is 15° to 25° C.

These areas are rich in solar and wind energy. The annual duration of sunshine amounts to 2500 to 3000 h, or even more, with yearly solar radiation of 136 to 160 kcal/cm^2. Accumulated temperature during the part of the year above 10° C ranges from 1700° to 5000° C.

B. Topography and Landscape

Pronounced differential uplifting during recent geologic time has created prominent and varied terrain features. These include inland basins (e.g., Junggar, Tarim), high plateaus (e.g., Inner Mongonia, Ordos, Qinghai-Tibet, Loess), and very high mountain ranges (e.g., Tianshan, Altai, Kunlun, Karakoram, Ximalaya or Himalaya). Depressions have also been created within the plateaus and basins, of these the Turfan is the lowest point in inland China. Ground surfaces vary from place to place. For instance, in the arid area, they are generally flat or undulating and subject to strong winds. Consequently, surface materials tend to be loose and coarse textured. As a result, shamo (sandy deserts) and gobi (stony deserts) land types are widely formed. In contrast, in the Loess Plateau, the ground surfaces are hilly and noted for their gullies, which were formed by serious water erosion.

Surface water flow in the inland river system is rather scarce. The distribution of groundwater resources is varied, mainly concentrated along river channels. Certainly, the underground water resources have potential for exploitation. In addition, there are large glaciers and accumulated snow packs on the large mountains, which provide a reliable water resource for developing irrigated agriculture, especially for the desert region.

C. Vegetation

Because of the varying degrees of influence by ocean monsoons, the climate changes from subhumid to semiarid and finally to arid, resulting in three corresponding regions of natural vegetation: forest, grassland, and desert. Except for some places where the natural vegetation has remained, most of these areas have sparse vegetation. Much of the rangeland has been overgrazed. Some of the mountains and desert ranges no longer support vegetation, and large tracts are entirely barren due to natural conditions as well as human activities such as overgrazing and other abusive practices.

D. Agriculture

There are many conditions suitable for developing dryland farming. The temperature and plant growth period support one or two harvests a year. Rainfall during the growing season is also favorable for most crops and therefore for agriculture and livestock production. When water is provided or soil water is effectively managed, the high temperatures and abundant sunlight become assets for plant growth. These factors contribute to the excellent yields and good quality of grain and cotton (*Gossypium*) crops in this region. Fruits, such as Xinjiang's melon from Hami and seedless grapes from Turfan, are famous for their taste because of their high sugar content. In addition, there are fine-food crops, vegetables, forages, forest and plant resources, cattle, horses, goats and sheep as well as wildlife resources; the Yak Tibetan sheep, Tan sheep, Xinjiang fine wool sheep, and Yili horse are well known.

These dryland areas are rich in land resources. The average arable land per capita is 0.21 ha, twice that of China as a whole. The soils differ greatly, ranging from very well-developed, highly fertile areas to poorly developed tracts with low fertility. In general, the fertility is low as is the content of organic matter. However, most of the soils are very deep, have a high water-holding capacity, and can therefore be ameliorated to high fertility.

The population density in dryland areas is much lower than that in irrigated areas or in rainfed areas, where the precipitation is sufficient for crop production. For instance, population density is 150 to 300 persons/km^2 for dryland areas in the Guanzhong Plain, Shaanxi Province, with rainfall of 500 to 700 mm, while the population density on irrigated lands in the same area is 500 to 800 persons/km^2. In Gansu Province, with rainfall ranging from 50 to 450 mm, an even greater difference occurs in population between rainfed and irrigated areas. For instance, in the middle part of the province where drylands dominate, the population is 60 to 100 persons/km^2, while that in the irrigated area is 79 to 144 persons/km^2. Both more land and low population density exert a strong influence on land use.

When more land is available for food production, fewer inputs are utilized by farmers. Taking fertilizers as an example, organic manure has long been used in China to provide plant nutrents and to improve soil structure. This is also true for the drylands. The amount of chemical fertilizers used in China is relatively small, and for a long time, very little chemical fertilizer was applied in the drylands. In recent years, chemical fertilizer usage has increased to some extent as the government has given attention to dryland agricultural production. Still, inputs for chemical fertilizers are far less than those on irrigated lands. For example, chemical fertilizer inputs on Weibei dryland are 225 to 300 kg/ha, corresponding to 45 to 60 kg N/ha, while those in the irrigated area in the Guanzhong Plain to 600 to 750 kg/ha. This discussion illustrates that these areas have great potential for adopting advances in modern science and technology to develep their agriculture and overall economies. However, as already mentioned, there is less rainfall in the dryland areas compared with the other regions, and this is a major limitation.

Although there are large glaciers and accumulated snow packs on the large mountains as well as some groundwater in some places, shortage of available water is still a significant problem for agricultural production, accounting for cropping intensities, crop yields, and income levels in the drylands. For instance, intensity in the irrigated Guanzhong Plain is 1.8 crops or so, while that in Weibei rainfed highland is 1.2 to 1.3 crops.

Generally speaking, grain crop yields on drylands are also much lower than those in irrigated ones. In 1987, grain yield in the dryland areas was only 0.57 to 2.69 Mg/ha, while tha national level was 3.53 Mg/ha. In addition, there are great differences in grain yields between wet and dry years. Taking Guanzhong Plain as an example, both dryland farming and irrigated farming are conducted. For dryland farming, grain yields generally are 1500 to 2250 kg/ha. However, in dry years, yields can be only a few hundred kg/ha, or none. For irrigated farming, the reverse is true: the grain yields are 3000 to 3750 kg/ha, and there is not much difference between dry and wet years. Grain obtained per capita depends on many factors, but per capita land area, total population, and crop yields are the main ones. Although there is more land and less population on the drylands, the grain obtained per capita is much lower than that in irrigated areas because the crop yields and cropping intensity are very low. For instance, in 1987, per capita grain obtained in the dryland areas was 207 to 355 kg, while the national level was 460 kg; grain obtained per capita in Weibei rainfed highland was only 300 to 400 kg, while that in the irrigated area was 500 to 700 kg. In the middle part of Gansu Province, where rainfall is less and temperature is low, per capita grain is 230 kg, while that in the irrigated area, west of the Yellow River was 614 kg. In some areas where rainfall is sufficient for crop production and the population is relatively low, per capita grain obtained is nearly the same as or even higher than that in irrigated areas. For example, per capita grain in Qianyang County on Weibei rainfed highland, Shaanxi Province, is 443 kg, which is 70 to 90 kg higher than that in irrigated areas such as Changan County, Qishan County, where population density is high.

The net per capita income in the dryland areas is lower than that in irrigated areas. In 1987, per capita income was only 122 to 354 yuan (3.75 yuan = 1 $U.S.), while the national per capita income was 460 yuan. The net per capita income on Weibei rainfed highland in Shaanxi Province was 180 to 230 yuan, while that in the irrigated area of the Guanzhong Plain was 300 to 400 yuan. In the middle part of Gansu Province where there is a large population and less land, net per capita income was 270 yuan, while that in the irrigated area west of the Yellow River was 440 yuan, higher than in Shaanxi Province. Because of the dry climate and lack of water resources, the six provinces and autonomous regions in northwestern China, which have more than 13.3 million ha of rainfed farmlands and a population of nearly 70 million remain the poor areas of China. The low crop yield and low per capita income result in less food and clothing. Although dryland farming has been significantly developed since the establishment of "New China," it is clear that the low production levels have not been changed greatly; the ecological environment of dryland farming desperately needs to be improved.

From the above discussion it is clear that the main agricultural characteristics of dryland areas in China are sparse population, large land areas, low inputs, low soil fertility, poor grain yields, and poor living conditions; but the production potential for agricultural as well as other industries is high.

III. Distribution of Different Drylands

The drylands have been scientifically differentiated into areas and natural regions, even belts within the same region. This information is necessary so that state and government authorities at all levels can determine the development strategies for agriculture, animal husbandry, water conservation, transportation, and communication, and prepare medium- and long-term plans for instructions regarding agricultural production of different types with different guidelines, and introduce various kinds of advanced techniques and experiences. The Chinese government has given great attention to this kind of work. Since the establishment of the New China much research has been done on natural and agricultural regionalization, in which the extent, boundaries, and types of dryland areas have been delineated. The economic purpose and strategic measures for the development of agricultural production have been fully discussed and carefully stated (Working Committee of Natural Regionalization, Academia Sinica, 1959; Central Meteorological Bureau, 1979; Contributing Group of China's Comprehensive Agricultural Regionalization, National Agricultural Regionalization Committee, 1981; Li Maosong, 1987). Researchers have carried out similar work for each region, and detailed information has been summarized (Zhu Junfeng, 1985; Liu Xunhao and Nan Xianglin, 1987). Since there are different views and purposes on this work, some differences exist on the regionalization of the drylands.

For distinguishing different areas, the moisture patterns and their characteristics are the main criteria. In most cases, the precipitation line of 400 mm is commonly used to separate wet and dry regions (Economic Geography Research Group, Geography Institute, Academia Sinica, 1985; Contributing Group of China's Comprehensive Agricultural Regionalization, National Agricultural Regionalization Committee, 1981). In the dry areas, there is no cropping without irrigation when the annual precipitation is below 250 mm, and the crop yields are risky and unstable when the annual rainfall is 250 to 400 mm. As a whole, these areas are short of available water, one of the main constraints, so crop production fails to be a major enterprise of agriculture, but animal husbandry is dominant with grazing as a priority; however, in areas above 400 mm, agriculture (crop plantation) is the priority. More than 90% of cultivated lands in China are concentrated in the southeast area, and dryland farming can be assured in the regions where the annual rainfall is 400 to 800 mm. In addition, the 400-mm rainfall line is also the crit-

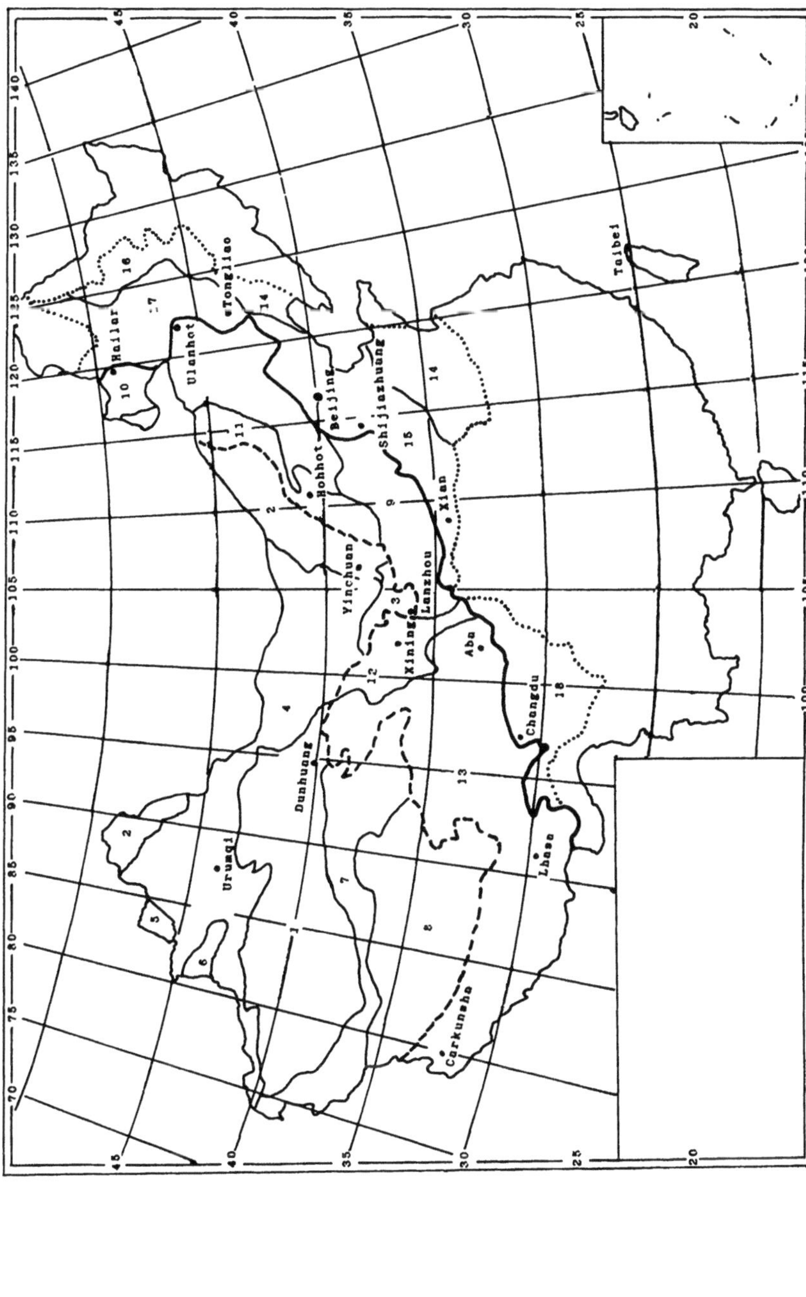

Figure 1. Distribution of different drylands in China. – – – –, Arid and semiarid area boundary; ———, dry and wet area boundary; · · · ·, subhumid and humid area boundary; ———, natural regionalization boundary. Source: Working Committee of Natural Regionalization, Academia Sinica (1959).

Table 1. Natural regions and belts in arid, semiarid, and subhumid areas

Area	Natural regions	Natural belts
Arid area	Warm-temperate MXPR	1. Desert-brown desert soil
	East of warm-temperate MXPR	2. Desert-brown soil
		3. Desert steppe premountains-sierozem
		4. Desert-gray brown desert soil
	West of warm-temperate MXPR	5. Desert steppe-brown soil
		6. Desert steppe premountains-sierozem
	Qinghai-Tibet Plateau	7. Dry desert
		8. Alpine frozen desert
Semiarid	Warm-temperate east monsoon	9. Steppe-manural loessial soil
	Temperate MXPR	10. Steppe-chestnut
		11. Steppe-light chestnut soil
	Qinghai-Tibet Plateau	12. Forest with meadowland meadow steppe
		13. Meadow steppe and meadow
Subhumid	Warm-temperate east monsoon	14. Semi-arid deciduous broadleaf forest-luvic drab soil
		15. Semi-arid deciduous broadleaf forest and forest steppe-drab soil
	Temperate east monsoon	16. Forest steppe-luvic black soil
		17. Steppe-black soil
	Qinghai-Tibet Plateau	18. Meadowand conifer forest

MXPR, Mongolia-Xinjiang plateau area.
Source: Working Committee of Natural Regionalization, Academia Sinica (1959).

ical line for forest growth. The natural distribution of forests with predominantly arbor species reaches roughly to the 400-mm rainfall line, and over 90% of the forest lands, including some other forest resources, are concentrated in most of the mountain ranges to the west of this line.

The dry areas are further classified into arid and semiarid, and the wet areas into humid and subhumid, the latter being in the south just below the 400-mm rainfall line (Fig. 1). As mentioned at the beginning, the drylands in the review include arid, semiarid, and subhumid areas, so the drylands differ from dry areas. In addition to dividing the areas according to precipitation or aridity, the drylands have been further differentiated into 10 natural regions and 18 natural belts (Fig. 1, Table 1), of which the arid and semiarid areas have 7 regions and 13 belts, and the subhumid areas 3 regions and 5 belts, based on the availability of heat and water, and the common characteristics of soil, vegetation, and land use (Working Committee of Natural Regionalization, Academia Sinica, 1959;

Contributing Committee of Natural Geography of China, Academia Sinica, 1981). From the above zonification, it is evident that the vast drylands in China have diversified types of environments. Evaluation and study of the natural resources for the different areas is very important for developing their economies.

A. Arid Areas

Arid areas are defined as those with less than 250 mm annual precipitation. They are mainly situated in the Xinjiang Uygur Autonomous Retion, the Hexi Corridor in Gansu Province, the middle parts of Ningxia Hui Autonomous Region, and the northwest parts of Qinghai Province and Inner Mongolia. There are 5.158 million ha arable land with a population of 17.48 million. The average arable land per capita is 0.29 ha (Wan Jiangzhong et al., 1987).

The climate in arid areas is mostly desert and dry, with scanty and sporadic precipitation decreasing from east to west. East of the Ondormiao-Baillingmiao-Otog-Yanchi line, annual precipitation amounts to 300 to 400 mm, with an aridity index (evaporation/precipitation) of 1.5 to 2.0. Between that line and the Helan Mountains, annual precipitation decreases to 200 to 300 mm with an aridity index of 2.0 to 4.0. West of the Helan Mountains, precipitation drops below 200 mm with an aridity index greater than 4.0. Toksun in the Turfan Basin has the lowest average annual precipitation record (3.9 mm) in China.

There is great variation in rainfall both seasonally and annually. In parts of southeastern Xinjiang there may be no rain for 3 successive years. Yet, a single rainstorm might cause serious flood damage. Generally, the dry season lasts more than 6 to 8 months annually. As both warm temperate and temperate zones exist in the areas, there is also great variation in temperature among regions and seasons. Pronounced dissimilarities in temperature are found in a basin and the areas outside. Because of the lower altitude, temperature inside the basins is generally higher than that in the mountains around them. The Turfan Depression has the hottest summer in China because it is a comparatively small enclosed basin located below sea level. The lowest temperature in the country has also been recorded in these areas. Accumulated temperature above 10 °C ranges from 1700° to 5000 °C annually. The frost-free period is too short in most parts of the areas for crop production.

The areas are very susceptible to wind erosion and the damage of wind and drift-sand is serious. For example, the Junggar (i.e., Dzungaria or Zungaria) Gate has 155 days of strong wind annually, and wind velocities in the western margin of the Turfan Basin exceed 50 m/s. In China, no other regions witness as much gale as does Qinghai. The wind with a speed equal to or greater than 27 km/h prevails over the province. Most places have over 50 windy days annually on average with more than 100 windy days in the west basin and the west of Qinnan Plateau, and over 150 windy days in the Tanggula mountain range. The wind constantly blows away the topsoil. Sometimes it is strong enough to blow off the soil as thick as several mm in a few hours.

The annual amount of wind erosion on cultivated lands is about 3 mm, with 45 Mg/ha of sands blown away and more than 15 mm on fallow lands. With serious wind erosion, fertile topsoil is scraped away, organic matter reduced, nutrients lost, and soil aggregates destroyed, and physical and chemical properties of the soil worsened. It is estimated according to the second soil survey that the organic matter content in the cultivated land is 20% to 30% lower than it was 20 years ago. Wind threatens all stages of plant growth and decreases its output. A fierce wind in Caidam Xiangde farm in 1959 left 93 ha of crops covered by shifting sand. It was determined that the yield loss per ha was 150 to 225 kg, and the maximum yield loss reached 952 kg.

Wind has also been disastrous to animal husbandry. In 1979, one sandstorm on the Qiliantuoli livestock farm buried 48 producing sheep and killed 2802 sheep. In spring, when the lambs are born and reared, a 7-point wind blowing continuously for 2 h can make the ewes give birth early. As a direct result of wind erosion, the land is being desertified. The average expansion of the sandy area is 65,000 ha/year. In Caidam, sands mounded by wind are pushed 20 to 50 m/year. The sandy area in Gonghe County has doubled in 20 years. As in the arable land, wind also reduces the productivity of grassland and speeds up its degeneration. There are 70,000 ha of grassland in need of water; 26,000 ha of such land are degenerated and 26,000 ha have become desertified. In some parts of Haiyan County, the grass yield gradually declines by 3.5% each year in succession. The degenerated area covers 32.3% (Wang Shaoming, 1987). No doubt, wind erosion has a great effect on reservoirs, like adding fuel to flames, and reduces the holding capacity remarkably.

Prominent and varied terrain features were created in the areas, such as inland basins (e.g., Tarim, Junggar), high plateaus (e.g., Inner Mongolia, Ordos), and very high mountains (e.g., Tianshan, Altai, Karakoram), as well as a famous depression (Turfan). The Tianshan Mountains run across the middle of Xinjiang and divide it into the Junggar and Tarim basins, and the eastern part of the mountains surround the Turfan Depression on all sides by their branch ranges. At an average elevation of about 1000 m above sea level, the terrain in the Inner Mongolia Plateau is flat and broad with no high mountains or deep valleys. It is a typical example of rolling tableland, where wide but shallow basins surrounded by low hills are common. As a whole, the ground surfaces are flat or undulating. Subject to strong winds, loose and coarse-textured materials are left on the ground surfaces, thus forming sandy desert and stony desert lands, which occupy areas of 562,500 and 569,500 km^2 respectively. Few perennial streams exist in the region. Some large rivers originate in the neighboring high mountains and are fed by melting snow. However, with the major exception of the Yellow and Erux rivers, all rivers terminate in basins within the region. Lakes are common but are mostly saline. In addition to some groundwater mainly along river channels and piedmont plains, accumulated snow on mountains is a reliable water resource for irrigation. Misuse and waste of the water resources as well as lack of groundwater resources are still major problems.

Sparse vegetation is typical in the areas. The sandy and stony deserts support few plants such as *Saxaul*, wormwood, *Zygophyllum*, and *Anabasis*, and the vegetative cover of shrubs is usually less than 30%; there are stretches with no vegetation. However, there are steppe zones where the cover generally exceeds 50%. On the lower slopes of the high mountains and in the Inner Mongolia Plateau, the vast grasslands are good grazing grounds. It is estimated that there are more than 19 million ha of fixed desert lands suitable for grazing. In addition, there are big stretches of timber on the upper slopes of the mountains.

In these areas agriculture has been developed only in specific locations, generally for one-crop systems. Since most crops are grown with irrigation, the crop production there is called "irrigated agriculture." Crop intensity is low, even in irrigated lands, only 0.81 to 1.14 crops per year. Animal husbandry can be managed in most of these areas except on the deserts and in the wilderness, but the stock-carrying capacity per ha is very low. Desert soil sequences are dominant with some scattered oasis soil sequences. Most of the soils are poorly developed with low organic matter and fertility, but high in soluble salts. They are generally unfavorable for agriculture, but large tracts of fertile soil are found in some places.

B. Semiarid Areas

These areas generally receive between 250 and 400 mm annual precipitation (Working Committee of Natural Regionalization, Academia Sinica, 1959; Central Meteorological Bureau, 1979), but for practical use, some places located within or near the areas, receiving less than 500 mm annual precipitation, are also included in this category (Contributing Group of China's Comprehensive Agricultural Regionalization, National Agricultural Regionalization Committee, 1981; Wang Qian, 1983; Li Maosong 1987; Xin Naiquan and Zhang Shaoli, 1987). These areas include 250 counties, mainly located in the west of three northeastern provinces, the southeast of Inner Mongolia Autonomous Region, the northern and middle of Shaanxi Province, most parts of Shanxi Province, the south of Ningxia Autonomous Region, the middle of Gansu Province, the east of Qinghai Province, and Yili Prefecture which is in the northwest of

Xinjiang Autonomous Region. There are 14.67 million ha of arable land, and a population of 52.13 million (Wan Jianzhong et al., 1987).

The climate is cold and cool in the west portion of the three northeast provinces, the Yin Mountain which is located to the east of Baoto in Inner Mongolia, and the counties along the Great Wall where there is a lot of strong wind during spring. The southern areas are temperate and warm. These areas might be divided into three zones based on temperature: temperate zone, warm temperate zone, and Qinghai-Tibet Plateau. In the temperate zone, the accumulated temperature above 10°C is between 1700° and 3200°C with 100 to 155 days above 10°C and 120 to 150 days frost-free. Average temperatures vary from −10° to −28°C in January and from 20° to 24°C in July. Large diurnal temperature differences exist, which is favorable for plant growth in summer, but easily causes frost damage in the fall. The annual precipitation varies mostly from 150 to 450 mm, but some places receive no rain at all. The aridity index is 1.3 to 2.0 except for in the mountainous regions. Less snow in winter and serious spring drought often occur. In contrast, the accumulated temperature above 10°C in the warm-temperate zone ranges from 3200° to 3600°C with a frost-free period as long as 5 to 5.5 months. The average temperature varies from −6° to 8°C in January and 18° to 20°C in July. In high altitude places, which occupy a large area in the zone, the temperature is lower. The annual precipitation varies from 350 to 450 mm, and the aridity falls to between 1.25 and 1.50. Spring and summer drought is often and serious. In the Qinghai-Tibet Plateau, the temperature and rainfall are highly variable. The accumulated temperature above 10°C ranges from 400° to 2000°C, and the July average temperature is below 20° but higher than 5°C. In most of the places annual precipitation varies between 300 and 500 mm, and the aridity is between 1.0 and 2.0. Some places in the valley along the Yalutsangpo River have high precipitation. Lhasa is an example where the precipitation is as high as 918 mm annually, but highly variable. In addition, the semiarid areas in the plateau also have the same climatic features as in other areas in the plateau: high solar radiation, long days of sunshine, large differences between day and night temperatures, and strong winds.

The Qiangtang Desert and Caidam Basin in the Qinghai-Tibet Plateau are located in the west, and the Loess Plateau is located in the east of the semiarid areas, and some valleys and basins between mountains as well as hills are scattered in these areas in addition to parts of the Inner Mongolia Plateau and the Northeast Plain. The Caidam Basin is enclosed by the Qilian Mountains to the north and northeast, the Kunlun Mountains to the south, and Altyn Mountains to the northwest. The basin measures 850 km from east to west and 250 km from north to south at its widest part. It is a depression in a high plateau, the elevation of its floor being between 2500 and 3000 m above sea level; it is characterized by plains, hilly regions, and deserts in the northwest and level land in the southeast.

The Loess Plateau is bounded by the Great Wall to the north, Taihang Mountains to the east, Qilian Mountains to the west, and Qinling Mountains to the south. The first three boundaries are all in the semiarid area, but the last reaches the subhumid area. Averaging 1000 m or more in elevation and covering 600,000 km^2, it is the world's largest loess plateau. The Yellow River and its tributaries—the Taohe, Weihe, Luohe, and Fenhe rivers—all flow across it. The plateau is largely covered by a layer of loess which generally varies in depth from 50 to 80 m but exceeds 100 m in some places. The yellowish soil is a loose, loamy deposit blown in over the centuries all the way from the highland deserts of Inner Mongolia. Sparse vegetation and concentrated rainfall in summer have caused serious losses of water and soil, and over the ages have resulted in the formation of numerous gullies. The complex terrain of the plateau is typical of loess land formation. Loess land forms are either the highland plain or hilly region type. Loess land of the plain is relatively intact, with a level center and gullies on the margins. Typical examples of this type are the well-known Dongzhiyuan in eastern Gansu and Luochuanyuan in northern Shaanxi, both of which cover extensive areas. The loess hilly regions, whose variation in height was brought about by the constant flowing of water, are noted for their gullies. This land form comprises two distinct types of terrain: the liang type, which is elongated in shape and has an arched top; and the mao type, which has an oval outline or a round top.

Depending on the degree of erosion, a loess highland plain can be turned into a liang, then a mao, and finally a rolling plain when the marginal gullies become increasingly obvious. Large tracts of loess hilly land, which are distributed widely, can be found in western Shanxi and northern Shaanxi. In addition to serious water erosion, wind erosion is a problem in some places. Groundwater in most of the areas is sparse. In some places it occurs too deep to use for irrigation, and in others it is not of sufficient quality. Also, where dainage is poor, salinization may occur.

With a temperate climate, the semiarid area in the Northeast Plain and southeastern portion of Inner Mongolia is a vast prairie or dry pasture covered mainly with perennial grasses such as goat grass (*Aneurolepidium chinense*) and feather grass (*Imperata cylindrica*). These grasses have long, narrow blades that are woolly and curl towards the midrib, preventing excessive loss of moisture. Some plants on the grasslands have long roots to tap moisture deep down in the soil. Although this is a major livestock-raising area, grass production per unit area is low, and a number of places have been turned into agricultural lands.

The Loess Plateau with warm-temperate climate also belongs to the steppe vegetation pattern, but the grasslands have almost all been turned into farmland. The semiarid area in the east-central and southern portions of the Qinghai-Tibet Plateau has quite different vegetation. Situated at 4000 m above sea level with no summer, this area is low both in temperature and precipitation, and thus alpine meadow-pasture and pasture dominate. The sunny slopes of mountains provide ample grazing grounds for the yake, Tibetan goats and sheep, which are the main domestic animals. A wide range of agriculture and production systems is found in the semiarid areas. In the temperate semiarid area, most of the lands are used for grazing, and overgrazing has led to degradation of pasture and promoted desertification. Some pastures have not been used as grazing fields due to lack of drinking water for animals. A number of places in the area have been turned into agricultural centers. Many state farms which practice both crop-raising and stock-breeding have been established on the great grasslands of the northeast. Their main crops are soybeans (*Glycine max*), sorghum (*Sorghum bicolor*; *S vulgare*), and sugar beets (*Saccharum officinarum*). Irrigation from the Yellow River has turned the Hetao area into a granary, north of the Great Wall. However, in most of the area with no irrigation, crop yields are low and unstable.

The semiarid valleys of the Qinghai-Tibet Plateau grow such cold-resistant crops as spring qingke, a variety of highland barley (*Hordeum vulgare* sp.), spring barley (*Hordeum vulgare*), and peas (*Pisum sativum*). In recent years, winter wheat (*Triticum aestivum*; *Tr. durum*; *Tr. compactum*) has been planted successfully in some places and the yields have been high.

The Loess Plateau is a major region for crop production together with animal husbandry in its semiarid areas. The farming is extensive with low productivity and frequent natural disasters. The competition between crop plantation and animal husbandry is keen and grasslands are becoming deserts, especially in most parts of Shanxi, the nothern part of Shaanxi, the middle part of Gansu, the south of Ningxia, and the east of Qinghai. All these areas in the plateau have deep soil layers, and there are many mountains. Unfortunately, because trees have been cut arbitrarily, leaving lands barren, runoff and erosion are serious. The frost-free period in the plateau is 100 to 180 days, and most areas adopt one-crop systems with a few places having three crops in 2 years. The crop intensity is 0.85 to 1.07 crops per year.

The main soil sequences in the semiarid areas are aeolian sand, chestnut, brown, sierozem, and loessial. As these soils have long been cultivated and suffered from serious water and wind erosion, the natural fertility has been depleted, and therefore most of the soils are low in organic matter and deficient in N and P as well as some micronutrient elements. Also, some soils are low in water-holding capacity.

C. Subhumid Areas Susceptible to Drought

Areas with 500 to 800 mm annual precipitation are defined as subhumid areas. These areas embrace 243 counties and cities, mainly situated in Beijing, Tianjing, the middle and southern parts of Hebei Province, Shandong Province, the northern parts of Henan Province, the southeast of Gansu Province, and some of the three northeast provinces. In these areas, there are 10 million ha of arable land and 70 million people (Wan Jianzhong et al., 1987).

Climates in these areas are also different. The subhumid region in the three northeast provinces is temperate with a frost-free season of less than 150 to 160 days, accumulated temperatures of 2000° to 3000° C, and 110 to 138 days with temperature above 10° C. The average temperature is between −27° and −16° C in January and 18° and 24° C in July. Annual precipitation ranges from 400 to 700 mm with aridity 1.0 to 1.5, except for in a few places where the precipitation is less than 350 mm. Excluding the subhumid area in Qinghai-Tibet Plateau, the other areas are warm-temperate with a relatively longer frost-free period (up to 170 to 220 days). Precipitation is relatively abundant. However, as the monsoon in this region has a great effect on the climate, there is a sharp variation in annual amounts of precipitation. Also, the rainfall is uneven among seasons: rainfall is often deficient during spring and fall, and drought in spring is more serious here than that in the three northeast provinces, making it difficult to obtain a good stand of crops in the spring and to produce summer crops.

The subhumid region in the Qinghai-Tibet Plateau is another world. There is great variation both in temperature and rainfall. The total accumulated temperature is from 1000° to 2000° C with frost-free periods of 2 to 7 months and less than 130 days above 10° C. January mean temperature generally varies between 0° C and −5° C. Precipitation ranges from 500 to 1000 mm with aridity values from 0.5 to 1.5, or even more. This region also has some common features of the plateau such as strong radiation, long days of sunshine, strong winds, and large diurnal temperature differences.

Plains, low plains, and plateaus are the common landscape in these areas. The low plains here occupy more than half of the total low plains in the country. Most of the areas have been turned into farmland except for in a few places in the northeast mountains and large tracts on the Qinghai-Tibet Plateau. Water resources are much more abundant. Besides groundwater, there is river water and numerous reservoirs have been built to collect water for supplemental irrigation. In some places, however, poor irrigation practices and inadequate drainage have resulted in salinization of large tracts. The natural vegetation is steppe and forest-steppe types in the temperate regions, forest, or forest-steppe types in the warm-temperate regions, and forest, meadow, as well as shrubs, in the Qinghai-Tibet Plateau. The forests in the first two regions are composed of temperate deciduous broad-leaf trees, sometimes called summer-green forests because they grow thick foliage in summer and lose their leaves in winter. The two regions have many broad-leaf forests similar to those in western Europe, northeastern Asia, northern Japan, the Korean Peninsula, and eastern North America. The forests in the plateau consist of cold-temperate deciduous needle-leaf trees such as spruce and pine, which are particularly suited to the severe cold because the small area of their leaves reduces water loss by transpiration to a very low level during winter when water is unobtainable from the frozen soil. As most of the area has been turned into farm land, the forest and steppe vegetation disappeared long ago in most parts except in the three northeast provinces and the Qinghai-Tibet Plateau, where large tracts of forests still remain.

The main soils are black, meadow, alkali-saline, and the drab soil sequence, which is deep, fertile, and very well developed.

These areas are very important for agricultural production. Except for the mountains and the plateau, the plains and low plains have long been used to grow wheat, maize (*Zea mays*), and rice (*Oryza sativa*), and are one of the major areas for cotton (*Gossypium arboreum*; *G. herbaccum*; *G. hirsutum*; *G. barbadense*), which likes warm climates. Since precipitation is more abundant and the frost-free period longer, crop yields are higher and more stable, and two crops a year or three crops in 2 years are dominant systems. Cultivation is extensive and the crop intensity is 1.3 to 1.5 crops per year. However, due to the uneven distribution of precipitation often resulting in spring drought, agricutural production is greatly influenced by water stress and therefore the areas are included here as drylands.

D. Humid Rainfed Areas Suffering from Water Stress

There are many drylands in moist areas in the south of China that are defined as humid areas. There is more than 500 million ha of arable land in the humid areas, of which rainfed lands account for 38%. Since water stress often affects crop production, these drylands may be called humid rainfed areas suffering from water stress.

These drylands are distributed to the south of the Kunlun Mountains, Qinling Mountains, and Huai River basin with an annual precipitation of over 800 mm. The drylands constitute 13.33 million ha distributed in 15 provinces, municipalities, and autonomous regions in the south of China (Taiwan Province is not included). Water resources are plentiful and the climate is mild and warm. The crop-growing seasons are long, with two or three crops a year in most of the drylands. The natural conditions are favorable for crop production, and crop yields are high, but on the hilly drylands in some places the field capital construction is poor and loss of water and erosion of soil are severe; thus, the soil is not fertile and crop production is low.

The main crop is rice, but wheat, maize, sweet potatoes (*Ipomoea batatas*), broad beans (*Vicia faba*), peas, and others are cultivated in hilly mountain areas. Oil-bearing crops, hemp (*Cannabis sativa*), silks, tea, sugar crops, vegetables, tobacco (*Nicotiana tobacum*), fruits, and medical herbs are also grown on these rainfed lands. Obviously, the rainfed agriculture in south China plays an important role in stabilizing grain production, developing cash crops, and raising commodity production rates. It also has considerable potential for increasing farmers' incomes and improving their living standards. As most attention in the past was given to the irrigated lands, grain yields on the rainfed lands here have been relatively low, generally 2.25 to 4.50 Mg/ha. Recently much progress has been made in improving farming systems on these lands and crop yields have increased. For instance, the multiple intercropping system for rainfed agriculture has been carried out in the hilly areas in the southwest part of Hunan Province. The system of three crops a year in the stripping intercropping of wheat, corn, and grasses has been extended on rainfed lands in Sichuan Province. The apparent yield increase benefits have been obtained in improving and extending the stripping intercropping rotation in Zhejiang Province. Improved cultivars, increased soil fertility, and the improvement of drought-resistant cultivation practices have raised grain yield from 6.0 to 11.25 Mg/ha. South China, thus, has great potential.

The main constraints in the arid, semiarid, and subhumid areas, together with hilly drylands in humid areas, are very clear from the facts illustrated above: The natural conditions are poor with scarce rainfall and vegetation; the loss of water and soil erosion are serious; the soil is barren and cultivation is extensive; and there is a large area for cropping but production is low. Because of these problems, agricultural production is often backward and the living standard comparatively low. Nevertheless, these areas have great possibilities, since there are abundant soil resources, and there is not only plentiful arable land per capita with a capacity for increasing yield, but also a lot of uncultivated land for exploitation. In addition, there are ample resources of light and heat. These advantageous conditions provide a basis for the comprehensive development of agriculture, forestry, and animal husbandry.

IV. Soils in the Dryland Areas

The varied patterns of climate, rock formation, topography, and vegetation, as well as the long history of agricultural development in the dryland areas, have resulted in many kinds of soil both in horizontal and vertical zonality. A typical example showing the soil vertical distribution can be seen in the Helan Mountains with the highest peak 3554 m above sea level. Every rise of 200 m in elevation will decrease the average annual temperature by 1° C and increase the precipitation by 23 mm. The vegetation from the foot upwards is desert steppe, mixed forests of conifers and deciduous trees, conifer forests, and mountain meadows, in succession. As a result, there is a regularity in the variation of soil type: sierozems below 2000 m; gray-drab forest soils from 2000 to 3100 m; and mountain meadow soils above 3100 m. Liupan Mountain is in the low latitude, where precipitation is high, thus the soils are burozems (Wang Jizhi, 1986).

The horizontal soil distribution is more clear and occurs on a large scale. Travelling from arid, semiarid in the northwest, to subhumid in the middle and east of the country, one can see the desert and semidesert soil (desert soil sequence with oasis soil sequence scattered through it, and aeolian sand soil sequences), then steppe soil (chestnut, brown, and sierozem soil sequences), and finally the forest-steppe soil (loessial soil and drab soil sequences). In addition, some alkali-saline soils formed on different zonal soils are found extensively in low-lying areas such as the plains, basins, and valleys of arid and semiarid inland regions. Because the goundwater table is high and there is great evaporation of surface water, easily soluble salts in the subsoil are drawn up into the topsoil, making the land unsuitable for growing crops. With soil amelioration, however, such land can become highly productive.

It should be noted that in the past the soil classification system used in China was based on the soil zonal distribution. Recently, a new soil classification system based on diagnostic soil horizons (Soil Survey Staff, USDA, 1975) has been proposed. According to this system, the Chinese soils are classified into 12 soil orders. However, most of soils in the arid, semiarid, and subhumid areas fall in four orders. The aoelian and oasis soil sequences and cultivated loessial soil belong to the Entisol; the desert, brown sequences and sierozem to the Aridsol; the chestnut, Chernozem, black to the Mollisol, and the drab, manural loessial to the siallitite. For details of the classification system see Soil Classification Research Group, Institute of Soil Science, Academia Sinica, Nanjing (1985) and Gong Zitong et al. (1988). In this review we will describe the principle soils individually.

A. Oasis Soil Sequence

Mainly formed under desert conditions, the oasis soils are widely distributed in the arid areas of Xinjiang and Gansu Provinces. In desert areas, this soil sequence covers more than 10,000 km^2 and has been used as farmland for many years. The oasis soils in the south of Xinjiang and the Hexi Corridor in Gansu Province have been used for agriculture since the first or second century B.C., and most of the remainder of the oasis soils have been used since the sixteenth to seventeenth centuries (Zhu Maoshun, 1965). Formed by irrigation with muddy and sandy water, as well as by cultivation and fertilization, this soil has deep, deposited layers differing in structure and physical, chemical, and biological properties. The deposited horizon is comparatively homogeneous, loose, and porous, and comprised mainly of loam or sandy loam generally with a bright gray color. In the deposited layer, broken tiles and coal cinders brought in during the course of manuring are often found. The oasis soil at the southern foot of Tianshan Mountain has a strong lime reaction. Silt is the main soil particle, but there is a great variation in particle composition. For instance, the oasis soil in Akisu contains 60% to 70% coarse sand, 10% to 15% clay, and 10% to 20% fine sand, while the one at the foot of Kunlun Mountain contains less limestone and more fine sands: fine sands, 40% to 60%; coarse silt, 15% to 20%; fine silt, 12% to 18%, and clay 10% or so, depending on the irrigation water. As the amount of water added from irrigation exceeds many times that from precipitation, the soil water status has been changed and the soluble salts and gypsum have been leached to 3 to 4 m. As a result, the top layer has less than 0.2% salt content. In places with a high water table, the potential for secondary salinization may occur. However, there is no salt accumulation in the top layer following many years of irrigation. Although the leaching process has resulted in loss of gypsum, lime content is still high because of its lower solubility and its addition by biomass and irrigation water. The lime may be as high as 25% to 30% in some places, and it is uniformly distributed in the topsoil profile with a little higher content in the subsoil. Clay has also moved into the subsoil so the subsoil is a little heavy in texture and somewhat compacted (Figs. 2, 3).

The radical difference between oasis soil and desert soil lies in that the former has higher biological activity. As irrigation has changed the heat and water status, the oasis soils have gained population and diversity of microorganisms (Table 2). For example, the soils cropped with cotton in the Tarim Basin have much more bacteria for nitrogen fixation, ammonification, and nitrification, as well as for decomposition of cellulose and mineralization of phosphorus. Large populations of N-fixing bacteria is a significant property for this soil, and earthworm activity is great. The intensive biological activity benefits humus formation and nutrient accumulation so that organic matter reaches 1% to 2% in the top deposited layer and 0.5% to 0.7% in the layers below, which is 1 to 2 times higher than that in the desert soil. The only exception is found in the Tarim Basin and the Hexi Corridor where the organic matter is low. Also, the soil is rich in potassium but lower in nitrogen and phosphorus. Even so, the latter two elements are still considerably higher than found in desert soil.

Three types of this soil sequence have been classified. They are grayish oasis soil, whitish oasis soil, and moistened oasis soil.

The grayish oasis soil occurs in middle and top alluvial fans and also in the upland of the alluvial plain, 300 to 800 m above sea level. Since the water table is low, about 10 to 20 m below the surface, groundwater does not influence the soil formation. There is a deep deposited layer ranging from 50 to 100 cm at the surface, followed by a gray-brown, loose plowed layer, a compact, heavy plowed pan, and finally an old deposited layer with cultural things such as coal cinders and tiles, as well as earthworm

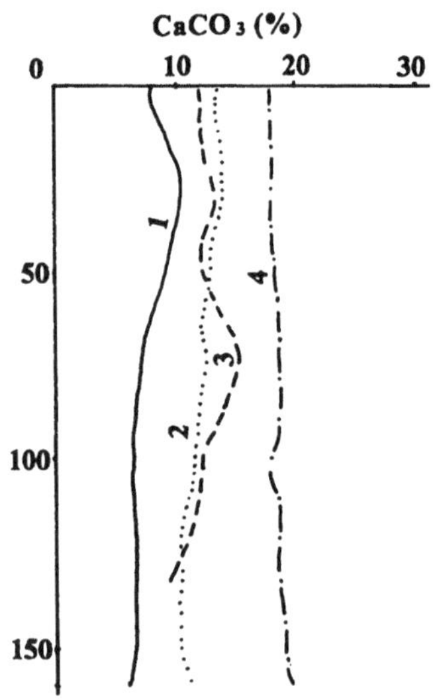

Figure 2. Distribution of $CaCO_3$ in Oasis soil profile. 1, Gray yellow soil; 2, white dry soil; 3, white yellow soil; and 4, black moistened soil. Source: Nanjing Institute of Soil Science, Academia Sinica (1978).

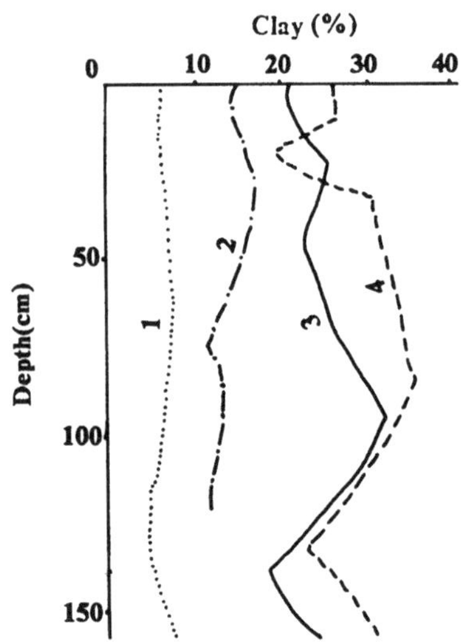

Figure 3. Distribution of clay particles in Oasis soil profile. 1, White dry soil; 2, black moistened soil; 3, gray yellow soil; and 4, white yellow soil. Source: Nanjing Institute of Soil Science, Academia Sinica (1978).

holes and dejections. The original soil layer underlies this profile and contains white spots of lime and gypsum. The soil varies from silt loam to clay loam. The clay content is slightly higher in the deposited layer, and clay and fine silt accumulate in the pan. Compared with the desert soil, the oasis soil contains more Ca, Mg, P, and S. The mineral composition is similar to the desert soil except that there is montmorilonite. The humus layer reaches 40 to 80 mm in depth and organic matter ranges

Table 2. Diversities and numbers of microorganisms in Oasis soils (Organisms $\times 10^4$/g soil)

Depth (cm)	Total number of bacteria	*Azotabacter*	Nitrifiers	Ammonifiers	P-decaying bacteria	Cellulose-decaying bacteria
0–15	680.3	83.0	0.16	149.5	11.7	0.10
15–30	129.4	159.0	0.03	2.2	1.7	0.04
30–50	94.7	6.0	0.01	0.2	0.3	0.01

Source: Nanjing Institute of Soil Science, Academia Sinica (1978).

from 0.6% to 4.0% with an average of 1.5%. Total N ranges between 0.05% to 0.15%, and the CEC 90 to 180 mmol/kg soil. The soluble salts are low, mostly less than 0.1% in the top layer while calcium carbonate accumulates uniformly in this layer, generally 8% to 12%, higher than that in desert soils.

The whitish oasis soil is mostly distributed in the Tarim Basin, Turfan Basin, and Hami Basin, and the most westerly part of the Hexi Corridor. Generally it is located in the upper parts of the oasis with a warmer climate and less precipitation than the grayish oasis soil. Situated in an area with a long agricultural history and long-term irrigation, the soil has a deeply deposited layer without the original soil properties. The plower layer is gray-yellowish or bright gray-brown, or gray-whitish with poor structure. Under this layer, the dark, finer-textured compact pan and the older deposited layer occur in succession. The organic matter, ranging from 0.6% to 1.7% with an average 1%, is lower, but the organic matter layer is deeper. Less available N and P are coupled with sufficient available potassium, so the soil has good response to N and then to P fertilizer. There is not much difference in the distribution of lime in this soil profile and thus no accumulation in the topsoil, but the content differs from one place to another. For instance, the lime content in soil in the Turfan Basin is 6% in topsoil and 6.2% to 7.2% in subsoil, but that in Ruoqian County is as high as 30%, depending on the chemical composition of irrigation water and the deposited material. The higher lime content is one of the factors making the soil hard and impervious, which distinguishes it from the grayish oasis soil. Soluble salts are low in this soil, generally 0.1% to 0.2%, with the highest being less than 0.4%. Gypsum is also low with 0.15% as the highest, but this is still higher than the grayish oasis soil. This is a very important property from the genesis of the soil. The pH ranges from 8.0 to 9.2, and bicarbonate from 0.02% to 0.05%. The particle composition depends on the deposited materials. Soils on the northern part of Kunlun Mountain are silt loams with sand contents of 50% to 70%, while those on the southern part of the mountain are silt loams to clay loams with higher contents of clay. The

CEC contents are less than 80 mmol/kg soil for the former and more than 100 mmol/kg for the latter.

The moistened oasis soil developed from meadow soil, scrubby meadow soil and part of the bog soil during long-term irrigation together with cultivation (Zeng Baoling and Xu Zhikun, 1963). For this reason, it was strongly influenced by groundwater through moisture processing and salt processing in addition to its deposited layer. Situated in low-lying, flatlands with a high water table (generally 1.5 to 3.0 m, even 1 m), this soil appears wet, and has a low temperature. In spring and autumn, when groundwater rises, the moisture phenomenon becomes more serious. The typical characteristics of this soil is the rusty spots in the profile, which increase with depth. In some places, lime nodules and white crystallized gypsum are found. Secondary salinization occurs widely, and salt accumulates on the surface. Animal activities are frequent in the profile, and earthworms have made a lot of holes and left dejections, but are limited to the deposited layer. Organic matter is uniformly distributed, ranging from 1% to 3.3%, with an average of 1.6%, but it is lower in the cultivated soils. Soluble salts are also higher, often above 0.5%. Lime content is higher in the top deposited layer than in the parent material and is uniformly distributed in the top layer. The distribution curve for gypsum is the same as lime, and the gypsum was formed following irrigation. As the deposit process occurs, the soil depth is increased and the water table decreased so that desalinization begins and the soil is improved.

B. Desert Soil Sequence

The desert soil sequence was formed in the desert areas. Located inland far from the ocean, wet air masses cannot come in and, therefore, the areas are very dry with sparse vegetation. Wind erosion is intensive and the ground surface is mainly covered by coarse sand. Long days of sunshine, high solar radiation, and abundant water resources from melted snow and ice and ground water are favorable for developing agriculture. There are large plains, hills, low mountains, eroded highlands, and basins in the desert area. The elevation is highly variable, ranging from below sea level to more than 3000 m. Little rainfall and great temperature variation provide the conditions for desert soil formation. The parent materials are primarily accumulated remains and accumulated slope-remains in the hills and low mountains, and alluvial, warped, and loessial deposits in the plains. Apart from the loessial parent materials, the rest are mostly accumulated coarse sands, even gravels.

Based on the parent materials, three types of soil sequence are formed: gray desert soil, gray-brown desert soil, and brown desert soil. The first two are formed in temperate and warm-temperate zones respectively, while the latter occurs in the transition area between desert and semidesert areas in temperate zones.

All of these soils can be used for agriculture when irrigation is available. Low rainfall and intensive evaporation make the soil water movement upwards as the main direction, and the moist layer is only 30 to 40 cm even in fine-textured soil. This factor is responsible for the slow weathering and weak soil-forming process and the accumulation of weathered products near the soil surface. The small amounts of organic matter from higher plants and rapid decomposition result in low organic matter accumulation. Since leaching is negligible and water moves up to the surface after short-term rains, and calcium bicarbonate has been changed into calcium carbonate with rapid evaporation, lime formed during weathering and in soil formation is accumulated near the surface, as are gypsum and soluble salts in the soil profile. In addition to the gypsum layer, the brown desert soil has formed a salt pan, consisting of sodium chloride. The accumulation of salts and gypsum is also influenced by parent material and topography, and numerous gypsum layers and salt pans were formed during the geological period. In contrast to soils developed in loess, those developed on other parent materials have a shallow gravel layer of less than 1 m.

There are three layers in the soil profile: a porous crust with a sheet, fish-scale top layer, a brown and compact subtop layer, and a lime-gypsum accumulated layer. The genetic layers vary in depth and appearance, depending on parent materials and soil-formation ages. In general, the compact layer was formed by alkalization, especially in the gray desert soils, but the alkalization was produced under high lime and gypsum conditions with no leaching, differing from the general alkalization.

Despite the common features, there are some dissimilarities in the three types. First, organic matter is reduced and the ratio of humic acid to fulvic acid is increased from gray desert, gray-brown desert of brown desert in succession. Second, the depth and developed degree of porous crust and fish-scale layer is weakened in the same order, and the color and luster of the surface and subsurface layers becomes lighter, i.e., light brown in gray desert, red-brown in gray-brown desert, and rosy in brown desert. Third, the accumulation of lime on the surface and the accumulation of gypsum and soluble salts are increased, and alkalization decreased, from gray desert to brown desert soil.

Gray desert soil is distributed near the fringe of the temperate desert area, i.e., the narrow transition belt between brown soils and gray-brown desert soils, bordered by gray-brown desert soils to the west and sierozems to the east. It includes primarily the south of Junggar Basin, the sloping plainand old alluvial plain north of the Tianshan Mountains, the plains in the middle and west of the Hexi Corridor and front of the Helan Mountain, and south of the Alashan Plateau, covering about 1000 to 2000 km in length from east to west. Some soils in the category were once termed "gray desert calcium soils" (Ma Yongzhi, 1938, 1944, 1945), "sierozem" (Ma Yongzhi, 1945), "gray desert soils" (Xinjiang Comprehensive Survey Team and Soil Institute of Academia Sinica, 1965), "gray-brown desert soils" (Sand Control Working Team, Academia Sinica, 1958), or "brown soils" (Wen Zhenwang et al., 1957).

Recent investigations have demonstrated that all these soils should be put into this soil type as they possess similar properties. The soil distributed area has an average yearly temperature of 5° to 8°C and an annual rainfall between 100 and 200 mm with 200 days available for plant growth. The climate changes obviously from east to west. In the Junggar Basin, winter is longer with lower temperatures, even rainfall distribution and thick snow (10 to 18 cm), while in the Hexi Corridor, temperatures are a little higher, the snow layer thinner, and precipitation less and distributed mostly in summer. The vegetation is primarily desert small shrub and shrub pattern. In addition, some short-life plants appear in spring in the Junggar Basin due to more rainfall during this period. The vegetative cover is about 10% of the soil surface with the highest being 20% to 30%. Gray-black moss and algae grow between the plants. The heat-water condition and vegetation are favorable for both desert soil and prairie soil formation. The gray desert soil, although developed on different parent materials, is commonly characterized by its lime reaction and high fine particles (higher than the other two desert soils). The soil surface has irregular narrow fissures, sometimes being full of multiangular cracks. On the edges of the fissures mosses and algae often grow, forming rough, uneven, and black crusts. Some places, influenced by wind erosion, are dotted with gravel of different sizes. In the Junggar Basin, small soil heaps and holes formed by mice are a typical characteristic of the local landscapes.

Well-developed gray desert soil consists of the following layers: (1) dry, loose, and fragile crusts, 1 to 3 cm in depth, brightly gray or brown gray, with sponge fissures and vertical fissures; (2) a 5- to 10-cm layer beneath the crust, consisting of brown, fish-scale-shaped sheets with rare and small fissures between sheets; (3) a 10- to 15-cm brown or bright red-brown compact layer with lime deposits; (4) a transition layer found beneath, being brighter, with no structure or vein-shaped, ranging in depth from 0 to 20 cm or more containing deposits of soluble salts; (5) an accumulation layer of gypsum and soluble salts. Organic matter in the soil is highest in the 0- to 10-cm layer, similar to some subtypes of brown soil and sierozem. The humic acid contents are much higher than that in sierozem.

With the exception of some soils located in the flood-alluvial fan, the other gray desert soils are mostly salinized in the middle or deep layers. Soluble salts accumulate between 40 and 60 cm over a gypsum layer, with the total salt content more than 1%. The soil pH is often greater than 8 and exchangeable Na consists of 10% to 30% of the total CEC, the latter being very low, about 20 to 30 mmol/kg. Most of the soils are composed of coarse silt-fine sand or fine sand-coarse silt. The clay content is highest in the middle of the soil profile, particularly in the brown-gray compact layer or alkaline layer, where the clay content may reach 20% to 28%. Hydromicas are the dominant clay minerals with a little chlorite and montmorillonite. For 1- to 5-micron particles, in addition to hydromicas and quartz, chlorite and feldspar are found, which shows that in the formation of gray desert soil, mineral weathering is weak. Hematite and other ferric minerals are not found. Iron may exist in colloidal films on the surface of clay and minerals, resulting in a brown compacted layer or brown alkalized layer.

The gray-brown desert soil was once termed "brown desert soil" (Ma Yongzhi, 1938, 1945) and "gray-brown waste desert soil" (Ma Yongzhi, 1957). It is located in temperate-desert areas with very dry, continental-like climate. The parent material is mainly sand-gravel diluvial materials or diluvial-alluvial deposits. The climate is hot, dry in summer and cold in winter with less snow. Drought-enduring bush and semibush with deep roots and juice stems dominate the vegetation and cover 5% to 10% of the total lands. The surface layer is a well-developed crust, like dry bread, gray or bright gray, 3 to 4 cm deep with a 5- to 15-cm brown or red-brown compact layer below. Due to the coarse texture, there is no clear sheet-structured layer. The accumulated gypsum and soluble salt layer, although very common at depths of 10 to 40 cm below surface, varies with areas and topographies. Humus content is generally between the amounts found in gray desert soil and brown desert soils. Lime accumulation at the surface is obvious, with the amount being 7% to 9% in the surface layer, decreasing to 3% to 5% with depth. The amount of gypsum varies greatly, increasing from surface to deep layer, where it may exceed 20%. Soluble salt, sometimes in excess of 1%, also accumulates in the gypsum layer. Sulfate is the main component of the salts with low chloride and bicarbonate, and Ca constitutes the main cation. The soil has alkaline or strong alkaline reaction with pH between 8.0 and 9.5. The soil texture is generally gravel silt or silt loam. Clay content is increased in the middle of the profile. P and K contents are high in the soil.

The brown desert soil was once termed "brown desert calcareous soil" (Ma Yongzhi, 1938, 1944, 1945), mainly developed on eroded mountains, hills, and inclined plains in front of mountains (gravel gobi) with great variation in elevation. Large pieces of black gravel mulch appear in the soil surface, and therefore the profile is composed of gravels or stony or stony fragments, splitting up the profile clearly into some layers. In the top layer, there is a weakly developed porous crust less than 1 cm thick, light gray or milky yellow. The next is a red-brown or rosy layer dyed by ferric materials, without obvious structure but with more clay, only 3 to 4 cm in depth. Then the accumulated gypsum layer appears with different shapes and gypsum content. In very dry and low plains, a black-gray firm pan made of salts often exists beneath the gypsum layer. Below this is the sand gravel layer or broken parent rocks. The whole profile is less than 50 cm thick.

C. Aeolian Sandy Soil Sequence

Formed on aeolian sandy parent material in desert areas and desertified areas, this soil sequence is widely distributed in arid and semiarid aeolian areas in north China with latitude between 36° and 49° N. The soils, totaling 637,000 km^2, were formed under a dry climate, varying greatly in temperature. Through selection by wind, these soils have a very uniform texture with fine sand accounting for more than 90%. Clay content gradually increases with the age of soil formation. Sand dust blown by wind is mostly deposited on the soil, increasing clay content at the surface. This is very important for changing the physical properties of the soil and improving soil fertility. Clay and organic matter improve soil structure and increase soil bulk density and porosity in the process of soil formation. The soil layer under dry sand holds 2% to 3% water. During the growing season each year, precipitation in excess of 10 mm occurs more than ten times, which infiltrates deep into the sand layer, providing available water for plant growth. However, much of the water percolates below the root zone and flows along slopes into low-lying land, supplementing the groundwater or forming seep areas. The aeolian soil in desert areas has 1-m dry sand layer with a water-holding capacity of less than 1.5% w/w. Because precipitation is rare, it is necessary to use supplemental irrigation for soil amelioration.

Plant growth in the aeolian soil results in rapid changes in the soil-water status. As plants take up water from the soil, the soil surface dries quickly. Furthermore, low precipitation and intensive evaporation from the top layer inhibit plant growth, and during summer droughts, plants often dry and die. Dry sand layers of 1- to 2-m depths can be found even in areas where precipitation is higher (Fig. 4). High heat conductivity and low heat content make the surface soil subject to large temperature changes between day and night. The surface temperature is as high as 61° C in daytime with a variation of 46° C between day and night. The high temperature and diurnal variation are unfavorable for plant growth but favorable to water condensation in the soil. Because of the special water and heat conductivity properties, low humus content, and presence of carbonate, a highly porous crust is formed on the surface. This is one of the chief features of the soil.

With soil development, microorganisms rapidly increase. From flowing aeolian soil to fixed aeolian soil, bacteria, fungi, and actinomyces multiply significantly, and the activities of N-fixing microbes are increased. This plays an important role in raising soil fertility and improving

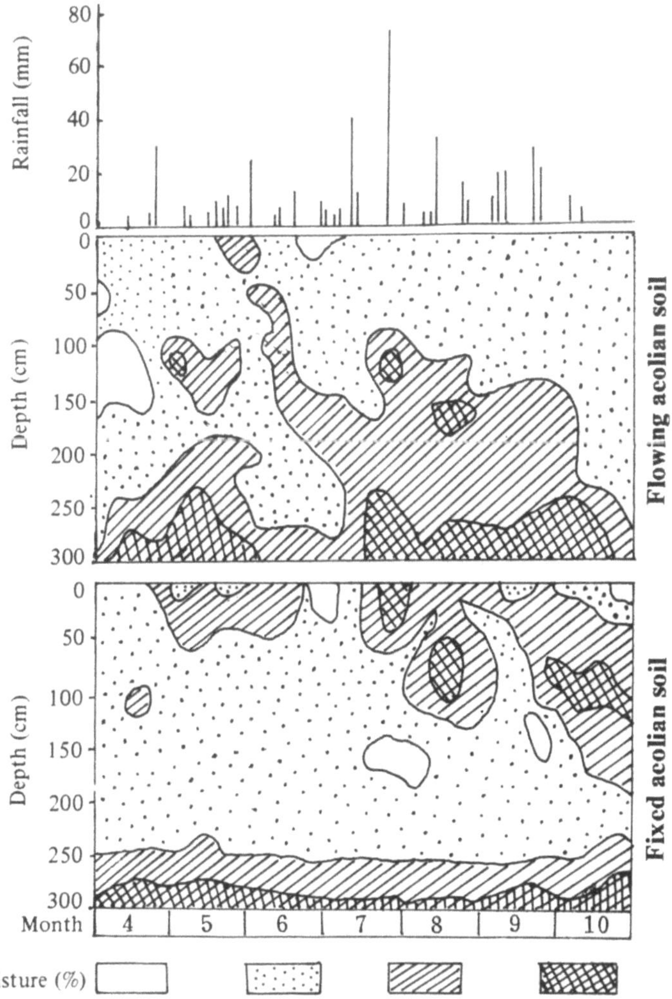

Figure 4. Seasonal variation of water in flowing and fixed aeolian soil. Source: Nanjing Institute of Soil Science, Academia Sinica (1978).

plant nutrition. In the process of sand stabilization, plant residues gradually accumulate in the top layer of the soil and thus result in increases of soil organic matter. The organic matter content is 0.1% to 0.3% in flowing aeolian soil, 0.2% to 0.8% in partially stabilized aeolian, and more than 1% in stabilized aeolian soil.

Aeolian soils are often saline and easily eroded. The salts move with wind and sand and are deposited on soil surfaces. Plants grown on aeolian soils contain high salts, especially salt-tolerant plants that contain high amounts of alkali and alkaline-earth metals, Cl and S. With the development of aeolian soil lime clearly accumulates in the profile. For example, lime content is less than 1% in flowing aeolian soil in semidesert areas, about 1.5% in partially stabilized soils, and 2% to 3% in the surface layer of stabilized soil. The flowing aeolian soil is concentrated in desert areas in large or small dunes, covering 2% to 3% and up to 70% to 80% of the total lands. In steppe areas, the flowing aeolian soil does not contain lime, while in semidesert and desert zones, lime often occurs and soluble salt content can be as high as 0.41% in a 2-m deep layer. The fixed aeolian soil in steppe areas contains higher organic matter, generally about 0.7% to 1.6%, and has no lime reaction and negligible soluble salts. In contrast, fixed aeolian soil in desert steppe areas contains 2% to 3% lime and less than 1% organic matter. Although the soluble salts are high, they do not cause soil alkalization. Saline aeolian soil also appears in the desert area, which not only contains higher gypsum, but also higher soluble salts, generally above 0.2%. Some areas have accumulated layers of salts and gypsum. In the low-lying and surrounded by hills with a high water table, meadow aeolian soil has been developed, and bog aeolian soil has been developed where the water table is about 50 cm or groundwater flows on the surface.

Generally speaking, because of different natural conditions, aeolian soils have different characteristics. The aeolian soils generally contain no lime in chernozem and chestnut zones, but large lime and low soluble salts in brown desert zones (Chen Longheng, 1963; Chen Longheng and Chen Wenrui, 1965); not only high lime but also clear salinization in desert areas; and are widely salinized in partially stabilized and stabilized aeolian areas (Chen Longheng, 1963; Liu Yumin, 1962).

D. Chestnut Soil, Brown Soil, and Sierozem Sequences

All of these sequences were once termed "chestnut soil sequence" (Thorp and Hou Guangjiong, 1935), and little study was made of all three sequences before the establishment of the New China. These soils are widely distributed in temperate and warm-temperate zones in the arid and semiarid areas, and constitute the main agricultural and husbandary base of north China. The distribution of the three sequences is in accordance with the bioclimatic zones.

Chestnut soil is located between chernozem in the northeast and brown soil in the southwest, in an arc-shaped belt from the northeast to the southwest (Wen Zhenwang, 1959). The climate is temperate, dry steppe with an annual average temperature $-0.5°$ to $-5°C$ and more or equal to $10°C$ accumulated active temperatures of $1600°$ to $3000°C$, which is similar to those of the chernozem, but precipitation is low and aridity is high. Brown soil sorrounds the desert area from the east, south, and west with a desert steppe climate. The annual average temperatute and

Table 3. Comparison of vegetation and climate among chestnut soil, brown soil, and sierozem

		Climate			
Soil type	Vegetation	≥ 10° C total accumulated temperature (° C)	Mean temperature (° C)	Annual rainfall (mm)	Aridity[a]
Chestnut	Steppe	1700–3000	3.3–5.7	250–450	1–2
Brown	Desert steppe	2200–3000	2–7	100–280 (150)	2–4
Sierozem	Desert steppe	2800–3100	6–9	200–300 (150) (350)	1.8–4

[a] Aridity = evaporation/precipitation.

Source: Nanjing Institute of Soil Science, Academia Sinica (1978).

accumulated temperatures are similar to the chestnut, but precipitation is less. The sierozem is in the transition zone from steppe to desert area, bordering with dark loessial soil in the east. Located between the temperate and warm-temperate zones, the annual average temperature and accumulated active temperatures the higher, but precipitation and aridity are in between (Table 3).

Although there is some difference in the soil formation, less humus accumulation, and more intensive calcification than in chernozem, a certain accumulation of gypsum and soluble salts are characteristic of all three soil sequences. The chestnut soil profile consists of a humus-accumulated layer, gray-white calcium-accumulated layer, and parent material. Seasonal leaching occurs so soluble salts have been leached from the middle part of the profile or the whole profile, while lime accumulates in a great quantity in the middle part of the profile resulting in a calcium-accumulated layer. Gypsum and soluble salts are only found in a few places. The soil is light in texture, mostly consisting of fine sands and silts. Sand and silt occupy 60% to 90%, fine sand and coarse silt occupy half or more than half, and clay 10% to 20% mostly. The clay distribution curve in the profile is in accordance with the lime accumulation curve, but clay accumulation is not clear in the profile.

Montmorillonite is the main clay mineral but depends on the parent materials. High montmorillonite content and little or no vermiculite are found in the deep soil layer developed from basalt. Soil developed on loess contains little montmorillonite and is high in hydromica and potassium oxide. Organic matter in the top layer ranges between 1.9% and 3.8% with a C:N ratio of 9–12:1. The CEC falls in the range of 100 to 250 mmol/kg, and soluble salts are almost leached. Also, gypsum is very low in the soil except the one found in Xinjiang containing 8% in the layer under 1 m.

The brown soil is located from temperate steppe to desert, between the chestnut and the desert soils, surrounding the desert from east, south, and west. In the Inner Mongolia Plateau the brown soil belt goes from northeast to southwest, while in Xinjiang it is distributed in an east-west direction. As this soil is located in the transition zone from desert steppe to steppe desert, it has a temperate, dry continental climate.

The eastern part of this soil belt is influenced by the southeast monsoon, two-thirds of the precipitation concentrated in summer with little rain or snow in spring. In contrast, the Xinjiang Uygur Autonomous Region in the western part is affected by west winds, and precipitation is distributed uniformly in four seasons. Since precipitation is so low that dryland farming is very limited, irrigated agriculture has been widely practiced. The profile of the soil consists of a light brown humus-accumulated layer, a gray-white calcium-accumulated layer, and parent materials. For most of the soil, gravel-desertification occurs. In areas without such a process, weakly developed multiangular fissures, and thin, false crusts are formed, where black lichen grows. This is a special feature that distinguishes brown soil from chestnut soil. Since leaching is minimal, calcium-accumulated layers appear 15 to 30 cm below the surface. The layer is compact and sheet-shaped with a depth of 20 to 30 cm, containing 10% to 40% lime. Also, gypsum and salt accumulations are widespread. Soluble salts vary from less than 0.1% in some subbrown soil to 0.3% to 1.5% in light brown soil. Sodium sulfhate is the main component. The soil layer is thin and consists mainly of sand-gravels.

The highest clay contents are found above the calcium-accumulated layer. The amount varies with parent materials, and generally ranges from 5% to 20%. The lower concentrations of clay in the surface layer are due to wind erosion. Hydromica is the main clay mineral, followed by montmorillonite and some iron oxide. The humus layer is thin with poor structure. Organic matter content, mainly in the top layer, is between 1.0% and 2.0%, and the C:N ratio is between 7 and 10. Due to the dry climate, the ratio of humic acid to fulvic acid in humus composition is less than 1. The soil has an alkaline reaction with a pH range of 8.0 to 9.5. The CEC is less than 100 mmol/kg in general, but it is as high as 200 to 250 mmol/kg in the soil developed on clay parent material.

Sierozems are also in the desert steppe zone, distributed in plateau-shaped hills, terraces in front of mountains, and diluvial-alluvial plains at the foot of mountains. These soils were derived mainly from loessial materials and also diluvial-alluvial materials. They can be used for agriculture and animal husbandry. Generally high production is obtained by irrigation. The soil profile is weakly developed: the humus layer and calcium-accumulated layer are not as clear as with the first two soil sequences, and the calcium-accumulated layer occurs at the bottom of the profile in some places. The soil surface is generally sandy, and if there is no covering sand, weak fissures and thin, false crusts at a depth of 0.5 to

2.0 cm appear, on which lower plants such as lichen and mosses grow, clearly distinguishing them from chestnut soil and manural loessial soil. The humus layer is extended to a thickness of 50 to 70 cm with poor structure. The organic matter varies from 0.9% to 2.5% in topsoil but less than 0.5% in the land without covering sands. The C:N ratio ranges from 8 to 12. The ratio of humic acid to fulvic acid in humus composition is less than 1. This is similar to that in brown soils, but differs from chestnut soils and manural loessial soils. As leaching occurs only seasonally, soluble salts, calcium carbonate, and gypsum are weakly leached. The Ca-accumulated layers are found in the upper layers, 15 to 30 cm, even 10 cm from the surface with a depth of 20 to 30 cm, in which lime varies between 15% and 25%, appearing as false hyphae and spots.

The particle composition varies with regions. In Xinjiang the soil texture is clay loam or sandy loam, while in the eastern part of this soil belt, sandy silt or silt soils predominate. Clay content, in accordance with calcium accumulation, is generally 8% to 12%, with at most 25%. Analyses show that the movement of minerals in the profile is not clear. Compared with the parent material, only Ca, K, and Na have moved. Si, Fe, and Al are relatively stable and seem to accumulate near the surface as other elements are leached. The ratio of sesquioxide falls between 5 to 10 as a whole, but is 2.8 to 3.4 in clay particles. Hydromica is the main clay mineral, along with small amounts of montmorillonite, chlorite, vermiculite, and kaolinite. This indicates a low degree of weathering. Although gypsum and soluble salt accumulation vary with regions, their accumulation generally occurs in deep layers of the profile. Due to the presence of carbonate, the soil solution is alkaline with a pH of 8.5 to 9.5. Cation exchange capacity is not high, generally ranging from 50 to 150 mmol/kg in the top layer.

E. Loessial Soil Sequence

In the famous Loess Plateau, a very important agricultural region with a long history of cultivation and also a region suffering from serious water erosion, many soils have developed on loess parent material. One of the soil sequences in the semiarid areas is the loessial soil sequence, which may be divided into three types: manural loessial soil, dark loessial soil, and loessial soil (Zhu Xianmo, 1957, 1959, 1964). The first two are old cultivated ones and the latter is a newly cultivated one.

The texture of loess is relatively uniform, mainly consisting of fine sand, silt, and clay, with coarse silt predominating. In the newly formed loess, the silt content is around 50%, and fine sand (more than 0.1 mm) comprises less than 2%. The old loess is similar to this, also dominated by silt but with a slightly higher clay content. Another feature of loess is that the texture varies with regions: fine sand decreases while silt and clay increase gradually from northwest to southeast (Liu Dongsheng, 1964).

The loess is column-structured, loose and porous, with a total porosity of 50% or so. However, as capillary pores occupy about 80% of the total porosity, infiltration of water is relatively poor. In the parent material, the total SiO_2 varies between 50% and 60%, Al_2O_3 between 8% and 15%, and FeO between about 5% and 6%. The loess is rich in lime, which concentrates in the silt particles and varies between 8% and 16%. Of the lime, $CaCO_3$ makes up more than 90% with a little MgO. The lime content also changes with regions and layers: it decreases from west to east and from north to south, and the content in the newly formed loess is more than that in the old. Organic matter and soluble salts are much less. More than 50 minerals are found in the parent material, of which, more than 70% appear in particles larger than 0.005 mm. Although the minerals in loess are complicated, their compositions in different layers are similar as a whole, mainly consisting of quartz, feldspar, and micas. Hydromicas are the main clay minerals, sometimes with a little montmorillonite and kaolinite because of weathering.

The properties of the parent material have a significant influence on soil formation and soil fertility in two ways. First, the loess layer is very deep, uniform in texture, loose and porous, and relatively good in the water conservation and fertility. Second, the loess is uniform in mineral composition with strong lime reaction, and under the bioclimatic influences the lime has been leached but the argillic processing is not obvious.

The features of soil formation may be characterized by the following:

1. Humus accumulation. As the loess is loose and porous, thus easy for root penetration, humus accumulation appears not only in surface soil but also in deep soil. For instance, in dark loessial soil crop and grass roots can penetrate into deep layers, and the humus formed by decomposition of the residues is combined with Ca, appearing as film around the surface of soil particles and small aggregates. Although the humus-accumulated layer is as deep as 1 m with a gray, dark color, the organic matter content is only 1% to 1.5% and even at its highest it is less than 2%.
2. Leaching and deposition of lime. The loess is rich in lime which leached and deposited to a certain extent during soil formation. In the loose, deep loess, the boundary line of the calcic horizon is not clear. Calcium carbonate appears mainly as hypha, and lime concretions occur only in the southern part of loess region.
3. Argillic characteristics. The argillic processing is weak. Therefore, the clay content of the soil is similar to that of the parent material.

During the long period of cultivation, two main measures, building terraces and applying manures, have been used to control erosion and raise soil fertility. As great amounts of manures have been applied to the soil, the soil depth has increased, and the original soil and cultivated layer were bruied, forming a fertile manural loessial layer, which is rich in organic matter and plant nutrients. The artificial layer in conjunction with the buried profile forms two humus layers in the soil profile. Humic acid occurs as the main component in humus composition, and the ratio of humic acid to fulvic acid is over 1.5, both of which are combined with calcium. As a result of the long-term application of manures, the lime content has increased, and the soil structure has improved.

Manural loessial soil is mainly distributed in the Guanzhong Plain, Shaanxi Province, and on terraces of the Fen-Wei valleys in the southwestern part of Shanxi Province, appearing as a long and narrow belt from east to west. Formed by a long period of cultivation, the profile structure and productivity of the soil is different from the drab soil underlying its deep and old cultivated layer (Zhu Xianmo, 1964; Liu Pengsheng, 1963). The soil profile may be divided into three parts: the upper part is formed by long-term cultivation and is about 50 cm deep, depending on the slope, distance from watershed, and use. This part includes the plow layer, plowpan, and old cultivated layer. Greatly influenced by cultivation, applications of manure containing large proportions of soil as bedding material, and autumn deep plowing, the plow layer—youngest one from the soil formation point of view—is continually increased, and appears as a loose and granular structure. The plowpan is compact, heavy, and about 10 cm deep. The old cultivated layer is buried under the first two layers, where pieces of coal, tiles, etc. brought in by manuring in ancient times may be found. In this layer, illuvial lime is also present.

Next comes the middle part, i.e., the gray-dark buried humus layer, which was formed from the original humus layer by cultivation. This part is loose with weak illuvial lime in the form of pseudomycelium.

The lower part is the original drab soil, which is less influenced by cultivation and has three distinguishable layers: argillic layer, Ca-accumulated layer, and the parent material layer.

All three parts constitute the special profile of the manural loessial soil. The cultivation property of the soil mainly depends on texture, while the other properties are greatly influenced by the plow layer. As a result, water penetration and holding capacity, water storage, and drought resistance are increased in the soil profile. The top layer is loose and porous and can easily decompose organic matter. The deep layer contains higher amounts of humus and exchangeable bases. Therefore, it has a higher supply of nutrients. For this reason the soil benefits plant growth for both young and old plants. The plow layer is light in texture, which makes cultivation easy. Also, the rich lime content in the soil benefits the formation of aggregates. Although the humus content in the profile is low, the depth of the humus is deep, and often appears in two layers. The amount of humic acid is higher than that of fulvic acid in the humus layer, and under the plow layer the ratio of the two falls from 1.2 to 1.5, which is similar to the ratio in the drab soil. Because the lime content in the soil is high, the humus combined with calcium is 4 to 10 times higher than that combined with Fe-Al. The lime content varies greatly with layers, ranging from 0.2% to more than 26%. Hydromicas are the main clay mineral, which is similar to loess parent material, while hydromicas and vermiculite dominate in the buried drab soil.

Specified by a deep and black loessial layer (humus layer), the dark loessial soil is termed an aged cultivated soil, widely distributed north of the manural loessial and drab soils. The annual average temperature in the area is 8° to 10°C with annual precipitation of 300 to 500 mm concentrated in July and August, and annual potential evaporation ranges between 1600 and 2400 mm. It is warm and rainy in summer and cold in winter with some snow. The dry season and wet season are very clear, and it is colder and drier than the area where manural loessial soil occurs. The organic horizon of the dark loessial soil contains 1% to 2% organic matter, and is as thick as 1 m or so, of which the organic matter comes mainly from the deep fibrous roots of the steppe grass. The extensive root systems are mainly a result of the dry climate, because roots must extend to the deep horizons to take up enough water for plant growth. Dust-falls have happened time and again in the past, thickening the original top layer of the soil. On the overlying new layer grass rooted and grew again. The ratio of humic acid to fulvic acid exceeds 2, and the humus combined with calcium is 4 to 10 times higher than that combined with Fe-Al, and 25% to 42% of the total humus is combined with minerals. The argillic process is weak and no distinct argillic layer is present, but calcium has leached and accumulated. The argillic part is in the humus layer, and the illuviating calcium carbonate is mainly in the form of pseudomycelium. Soluble salt and gypsum contents are low, showing that the dark loessial soil has undergone a marked eluvial process. The ratio of silica to sesquioxide falls between 2.6 and 2.8, with no significant change in the whole profile. Hydromica is the main clay mineral with small amounts of kaolinite and montmorillonite. The CEC acounts for 90 to 140 mmol/kg dry soil, similar to that in drab soil. The total N is 0.03% to 0.1%; total P, 0.065% to 0.047%; and total K, 1.33% to 1.67%. Because most of the P is present in an insoluble form, deficiencies of the element often occur. Coarse silt is more than 50%, and clay content is 15% to 20%. As a result, the soil texture is mainly light loam to medium loam.

The loessial soil is also termed "yellow soil," developed on the loess by cultivation. This soil is distributed widely in loess hill areas where soil erosion is very serious, and its profile consists of two layers: the plow layer and the bottom layer. The plow layer contains more nutrients and has a loose and definite structure. The bottom layer has the same properties as parent material. Transition from the plow layer to the bottom layer is not clear, and there is no distinguishable plowpan or illuvial layer. The texture and chemical compositions are similar to loess. Organic matter is less than 1%. Therefore, the total N is low. As the soil is composed of more than 60% fine sand and silt, it is easy to cultivate during much of the year.

F. Drab Soil Sequence

Located in the subhumid area, the drab soil sequence occurs mainly in temperate subhumid mountains and hills as well as some basins. Specific areas include the Yanshan, Taihang, Luliang, and Qinling mountains, and the Guanzhang Plain of Shaanxi Province, south of Shanxi Province, and west of the Henan basins. It was once termed "calcareous brown soil" (Hou Guangjiong et al., 1935), "Shandong brown" or "forest brown soil" (Zhu Lianqing et al., 1941; Ma Yongzhi, Wen Qixiao 1951). Later investigations demonstrated that this soil was different from the brown soil and it was classified separately (Xiong Yi and Xi Chengfan, 1965; Wen Zhenwang et al., 1957).

The climate where the soil forms is cold and dry in winter and warm and wet in summer. High temperatures during the rainy season promote rapid soil formation. Developed on a parent material rich in lime, leaching and illuviation of lime play a definite role in the soil's formation. Lime leaching in the soil occurs in three ways. The first is Ca-illuvial type, in which lime on the surface is significantly leached, but not absolutely lost, and therefore there is a clear Ca-illuviation. The second is a leaching-accumulation type of calcium, in which the argillic layer and surface layer have no lime and a neutral to weakly alkaline reaction, but the calcium-accumulated layer under them is very pronounced. The third is leaching type, in which there is no Ca-accumulated layer at a 1-m depth, but Ca appears in a saturated state and there is a small amount of free lime in the bottom layer. The surface has a neutral reaction, but the bottom layer is alkaline (Table 4).

The humus layer in drab soil is thin, and the argillic layer is clear, compact, and has a gum film illuviation. Both layers have a neutral reaction. The bottom soil has a clear calcium-accumulated layer with alkaline reaction. As the soil is suitable for earthworms, organic matter has been moved down to the Ca-alluvial layer. The clay content is less than 15% on the surface in most cases, less than 10% in the Ca-illuviated layer, but as high as 26% to 30% in the argillic layer, indicating that the argillic process is dominant. Hydromicas and vermiculite are the mineral indicators for the soil, and in the drab soil developed on loess, more than half of the clay is occupied by hydromicas, and then vermiculite with a small amount of kaolinite. The soil properties are different in different layers. The surface layer contains more organic matter and nitrogen than the argillic layer. The calcic horizon is rich in lime, which can fix P.

The drab soil sequence may be divided into four subtypes: ortho drab, luvic drab, cultivated drab, and cultivated meadow drab soil. The luvic

Table 4. Luvic and argillic drab soils

Calcium luvic type	Soil	Location	Depth (cm)	pH	$CaCO_3$ (%)	< 0.001 mm (%)
Luvic	Luvic drab	Shi county, Shanxi	6–31	7.2	< 0.2	6.6
			31–43	7.6	< 0.2	6.1
			43–68	7.4	< 0.2	21.4
			68–95	7.9	0	20.0
			95–148	7.6	—	9.3
Luvic and calcic	Drab	Xinyuan, Shanxi	0–5	6.8	0	7.1
			5–17	7.1	< 0.2	14.3
			17–50	7.1	< 0.2	28.9
			50–84	7.6	< 0.2	26.9
			84–130	8.2	16.3	8.1
			130–	8.2	19.2	11.0
Calcic	Calcarious drab	Jianpin, Liaonin	0–10	8.8	2.0	13.5
			30–40	8.9	3.8	15.5
			70–80	8.9	31.5	8.4

Source: Nanjing Institute of Soil Science, Academia Sinica (1978).

drab soil is situated in warm-temperate and wet areas and mountains of the drab soil zone with mixed forests of conifer and deciduous trees as the main natural vegetation. As leaching is high, the luvic drab soil has some differences from the ortho drab soil in the soil profile: lime in luvic drab soil has been leached and therefore its content is less than 0.2%; the pH is 6.5 to 7.5, and there is no calcic horizon. The particle composition in the soil is similar to that in the ortho drab soil, from silt to clay loam in texture.

Cultivation has turned the ortho drab soil into cultivated drab and cultivated meadow drab soils. The former was formed mostly on loessial diluvial and alluvial deposits. Over a long period of cultivation, the plow layer is loamy, loose, and about 15 to 25 cm deep with 1.0% organic matter. Underneath is a plowpan which is somewhat compact, followed by the compact and clayey bottom layer. Lime occurs throughout the whole profile with a pH range of 7.5 to 8.5. This soil is susceptible to spring drought; therefore, irrigation is important for agricultural production. Once termed "meadow drab soil" (Xiong Yi and Xi Chengfan, 1965), the latter occurs primarily in the plain surrounded by mountains in the North China Plain. Formed on diluvial and alluvial deposits, the soil surface is flat and has good water drainage. As the water table is high, the soil surface is often wet.

With a long history of cultivation, the soil has high fertility and a deep plow layer. The organic matter exceeds 1% and total nitrogen is greater than 0.05%, and K and P levels are high. The surface texture is silt to silt loam, but the bottom layer is clay loam or silty-clay loam. The high water-storage capacity and rich nutrients make the soil suitable for plant growth. In addition, rich ground water resources are available for irrigation. The soil is suitable for all crops grown in drylands with high and stable production, and two crops a year can be grown.

V. Management of the Drylands

The main purpose of managing drylands is to transform desert and desertified lands, control erosion, and make full use of precipitation. The people living in these areas have had a long history of taking effective measures for doing so and developing agriculture.

A. Transformation of the Desert and Desertified lands

It is estimated that in China the total desert and desertified lands occupy 1,308,000 km^2, covering 13.6% of the total area. Of this, the deserts account for 1,132,000 km^2 and the desertified lands 176,000 km^2 (Zhu Zhenda et al., 1980, 1989).

Most of the desert areas were initiated during the last period of the Mesozoic Era and the early appearance of the current ones occurred at the beginning of the Quaternary Period (Zhou Tingru, 1965). The deserts can be divided into two kinds: stone (gobi), consisting of gravel and pebbles, and sand (shamo). The former account for 569,500 km^2, or 50.3% of the total deserts, and the latter 562,500 km^2, or 49.7% of the total deserts (Zhu Zhenda et al., 1980). They include most parts of Xinjiang, Gansu, Inner Mongolia, Qinghai, and Ningxia, running from the western part of Xinjiang Autonomous Region in the northwest of the country to the western part of Heilongjiang Province in the northeast. The Taklamakan (Takelamagan, in new spelling) and Kurban-Tungut (Gurban-Tonggute) Deserts, the deserts in the Tsaidam (Caidam) Basin, and the Badan Jiryn (Baden Jiling), Tyngeri (Tenggeli) and Maowusu deserts, as well as the deserts in Alashan, lying from west to east, are well known in China. Taklamakan, the largest of those, formed in the Mid-Pleistocene and Holocene Eras (Zhu Zhenda, 1987), occupies an area of 330,000 km^2. Dunes are big and high and vary in shape. Some are shaped like pyramids, some are more than 150 m high, and others extend for more than 30 km. In the Baden Jiryn Desert, huge sand hills are 300 to 500 m high. The stone deserts are situated mainly in eastern Xinjiang, on the fringes of the Tarim (Talim) and Junggar or Dzungaria (Zhunger) basins in Xinjiang, the Caidam Basin in Qinghai, and in northwestern Gansu and northern Inner Mongolia.

The formation of China's deserts depends on geographical location and climate. They are located deep in the hinterland, far from the sea. The Qinghai-Tibet Plateau and the surrounding mountains, which were uplifted during the Mesozoic Era (Zhou Tingru, 1965), block the moisture-laden monsoon coming from the sea, so the hinterland is dry and receives little rainfall. Moreover, most of the deserts are in the interior basins where there are thick layers of loose alluvial deposits and sand sediments, which were formed during the Quaternary Period (Guangzhou Earthquake Brigade, Institute of National Earthquake Burean, 1977) and are easily picked up by strong winds in dry weather. All this provides conditions for the formation of deserts, though some are formed by sand from weathered bedrock.

The desertified areas originate from the desert-like landscape consisting of blown sands and undulating sand sheets and mobile dunes on the aboriginal nondesert territory. Since the fragile ecosystem was destroyed by excessive human utilization of the land resources under conditions of drought, frequent winds, and loose sand surface, the desertified lands have appeared in arid, semiarid, and even some parts of the subhumid zones. Desertification is a degradational process of the environment, causing the decline of biomass productivity and the reduction of arable land resources. By this process, an area of 176,000 km^2 has already been desertified. In addition, there are 158,000 km^2 at risk of desertification, which may be defined as desertification-prone areas. The desertified and desertification-prone areas, totaling 334,000 km^2, involve 212 counties in 11 provinces. It is estimated that 29% of the desertified and desertification-prone land in China is distributed in the western parts of Heilongjiang, Jilin, and Liaoning provinces, in the northern part of Hebei Province, and in the Hulun Buir, Hinggan, and Jirem prefectures, as well as Chifeng City of Inner Mongolia. Desertification in the above-mentioned regions is manifested by its quick growth in the marginal areas of dry-farming and steppe and by the damage from sand sheets around dry-farming systems. Of these, 44% are distributed in central Inner Mongolia, the north and northwest of Shaanxi Province, and the southeast of Ningxia Hui Autonomous Region, manifested by rangeland desertification and its acceleration of the spread to the marginal areas. Twenty-seven percent is distributed in western Inner Mongolia, Gansu and Qinghai provinces, and Xinjiang Uygur Autonomous Region, characterized by dune encroachment at the oasis periphery and by the harm caused from reactivation of vegetated dunes. Of the desertified land 42.2% was caused by the spread of rangeland desertification on the steppes, 52.3% by reactivation of vegetated dunes, and 5.5% by dune invasion.

Multipurpose surveys by the Chinese Academy of Science and other departments has shown that China's desert, and desertified and desertification-prone lands are rich in natural resources with long hours of sunshine and high temperatures. If water is available for irrigation, various temperate-zone crops can be grown, and they may even give higher yields than in some areas with a temperate climate. For example, the oases in the Turfan (Tulufan) Depression and the Tarim Basin in Xinjiang are noted for their good wheat and cotton harvests and are the country's principal sources of long-staple cotton. The eastern part of China's desert area, with an average annual precipitation of 250 to 400 mm, has more rainfall than the western part, and plants grow well there. Large and small lake basins, known as "desert pearls," have been formed in some places. The Tyngeri Desert has more than 400 such lake basins, and the Maowusu and Lesser Tyngeri deserts also have them in varying numbers. These lake basins are surrounded by excellent pastures.

The western part of the desert area has little rainfall. Huge quantities of melting snow from the high mountains around the deserts, and the ample groundwater in the piedmont plains, however, make the growth of vegetation possible. Large tracts of dense Euphrates poplar forest occur on the margins and in the river valleys deep in the deserts. The Taklamakan alone has a total of 420,000 ha of such forest along the banks of some of its rivers. These forested sections can be used as pastures, while nearly 13.3 million ha of fertile wasteland is available for reclamation. A great variety of wind plants grow in China's deserts. Among the 1000 or more species known, 350 are medicinal herbs such as *Ephedra*, *Liquorice*, and *Cistanche*. *Lop Nor hemp* (a wild variety), *Achnatherum splendens*, and reeds are fiber sources; *Lop Nor hemp*, in particular, is noted for its fine, long staple.

The natural resources, however, were neglected, even damaged in the past. In the arid and semiarid areas of northwest China, human activities have had a significant impact on the environment, particularly during the past 200 years. In many places, vegetation has been destroyed and the fragile ecosystem has deteriorated; persistent hazards, such as drought, salinization, and wind erosion and deposition have intensified the problems that accompany human activity. In fact, the already desertified area of 176,000 km^2 was mainly caused by human activities during historical time.

The Maowusu sandy land on the southeastern margin of the Ordos Plateau may serve as the worst case of desertification in Chinese history. Located in the semiarid steppe, the conditions in the Maowusu sandy land were more favorable in early Chinese history. Even now, extensive lignitic strata, several meters thick, are found in depressions between sand dunes. The tops of flat sandy ridges frequently have thick well-developed soils, while small stands of *Rhamnus erythroxylon* and *Salix monogolica* can still be found in the lowlands and on lower slopes. It might be concluded that the region was more humid during early historic time, supporting a vast grassland spotted with large patches of marshy areas.

Traces of human activities can be found up to Neolithic time, and their distribution pattern shows a pronounced temporal sequence. From southeast to northwest, ruins of the West Han Dynasty (206 B.C. to 8 A.D.) extend farthest into the sandy lands, followed by those of the Tang

Dynasty (618 to 907 A.D.), and then those of the Song Dynasty (960 to 1276 A.D.), while ruins of the Ming Dynasty (1368 to 1644 A.D.) are restricted to the southeastern margin of the sandy lands. Such a temporal sequence is well correlated both with spheres of political influence in different dynasties and also with different stages of desertification. According to historical records, the Maowusu sandy land was a lush grassland during the West Han Dynasty with a well-developed and prosperous agriculture. In 413 A.D., the Shi Kingdom founded Tungfan City as a capital (the current "White City" in northern Chenpian District, Shanxi Province). It had a population of 200,000, contained extensive grasslands and meadows within its bondaries, and its rivers were clear. In 828 A.D., wind-borne sands began to impinge on the city and, when it was finally destroyed in 904 A.D., the city was described as already "deep in the shamo (desert)." In 1473 A.D., the Great Wall was built along the southern margin of the Maowusu sandy land. At that time, it served as a boundary between pastoral and agricultural areas, but shifting sands began to intrude in agricultural areas. Again, in the midnineteenth century the Qing government (1644 to 1911 A.D.) began "to open up the Mongolian wasteland." Large tracts of sandy lands in the southeastern part of the Ordos Plateau were cultivated indiscriminately. This further destroyed the grasslands and extended the shifting sands. The boundary between pastoral and agricultural areas at the time moved northwestward to the present provincial boundary between Shaanxi and Inner Mongolia. The once-prosperous Chenpian District became a barren area with shifting sand dunes, half-fixed sand dunes, and saline lowlands that occupied more than 70% of the total area. From this point on, the process of desertification has been critically intensified by human activity.

To summarize, the process of desertification in the Maowusu sandy land has endured more than 1000 years since the Tang Dynasty, and it has pushed sand southeastward. Generally, northwest of the Great Wall critical desertification occurred between the ninth and fifteenth centuries; southeast of the Great Wall, a belt of shifting sand 60 km wide was formed during a 300-year period, from the middle Ming Dynasty to 1949 A.D.

Another unfortunate example is the Kolshin sandy land in the West Liao River basin. This area is also located in semiarid steppe with as good climatic and other conditions as those in the Maowusu sandy land, both being better than other parts of arid northwest China. Yet, due to misuse of the land, including both overgrazing and imprudent dry farming, desertification processes have dominated the area during historic time. Other notable areas of desertification include the steppe zone of eastern Inner Mongolia, the northern Ulan Buh Desert, and the northern and southern margins of the Taklamakan Desert. Desertification also occurs in some old and new oases, such as the lower reaches of the Shiyang River. This results chiefly from the overuse of surface and groundwater resources, leading to serious degradation of land productivity. Many unfavorable conditions continue to exist in arid and semiarid northwest China that are both natural and social in nature. Natural hazards such as drought, salinization, and mobile sands continue to threaten agricultural and other industries. The unstabilized sand dunes shift their locations not only to encroach over the roads, railway lines and farms, settlements and villages, and clog the irrigation canals, but, occasionally, the drifting sands also cover vast stretches of growing food crops. Land that is poorly managed or misused is also degraded.

After the founding of the People's Republic of China in 1949, through hard struggle the people of various nationalities in the desert regions began transforming the desert and desertified areas. A lot of work has been done, and a lot of methods have been implemented to control the desertification and to stabilize both the deserts and desertified areas.

1. Protection of Natural Vegetation

Although the vegetation in the desert and desertified areas is sparse, there still are some plants that grow and adapt to the conditions well, such as saxaul, wormwood, and tamarisk on the sand deserts, and *Zygophyllum*, *Nitraria*, Anabasis, and *Ephedra* on the stone deserts. Protection of the natural vegetation by fencing areas with such plants and prevention of their damage by animals or human activities for a certain period can rapidly rehabilitate the natural vegetation, which in turn covers the surface and stabilizes sands. These measures are widely practiced in such areas, especially on the boundaries between oases and deserts, to form protective screen on the periphery of the oasis. Numerous grass-protective belts around oases have been established from natural vegetation. Practices and experiments have shown that adoption of this approach, especially combined with irrigation, over a few years, has rapidly recovered the natural vegetation. For example, a vegetative cover in some dunes in the Turfan Basin was increased to 60% to 80% through several years of fencing (Jiang Yuan, 1980). Wind erosion can be somewhat controlled by natural vegetation belts more than 300 m in width and vegetation covers more than 30% (Zhu Zhenda et al., 1980).

2. Artificial Grass Planting to Stabilize Sands

Planting grass can rapidly rehabilitate vegetation cover and stabilize shifting sand in arid areas. Two ways of plantation have been utilized in China: planting by airplanes and by manual work. The former is used in areas with large pieces of shifting sand and less population. The grasses sown by airplanes are those which have high sand-holding capacity and can produce a large quantity of small seeds that are easily spread by planes (Yulin Prefecture Office for Directing Airplane Seeding, 1977; Song Wanlin and Chen Ailian, 1988; Xing Xuexiang et al., 1987).

Artemisia sphaeroce phala and *Artemisia ordosicaare* are examples of such species. The best time for seeding is before the rainy season and after a period of strong wind. Using *Rhizobium*-containing material to wrap legume seeds, forming "granular clothing," has been suggested for airplane sowing (Li Yuanfang et al., 1984). Seeding grass by airplane has brought good results in a short time. Yulin Prefecture, Shaanxi Province, extended its grassland to 11,600,000 ha through airplane seeding as well as manual work by the end of 1985, of which, Jingbian County seeded 6,400,000 ha of grass in 1985 (Yang Xiaoyin, 1987).

The latter is used in areas with less per capita land and more population and involves broadcasting or banding with sowing widths 1 m and 2 m apart and transplanting wild *Artemisia ordosica* seedlings after rains. The *Artemisia sphaeroce phala* and *Ar. ordosica* sown during the rainy season can grow to a height of 30 cm in the seeding year (Jiang Yuan, 1980), and the humus content in the 0- to 50-cm layer of the sandy soil has been reported to reach more than 1.4% and 2.0% or so after transplantation of these species for 8 to 10 and for more than 10 years respectively, while the sandy soil nearby without transplanted grasses had no humus (Yulin Prefecture of Shaanxi Province and Shaanxi Water and Soil Conservation Bureau, 1975).

3. Controlling Wind Erosion by Afforestation

Although wind erosion also occurs to some extent in humid climates, it is most common in arid and semiarid areas. It is essentially a dry weather phenomenon and hence creates a moisture constraint. All kinds of soils and soil materials are affected and at times their more finelly divided portions are carried to great heights and for hundreds of miles under dry conditions (Brady, 1974).

Wind erosion is serious and is one of the foremost problems in desert and desertified areas, because both the areas from where the sand blows and where it is deposited can be turned into wastelands. Factors such as vegetation or other cover of the land surface, crops, soil texture, and moisture content greatly influence wind erosion. Shao Liyie and Chen Guishang (1987) reported that within a range of 20 times the height of trees, wind velocity was reduced by 30%, and grain yield was increased by 38%. Undisturbed soil with 5% vegetation covering reduced 50% of the soil erosion, whereas, for the most part, land with vegetation covering over 15% was not subject to wind erosion. Prevention of wind erosion by cropping was more or less seasonal in character, the effect being dependent on crop types. Soil texture and the particular structure of topsoil were closely related to the extent of wind erosion. There was no wind erosion with wind velocity of 4.7 m/s in moist soil with a surface layer over 1 cm in thickness. However, the reverse was true for dry and loose topsoil.

In order to prevent wind erosion and curb the spread of desert conditions, various techniques have been adopted (Liu Xinmin et al., 1982; Li Mingang, 1980; Ling Yuquan, 1980; Liao Ciyuan et al., 1980; Liu Shu and Shi Qinghui, 1980; Liu Yingxin and Yang Xilin, 1980; Wang Fukong, 1980). An important one is afforestation, including the establishment of windbreaks and shelter belts to provide mechanical obstacles to the free sweep of winds. This technique is basic to any management program in these regions, because wind action on the soil surface has to be controlled before engaging in other practices, including those involving agricultural activity (Central Arid Zone Research Institute, Jodhpur, 1988). The windbreaks and shelter belts can reduce wind velocities, and therefore effectively stabilize shifting sand dunes and soil in situ when their roots are well established, and they can function as barries to trap drifting sand and blowing soil. In addition, such obstacles can also reduce evaporation from soil by minimizing the desiccating effect of winds (Chen Wenrui and Zhang Jixian, 1962). Since winds and shifting sands have a cause-effect relationship in the desert and desertified areas, the wind shelter belts and windbreaks are also termed "sand shelter belts" and "sand breaks." Shelter belts should be included in windbreaks, but in this review, wind- or sand breaks imply tracts of woods in order to differentiate from the belts.

a. Shelter Belts and Windbreaks

Depending on the magnitude of the wind erosion problem and sand resource, the width of tree shelter belts varies from serveral to hundreds of meters with few to several rows of trees. In addition to arbors (trees), shrub species are also used in the establishment of the shelter belts with staggered planting and in a pyramidal shape. The most suitable compositions of trees and shrubs in various permutations and combinations have been tested in different regions to impart a pyramidal cross section and form mixed forest belts with long-living and stable species. In this way, the diseases and pests occurring in pure tree belts can be avoided and the ecosystems of the forest belts can be stabilized (Desert Research Laboratory, Lanzhou Institute of Glacier and Frozen Earth, Academia Sincica, 1976, 1977). Suitable tree and shrub species, both indigenous and exotic ones, for this purpose vary from region to region, even from place to place. However, trees like *Populus tomentosa*, *P. euphratica*, *P. bolleana*, *P. cathayana*, *P. nigravar. italica*, *P. davidiana*, *P. dakuansis*, *P. alba*, *P. simonii*, *Elaeagnus angustifolia*, *Haloxylon ammodendron*, *H. persicum*, *Caragana microphylla*, *C. korshinsku*, *Ulmus pumila*, *Salix matsudana*, *Pinus sylvestrisvar. mongolica*, and shrubs like *Tamarix chinensis*, *Amorpha fruticosa*, *Hedysarum scoparium*, *H. laevc*, *Calligonum mongolicum*, *Lespedeza bicolor*, are commonly used in these areas because they are very well adapted to the environments. Furthermore, some fruit trees such as *Prunus armeniacavar. ansu*, *Zizyphus jujubavar. inermis*, *Juglans regia*, *Morus alba*, and *Prunus persica–Amygdalus persica*, which are indigenously widespread in the areas, have also been employed for this purpose as well as for economical benefits. Among those, *Pinus sylvestrisvar. mongolica*, *Populus bolleana*, and *Haloxylon ammodendron*, as well as *Haloxylon persicum* are more suitable and are recommended for extension because of their fast growth, good quality timber, high adaptation to the conditions, tolerance to poor soils, very well-developed roots, lush branches and leaves, and good ability to stabilize sands and prevent winds (Zhou Peizi et al., 1988; Yang Wenbing, 1988; Wang Kangfu and Dong Shudi, 1988; Pan Borong, 1987; Zhang Liyuan et al., 1988).

In addition to shelter belts, other tree windbreaks are also utilized. They include scattered woods, tree patches in the interdune areas, tree patches around dunes scattered in oases, and tree patches around cropping fields. In general, the tree patches in the interdune areas are often combined with grasses and shrubs planted on dunes to stabilize the shifting sands, and they are widely used in desert and desertification-prone steppes.

A lot of measures have been adopted for afforestation (Forest Institute of Inner Mongolia Autonomous Region, 1972). One of them is combining trees with shrubs and grasses. In moving aeolian sand soils, planted trees are easily blown up by winds or buried by sands, and thus have difficulty in surviving. Even so, they do not grow very well. Adoption of trees together with shrubs as well as grasses, i.e., planting grasses and shrubs first followed by trees can ensure young seedlings of trees to be grown successfully (Jiang Yuan, 1980; Qi Jiangzhong, 1988). Another method is called "afforestation for blocking and dragging sand dunes," adopted for shifting sand dunes. This is to plant shrubs on the lower part of the windward slope of a sand dune to "drag" sand moving ahead and to plant trees on the leeward side of the sand dune as well as interdune area to "block" sands (Fig. 5a). In this way, however, the trees on the leeward side and interdune lowland may be gradually buried by sands, and the shrubs at the low part of the windward slope may suffer from wind erosion (Maowusu Survey Team, Geology Department, Beijing University, 1976). For this reason, fast-growing trees, which are tolerant of being buried by sand, such as *Populus simonii* and *Salix matsudana*, have been employed on the leeward slopes and interdune areas. Shrubs having a high ability to stabilize sands and tolerate wind erosion, such as *Amorpha fruticosa*, *Salix mongolica*, and *Hedysarum scoparium*, have been adopted on the windward slopes. An alternative is transplanting 2- to 2.5-m tall trees on the lower part of the leeward slope and on the interdune lowland. The latter is called "blocking sands with tall trees in front (of dunes) and dragging sands with low shrubs behind." After sand dunes are leveled, the lands can be used for fuel or commercial forest or

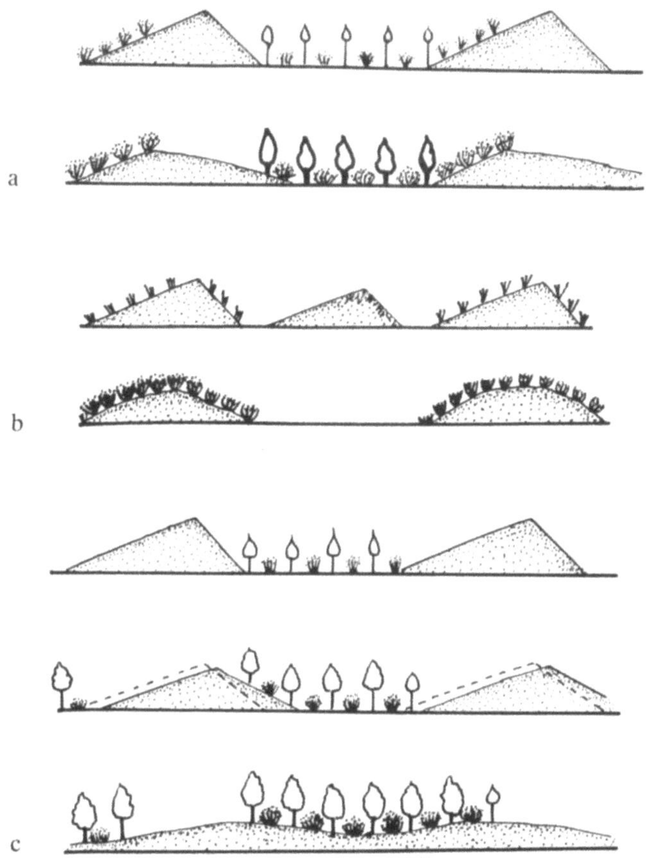

Figure 5. Methods for afforestation. a. Blocking and draging sand dunes. b. Releasing and blocking sand dunes. c. Afforestation on inter lowlands among check dunes and crescent dune chains. Sources: Jiang Yuan (1980); Qi Jiangzhong (1988); Maowusu Survey Team, Geology Department, Beijing University (1976); Yulin Prefecture of Shaanxi Province and Shaanxi Water and Soil Conservation Bureau (1975).

agriculture. The Xiaotanzhi Productive Brigade, Dingbian County, Shaanxi Province, leveled 6- to 7-m high sand dunes over 4 years using this method, and in the 5th year, began sowing proso (*Panicum milioceum*) and under taking land construction. In less than 10 years, the aeolian sand land was turned into highly yielding arable land (Yulin Prefecture of Shaanxi Province and Shaanxi Water and Soil Conservation Bureau, 1975). Stabilizing some shifting sand dunes by afforestation and releasing some by clearing natural vegetation and/or by artificially kicking up sands to let them move when strong winds occur is adopted primarily in the peripheries of lake beaches with smaller, fast-moving sand dunes

(Fig. 5b). In general, rows of odd- or even-numbered sand dunes are chosen in a piece of field to build up sand obstacles, and then they are afforested with *Salix mongolica*, *Amorpha fruticosa*, *Hedysarum scoparium*, *Artemisia sphaeroce phala*, *Ar. ordosica*, etc. The rest of the rows of dunes are then released. In this way, over a few years, the released sand dunes are moved to the afforested ones and stabilized, and the interdune areas are extended into broad flat sand fields, which can be utilized as arable fields or fruit gardens (orchards). Using this method combined with drainage, irrigation, and improvement of soil fertility, the Haitan Productive Brigade has turned sand fields into stable, highly yielding arable lands (Yulin Prefecture of Shaanxi Province and Shaanxi Water and Soil Conservation Bureau, 1975).

Afforestation on interdune lowlands among check dunes and crescent dune chains may be different. This is to plant trees and shrubs on the interdune lowlands in the first year, and the plantation is restarted in the second year on the location left from dunes which had already moved away. The same procedure is repeated year by year until the shifting dunes have been stabilized and just about all dunes are leveled (Fig. 5c).

Different trees and shrubs may be chosen for this purpose according to the environmental conditions. On lowlands with a high water table, a mixed forest with trees and shrubs composed of *Salix matsudana*, *Populus simonii*, *Salix mongolica*, *Hippophae rhamnoids*, and *Amorpha fruticosa* is widely adopted whether there is frequent water logging or not. In contrast, *Tamarix chinensis* is used instead when the land is slightly saline. Furthermore, on the lowlands with a low water table, afforestation may be directly carried out with shrubs such as *Salix mongolica*, *Amorpha fruticosa*, *Hedysarum scoparium*, etc. when wind erosion is not serious; otherwise, it should be conducted after sand obstacles have been established (Li Mingang et al., 1963).

In order to afforest or plant grasses successfully, sand obstacles are often used as mechanical windbreaks to protect young seedlings of either trees or grasses against wind erosion and block existing shifting sand dunes. This prevents them being buried in areas where plant growth is very difficult due to extreme lack of water and serious wind erosion. Different materials such as stalks of *Linum usitatissimum* (Xu Chengxiang, 1987), crop straw, clay soil (Geng Kuanghong, 1961), railings (Lin Yuquan and Zou Bengong, 1984), laths, and gravels are adopted for the establishment of such obstacles. This can be done in two ways. One is to mulch parts of a field, generally in a belt shape, or the whole field so that sands cannot be picked up when winds blow through the surface. The other is to erect some materials, such as straw, tree branches, and laths in a field in the direction vertical to the main winds at certain intervals according to local conditions. The sand obstacles have certain effects on the protection of plants as well as fields, but they can only be kept for 3 to 5 years. So after establishment of sand obstacles, plantation of grass or trees should be carried out in the same year or the year next (Geng Kuanhong, 1961).

Soil moisture is a major factor affecting afforestation (Qi Jiangzhong, 1988). Zhang Kangjian et al. (1987) took 999 soil samples at a depth of 30 to 40 cm from some locations of the Loess Plateau and made various analyses. In comparison with other factors, they found that the soil water content greatly influenced the site quality for afforestation. Choice of the most suitable soil moisture or use of irrigation is of prime importance for successful afforestation. As surface soil is usually dry in the arid and semiarid areas, using large trees and deep planting are most appropriate measures (Qi Jiangzhong, 1988). To match the precipitation, four types of vegetation combinations are suggested for stabilization of sand dunes in the northwest, north, and northeast of China (Liu Yingxin, 1988). They are: tree-shrub-subshrub in sandy lands with more than 400 mm rainfall; shrub-subshrub in sandy deserts with 250 to 400 mm rainfall; mechanical obstacles combined with vegetation in sandy deserts with 200 to 250 mm rainfall; and afforestation by irrigation in sandy deserts with less than 200 mm rainfall.

b. Tree Networks

To mitigate the ill effects of high wind velocity and the erosion and sand shifting that it causes, shelter belts and windbreaks have been adopted, and farmland-protective tree networks have been planted. Combined with sand shelter belts and fenced grass belts, tree networks significantly reduce wind velocity, change field temperature and humidity, regulate field climate, and avoid hazards like winds, sands, drought, frost, etc., thus effectively blocking sands and protecting arable fields against winds (Xinjiang Forest Institute, 1976).

The tree networks consist of two parts: main belts and subbelts. The best arrangement of main belts is at right angles to the direction of the main local wind. If this is impossible, the belts may have a certain angle with the wind direction, but it should not be more than 30 degrees. In addition, irrigation canals and main roads should be considered in making such arrangements. In fact, a lot of main belts are located on both sides of irrigation canals. Keeping the principle of the two belts being at right angles to each other, the subbelts are arranged according to the field roads and field permutations and combinations. Such arrangements of both main and subbelts can reduce the occupancy of more arable lands and make full use of the permanent topography and surface features, which are usually man-made (Plate 1).

There are three kinds of belt structures for tree networks: dense (Fig. 6a), sparse (Fig. 6b), and ventilative (Fig. 6c). The dense arrangement is such that the vertical section is so dense that almost no airflow can pass through. This structure can bring about sand accumulation

Plate 1. Tree networks in turfan. Source: Zhu Zhenda et al. (1980).

around or in the belts and lead flat arable land into trough-shaped lands, as well as create a less windproof stretch of ground. The sparse structure is low in tree density and therefore there are some ventilative openings distributed uniformly in the whole vertical section. This structure has certain and uniform effects on the prevention of winds, but it is not as good as the ventilative arrangement in protecting surface soil against wind erosion. The ventilative structure is different in that the upper part of the vertical tree section is dense, but the lower part is sparse with large ventilative openings. This structure has a larger and uniform effect on the prevention of winds, and if the shrubs, a component part of the belts, are suitable and are arranged rationally, sand accumulation is not serious. Because of this, the ventilative structure is adopted in most cases.

Figure 6. Tree network structure for protection of farmlands. a. Dense structure. b. Sparse structure. c. Ventilative structure. Sources: Yulin Prefecture of Shaanxi Province and Shaanxi Water and Soil Conservation Bureau (1975).

The width of the belt and the distance between two belts are very important not only for their effect of protecting fields against winds, but also as regards occupancy of the lands. Both too wide (such as a 15-m width with nine rows of trees) and too narrow (such as a 5-m width with three tree rows) belts are not suitable as they may occupy more land or may have little effect. The belt width established recently is 8 to 11 m with five to six rows of trees for main belts and 6 to 8 m with three to four rows of trees for subbelts. Despite many factors that should be considered in determining the distance between main belts, tree height is the main one for making such a decision, because the height of "tree wall" is highly correlated with the function of reducing wind velocity and thus of the protective scope of the structure. In general, the distance is 15 to 25 times the tree height with a range of 120 to 250 m for the main belts, but for the subbelts, the distance is greater, generally 300 to 500 m apart. For the tree networks to have a constant effect on protecting arable lands, in most cases a small network with narrow belts is constructed. Such a structure is achieved by planting four to six rows of tall tree species such as *Populus* with a width of 10 to 12 m to form a narrow belt, and then small networks are formed by reducing the distances between two main belts to 200 to 300 m. Data from Turfan Basin have shown that the small networks increase the efficiency of reducing wind velocity by 7.4% to 26.7% compared with the large networks (Desert Research Laboratory, Lanzhou Glacier and Frozen Earth Institute, Academia Sinica, 1976). Observations from the Chepaizhi region in the western periphery of the Kurban-Tungut Desert also indicate that the efficiency of reducing wind velocity in the protected fields by the small network is 20.0% higher than that by the large networks.

It is clear that the criss-cross tree network can successively reduce wind velocity, and therefore its function in the protection of fields against winds is different from that of an isolated tree belt. For this reason, these kinds of tree networks are widely practiced in the desert and desertified areas. With the reduction of wind velocity, horizontal exchange of air among fields and vertical exchange of heat and water between soil and the atmosphere are also reduced, and thus a favorable microclimate is provided for crop growth. Observations from oases on desert peripheries in Xinjiang have revealed that with the protection afforded by small networks the temperature in the surface soil increased by 0.7° to 3.5° C, in 5- to 20-cm deep soil by 0.4° to 1.0° C in spring, and in surface soil covered with snow by 0.2° to 0.6° C in winter. In summer, the air temperature at 1 m above the soil was decreased by 0.1° to 2.0° C, and the evaporation was reduced by 14% to 41%, but the relative humidity was increased by 4.0% to 14.0% (Desert Research Laboratory, Lanzhou Institute of Glacier and Frozen Earth, Academia Sinica, 1976; Zhu Zhenda et al., 1980). Si Binglian (1987) reported that since the establishment of a forest network on 36% of the total land, conditions in Dangpudi

Village located in the suburbs of Chifeng City, Inner Mongolia, have changed, and the frequencies of wind, sand, frost, and hail disasters has been reduced. The relative field humidity has increased by 4% to 10%, the annual evaporation rate has decreased by 29.5%, the soil water content has increased by 3% to 5%, and the average temperature in winter and spring has increased by 0.5° to 1.5° C, while that in summer has decreased by 0.1° to 0.7° C. In the past, temperatures in spring went up very slowly, and wind was frequent, so the sowing time was delayed till June, and short-day and low-yielding crops were planted. Now the sowing time is 2 months earlier, and long-day and high-yielding crops have been adopted. Since the forest network provides leaves and branches for fuel, the crop straw is saved for raising animals. This, in turn, improves soil fertility. Yu Zhifu and Zhang Renming (1984) reported that the Huangheai region in Shandong Province solved the problem of shortages of fuel, forage, and fertilizer with the establishment of tree networks, which also led to increased crop yields and reduced soil erosion.

Suitable tree species for such networks are chosen in accordance with the requirement to achieve a rational structure of the tree network and the possible economic benefits. Long-living, fast-growing, and highly protective trees should be used as the main species with commercially highly valued trees and high protective shrubs in proper proportions. Experiments and investigations have shown that most of the trees used in stabilizing sands and dunes, as mentioned above, are also suitable for this purpose.

4. Use of Water to Level Dunes

Using water to flush sand away to where it is needed or to level sand dunes in situ is a new approach adopted for the construction of arable lands in the desertified areas (Ci Longjun, 1963; Shaanxi Water and Soil Conservation Bureau, 1974). This method can change the undulating sand dunes into flatlands suitable for irrigation. As plenty of water, generally 2 to 2.5 m^3 for flushing away 1 m^3 sand, is needed for such work, this technique can only be adopted in areas rich in water resources such as rivers, lakes, canals, and reservoirs. It can also be achieved by pumping water from wells, but this is energy-consuming. Using floodwater for this purpose is also undertaken in some places. Engineering work comes first, and it includes constructions channels for drawing water, reservoirs for storing and raising the water potential, moats in dunes for water to pass through and to flush the sands away, surrounding ridges in low places for blocking sand and water flushed away from the moats and thus raising the surrounded ground surface to the same level as the leveled dune place, and outlets for draining away the remaining water (Fig. 7a).

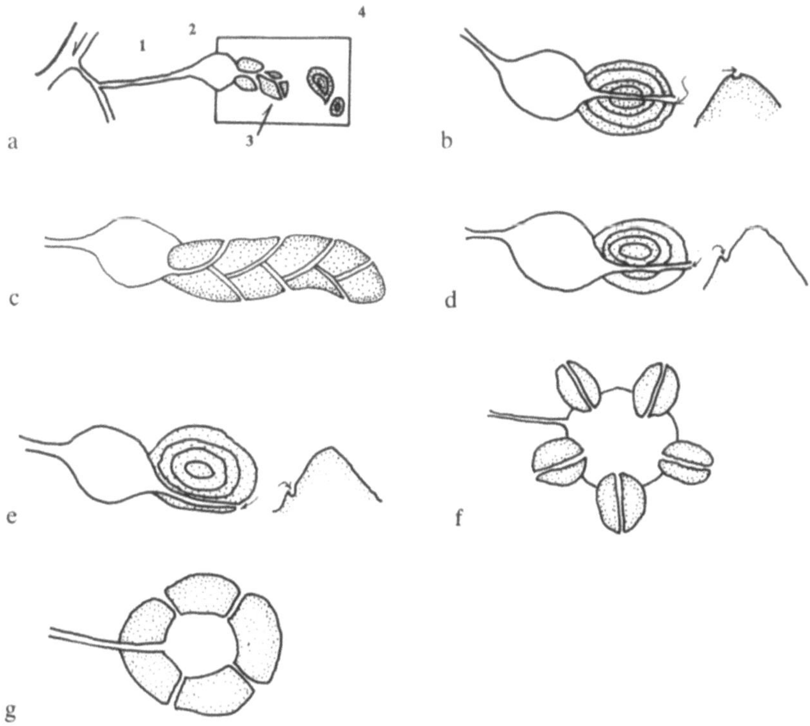

Figure 7. Engineering works for letting in water to flush away sand dunes. a. Layout of engineering work: 1, channels for drawing water; 2, reservoir for storing and raising water potential; 3, moats at dunes for water to pass through and to flush away sands; 4, surrounded ridges. b. Moat on top of a crescent or elliptic dune. c. Moats on both sides of a big crescent dune or dune chains. d. Moats at the waits of a check or crescent dune. e. One or two moats at the foot of a big dune. f. Moats of the dunes around. g. Moats on top of a big dune. Sources: Jiang Yuan (1980); Yulin Prefecture of Shaanxi Province and Shaanxi Water and Soil Conservation Bureau (1975).

Many methods have been employed for flushing away sands: letting in water from a channel in a high place into a moat on top of a crescent or elliptic dune to flush the dune outwards and level it (Fig. 7b); letting in water from a lower channel into moats on both sides of a big crescent dune or dune chains to wash away different parts of the dune or dunes gradually and separately (Fig. 7c); letting in water into moats at the waist of a check or crescent dune to flush the top away and to level it (Fig. 7d); letting in water into one or two moats at the foot of a big dune to flush it away from outside to inside gradually (Fig. 7e); making dams between small dunes and letting in water to form a reservoir, and then drawing water into the moats of the surrounding dunes to flush them away

Plate 2. Arable land made by flushing away sands with water in Mohe Tree Farm, Shenmu, Shaanxi. Source: Shaanxi Water and Soil Conservation Bureau (1971, 1972).

(Fig. 7f); letting in water into a reservoir formed on top of a big dune and digging four to five moats in all directions from the reservoir for water to flush away the sands (Fig. 7g); letting in water through a zigzag moat or straight moats to level the small undulating sand land or small dunes (Plate 2).

Flushing sand with water is often combined with flooding the land and irrigation using muddy water containing sediment. Flooding lands or irrigating with muddy water can significantly improve the properties of sandy soil. Over 10 years' flooding or irrigation with muddy water, fine sand was decreased from 82.2% to 88.8% to 7% to 17%, 0.05 to 0.001 mm silt was increased to 63% to 70%, and less than 0.001 mm clay was increased from 4.4% to 6.4% to 16% to 20% in the top 25-cm layer of sand soil (Desert Research Laboratory, Lanzhou Institute of Glacier and Frozen Earth, Academia Sinica, 1976).

5. Miscellaneous Measures

In addition to those mentioned above, other measures have also been adopted in the desert and desertified areas for the management of arable, range, and forest lands.

Improvement of the fertility and physical properties of the soil are also goals in managing arable lands in these areas. The measures used for this include turning over the deposited clay in the bottom of a soil to the surface, covering or mixing sandy soil with clay soil or sludge from rivers and lakes. leveling lands (Cui Wencai, 1975), regulating or limiting cultivated land for cropping (Chen Changyu, 1990), applying organic fertilizers, adopting drought-resistant tillage and cropping methods such as strip tillage, stubble mulch, planting green manure crops (Liu Zengxiang, 1965), interplanting crop with stubble (Luo Rongxi et al., 1987), increasing sowing rates, sowing in the deep layer, hoeing after periods of strong wind, hilling up and making ridges vetical to the wind direction, and the exploitation and full use of water resources for irrigation (Zheng Guoying and Liu Cuirong, 1991). Of these, fertilization, irrigation, and leveling lands are the most important (Zeng Baoling and Chen Zuoming, 1959). It should be noted that the gray desert soil contains more fine particles and has a higher value for agricultural use. However, it is often hardened and impervious after irrigation. If a compacted or alkaline layer exists, the problem becomes more serious (Cui Wencai, 1976). Thus, incorporation of clay soil into this soil has been used and advised for the improvement of its physical properties, in addition to the application of organic manures and summer deep plowing (Li Guihua, 1963, 1965; Cui Wencai, 1975; Soil and Fertilizer Institute, Gansu Academy of Agricultural Science, 1963). Planting rice is used for preventing wind erosion in the areas rich in water resources and also leaching salts in places where soil salinization is a problem.

Two things become important for managing rangelands: rehabilitation of the natural pastures and the construction of artificial grasslands using the water and soil resources. The first can be done by fencing some lands for a certain period and letting the natural vegetation recover. The most important practice is to overcome overgrazing and overstocking here by achieving a rational balance between numbers of animals and amount of herbage. The latter may be achieved by sowing suitable grasses, such as alfalfa (*Medicago sativa*), *Melilotus albus*, and *Astragalus adsurgens*, using irrigation, rationally applying fertilizers, and carefully grazing and harvesting.

Many examples of successful efforts have appeared in recent years. These include Wushenchao People's Commune of Ushan Banner in Inner Mongolia, noted for controlling shifting sand and building grassland; and the Yangqiaopan Production Brigade in Qingbian County, Shaanxi Province, where the people let in water to flush away sand so that they could build arable land and develop farm production. In the western part of northeastern China, much of the farmland was buried or destroyed by large quantities of sand carried by strong winds from the north in winter and spring. Since 1952, extensive shelter belts have been planted to anchor the sand over an area 1000 km from north to south and 500 km from east to west. In the northern parts of Shaanxi and the Hexi Corridor, as well as western Inner Mongolia, there are vast shifting deserts, which have moved steadily towards the southeast, encroaching on farmland and cutting off communications because of lack of vegetation. The people in these regions have enclosed desert land to allow the natural growth of grass. Grass and shelter belts have been planted over wide areas to stabilize the sand. In the Yulin region of Shaanxi Province, for example, shelter belts with an aggregate area of 0.23 million ha have effectively blocked the deserts from moving southward. Today, 44.2% of the total desert land in the region is being transformed. The local people brought in water to wash away sand; they opened irrigation channels, built dams, let in river-carried silt to help improve the soil, and put up shelter belts. All these measures combined have turned 20,000 ha of deserts into good farmland. Reservoirs have been built and irrigation channels dug to make use of the melting snow from the surrounding mountains. Shelter belts have been planted and various measures for soil improvement implemented; wasteland has been reclaimed and oases built.

The total area of arable land in the Xinjiang Autonomous Region has been increased by 150% in the past 40 years. Large tracts of desert there have become fertile oases. In recent years, efforts have been made to divert water from the Yellow River to the land dunes, to level the desert and to help develop shelter belts with man-made rain (sprinkler irrigation) and gravity irrigation. These measures have quickened the pace of transfomating China's deserts, and today, the work is in full swing.

Much effort has been expended, resulting in remarkable achievements in some places, and not a few desert and desertified areas have been transformed into fertile arable lands, however, the desertification problem is not yet completely under control. Economic and social problems such as shortages of food, fuel, and herbage, continue to be responsible for misuse, even abuse of the desertification-prone lands, and overgrazing is common. For example, during the Cultural Revolution (1966 to 1976), local pastoralists were compelled to cultivate sandy lands on a large scale. As a direct result, the southern margin of the Maowusu sandy land has encroached upon the northern boundary of the Loess Plateau, while its northern margin has advanced and merged with the Nobq Desert of the northern Ordos Plateau. That means the sands were extended further north- and southward (Zhao Songqiao, 1985).

It is estimated according to aerial photographs taken at the end of both the 1950s and 1970s (or aerially surveyed topographic maps) that 390,000 km^2 of desertification-prone land has been desertified, averaging 1560 km^2/per year. This occurred mainly on the steppe area with cultivation in vegetated and stabilized dune areas, collection of fuelwood, and overgrazing, disturbing the old vegetated dunes. The lower reaches of the Nenjiang River, the western part of Baicheng Prefecture, Horqin Steppe, Chengde Prefecture, the northern part of Zhangjiakou District, Duolun,

Qahar Steppe, Ulan Qarb Prefecture, and Ordos Steppe of Inner Mongolia are the areas with the most serious state of desertification. This situation will become worse in those regions if less attention is paid them and few preemptive measures are undertaken. The fragile ecological environment increases the risk of desertification in these areas, however, they do possess potential for animal husbandry and agricultural production and, from an ecological point of view, are characterized by a capacity for self-rehabilitation. These regions are active as regards both "further development" of desertification and "self-reversion," and, therefore, rehabilitating them should be given priority.

In accordance with the natural and economic characteristics, the presenting problems, and the pilot experiences, rehabilitation of the desert and desertification-prone areas should focus on two objectives: both the ecological and economic benefits. The ecological principle of appropriate utilization and multiple supplements should be realized; utilization and development of these areas should reside in their protection and rehabilitation. Considering this aspect, rehabilitation may be divided into two categories: in the latent and on-going desertification-prone land areas the rational utilization of resources should focus on a rehabilitation model of readjustment; in the severe and most severe desertification-prone areas, the rehabilitation model for reestablishing ecosystems through artificial supplementary energy and material inputs should be emphasized.

As for the strategy against desertification, the rehabilitation of desertification-prone land should be taken as complex systems, based on integrated conceptualization with the view of improving the ecosystems, and a master plan made. In moving fowards achieving a diversified economy, afforestation and animal raising should be emphasized. Simultaneously, population growth should be controlled to reduce pressure on the land, and the energy consumption system in these areas should be improved to ease up on the destruction of vegetation.

Regarding concrete measures to combat desertification, in the marginal areas of dry-farming and steppe in the semiarid zone, natural preservation should be adopted, dry-farming land use systems readjusted, land use rates for afforestation and forage farming enlarged, arable land with better water and soil conditions intensively managed, and shelter belts and tree networks built in combination with the planting of sand-holding species on dunes and the establishment of tree patches in the interdune areas to reduce wind velocity and stabilize mobile dunes. In the rangeland areas, rational carrying capacity should be determined, and proper rotation grazing, with grassland and forage farms established. In addition, sound management of drinking wells, ensuring a reasonable density of grazing spots, and construction of paved roads on pastureland are important steps. In the arid zone, continental river basins must be planned as overall ecological units. Water resources should be reasonably distributed from upper to lower reaches of a river. In an oasis, farmland protection networks should be established inside the oasis, with sand breaks at the oasis peripheries, and sand-stabilizing belts vegetated on the oasis outskirts to form a complex controlling system.

B. Controlling Water Erosion

Viewed from a global scale, the greatest danger that threatens soils is probably soil erosion (Hartge, 1986), a process which can lead to widespread soil degradation and desertification. In the most serious cases, soils that had been suitable for cropping have simply vanished. According to some estimates, the topsoil of the world is being depleted at the rate of 0.7% per year (Brown and Wolf, 1984). This is particularly true for China. In addition to wind erosion, which is more serious in desert and desertified soils, water erosion occurs widely in the country. On the drylands, the Loess Plateau region in central China is well-known internationally for its wide distribution of loess and severe water and soil erosion, which results in large-scale land loss and soil degradation. During the past few centuries, especially the past 100 years, natural vegetation has been extensively reduced by the pressures of increasing population and indiscriminate clearing, making the erosion problem more serious resulting in low and unstable agricultural productivity (Li Rui et al., 1990).

The Loess Plateau is located in the middle of the Yellow River region with altitudes varying between 400 and 2400 m above sea level. Soil conservationists estimate the area of the Loess Plateau to be 530,000 km^2. Excluding the exposed rocky hills and alluvial plains, it covers an area of 300,000 km^2. Most of the soil in this region possesses a high water-storage capacity. The soils developed on loess, in fact, are a big reservoir, and can store 550 to 600 mm rainfall, almost the annual average precipitation, in the top 2-m layer (Li Yushan, 1983; Zhu Xianmo, 1984c). The factor preventing soil storage of all of the rainfall is not that the precipitation exceeds the soil's water-holding capicity, but that the intense and concentrated rainfall surpasses the percolation rate of soil, resulting in runoff and serious erosion. Study of the Loess Plateau has shown that irrational production activities have made the problem more serious. Recently, average erosion has reached 2500 Mg/km^2 per year.

Three kinds of erosion are found in the plataeu: splash erosion, sheet erosion, and gully or linear erosion (Zhu Xianmo, 1956, 1986b).

Splash erosion is the result of soil particles being splashed and transported by the direct strike of raindrops. There is very sparse vegetation in the Loess Plateau, the land surface is usually bare and steep, and the loess is extremely loose. As a result, splash erosion is serious, causing stagnant surface water to be muddy, sealed soil pores, reduced soil permeability, promoted and converged runoff, and accelerated sheet and gully erosion. In addition, the rainfall in the loess region is usually of high intensity and short duration, often accompanied by hailstones; their impact at the soil surface leaves a great number of small hollows. Rainfall also makes the surface of the land muddy and crusted (Liu Zhi and Jiang Zhongshan, 1988). Dissipation of the impact of natural raindrops directly on the soil surface using double covering nets with 1.5-mm pore diameter, 15 cm above the surface of the ground, has a significant effect in reducing the degree of soil surface crusting and, therefore, soil erosion. However, loosening the soil surface or the crusted layer has no effect on the reduction of soil erosion, and it even decreases soil antierodibility, resulting in a rapid increase of erosion from heavy rainstorms, although it has the function of increasing the infiltration (Jiang Zhongshan and Liu Zhi, 1989). Field data show that the maximum intensity during a short period and the kinetic energy of rainfall are the main factors affecting splash erosion, and especially the combination of maximum intensity (in 30 min) with the kinetic energy. The intensity of downward splash erosion increases initially with the increase of steepness below 30 degrees, and then decreases when slope degree is more than 30. It has been explained that in the latter case, the area for accepting rainfall on steep slopes is relatively larger while there is less per unit area acceptance of rainfall; therefore, the kinetic energy of raindrops decreases for a unit of area (Jiang Zhongshan and Liu Zhi, 1990a).

Sheet erosion occurs on cultivated land consisting of loess parent materials. Profile sheet erosion is formed only on the table land and the top of saddle-shaped land. It may be found on sheep paths distributed over pasture slopes and is called squamous erosion because it often appears in the shape of squamules. Sometimes the runoff on a slope may be concentrated to such a degree that its kinetic energy is high enough to cut or score the surface, leaving clear line-shaped traces on it. This shows that gully erosion has begun. It develops from rill, then shallow gully, to gully (Wang Lixian, 1981; Zhu Xianmo, 1956, 1984c). Rill erosion is usually found on cultivated slopelands, gulch walls, and bare hillsides, where bedrock is exposed and strongly weathered. It does great damage to agricultural production. Sometimes rills may cover an area of more than 50% of the total cultivated land, and the eroded material may reach 72,000 Mg/km^2 per year.

Shallow gully is a transitional type of erosion between rill and gully. It can result from the convergence of rills or develop directly from furrows. It appears in branch-like, grating-like, radial, or parallel forms according to different shapes and human activities. Gully is the most distinct sign of erosion, often cutting more than 2 m deep, even several meters deep into the bedrock. Gully erosion caused by flushing runoff is the predominant type of soil erosion in the Loess Plateau. It forms many gullies, such as hanging gullies on steep slopes, vertical hollows washed by small waterfalls, sinkholes, loess caves, tunnelling gullies, and natural loess bridges. In addition to rain, overflow of irrigation often causes severe erosion. Not only does water cause severe gully erosion by itself, it often encourages other erosion, especially gravitational erosion in the loess region; this is mainly in the form of landslides, and a large-scale landslide having a thick layer can lead to more serious disasters. Also, this kind of erosion or mass movement in the region occurs in the form of avalanches, debris creeps, and various kinds of mudflows. According to Liu Bingzheng et al. (1990a), Chen Yongzong (1988), and Cao Yinzhen (1981), landslides occur on slopes of 35 to 55 degrees, and debris creeps on bare slopes of more than 35 degrees, while avalanches occur on steep slopes of more than 55 degrees.

Factors affecting soil erosion include climate, especially the intensity of rainfall (Liu Bingzheng and Chen Jiming, 1990; Zhu Xianmo, 1981; Wang Wanzhong, 1983; Jia Zhiwi et al., 1990), topography including slope degree, length, and shape (Jiang Zhongshan and Liu Zhi, 1990b), geology (rock properties and structure), soil, and vegetation (Zhu Xianmo, 1960; Institute of Forestry, Academy of Shaanxi Agriculture and Forestry Sciences, 1972). As far as the soil itself is concerned, the most significant soil characteristics influencing erosion are the infiltration capacity and structural stability. They are closely related. When rainfall exceeds infiltration capacity, runoff occurs and erosion follows. Under approximately equal conditions, the water infiltration rate is inversely correlated with the runoff discharge. These facts show the importance of the infiltration capacity to water erosion. The infiltration capacity is influenced greatly by structural stability, especially in the upper soil horizon. A study of the gray-drab forest soil indicated that when the soil had been under cultivation for several years, its water-stable aggregates decreased from 57.5% to 68.5% to 21.7% to 32.6%, and hence the total infiltration capacity within 2h decreased by 32.7% (Tian Jiying and Huang Yiduan, 1964). In addition, soil texture, pore space, organic matter content, the kind and amount of swelling clays, soil depth, and the presence of impervious soil layers all influence the infiltration capacity.

The structural stability influences the soil erosion in another way, determining to some extent the soil antierodibility or the soil resistance to water erosion. The antierodibility may be further divided into two types: soil antiwashability and soil antiscourability. The former is defined as a resistance to dispersion and suspension caused by runoff and is mainly dependent upon the affinity between soil particles and water. The greater the affinity, the more liable the soil is to disperse and suspend, and the more easily the structural and the microstructural bodies are disrupted and disintegrated. Consequently, the soil infiltration capacity decreases and the soil surface becomes muddy. Erosion then occurs due to suspension. This is also true even if the runoff is very slow and its mechanical force rather weak. Indices commonly used to express the soil's resistance to erosion, such as the soil dispersion ratio, erosion ratio, dispersion coefficient and aggregation, are mainly for the description of soil antiwashability. In loess regions, the erosion ratio is related to the soil's clay content (Tian Jiying and Huang Yiduan, 1964), and it has an inverse relationship with it (Zhu Xianmo, 1986b), especially when organic matter is high. Vegetation and land use also influence soil antiwashability. Both the greatest dispersion ratio and erosion ratio are found on farmlands, followed by grasslands, while the lowest ratio is on forest lands. Also, both the ratios are lower in the topsoil layer containing more organic matter than in the lower layer. However, the soil antiwashability cannot reflect the erosion problem in most cases in loess regions, because there is no apparent correlation between soil erosion magnitude and the antiwashability indices. For this reason, the concept of soil antiscourability has been developed; it refers to the soil's resistance to mechanical disruption and downward movement by runoff. It is mainly dependent on both the cementation force among soil particles and microstructural bodies as well as the resistant force of soil structural bodies to disruption. In the scouring process, erosion occurs when runoff is strong enough to push soil materials forward, even if the soil particles or soil lumps are not dispersed or suspended. If the cementing force among soil particles is strong, and the structural bodies are big or well cemented, the soil will possess a high resistance to scouring. The magnitude of soil disruption in static water, i.e., the stability of soil structural bodies, is suggested as an index for describing the antiscourability (Jiang Deqi and Zhu Xianmo, 1963; Zhu Xianmo, 1984a, 1984b). In general, the soil masses from loess regions, including parent materials, are easily disrupted, even when they are soaked in static water. The magnitude of disruption depends on the soil's water content, soil's expansion coefficient, binding types of the soil's mass microstructures (Zhu Haizhi, 1964), plant root quantities, and land use systems (Table 5).

Table 5. Relationship of soil expansion with disruption and erosion

Soil sampling	EC (%)	Time for full disruption in static water	Soil loss by flowing water conditions
Loess parent material	19.8	2 min 30 s	100% within 3 min
Loess parent material	31.1	2 min 1 s	100% within 3 min
Surface soil, farmland	13.9	Not disrupted (fixed by roots)	100% within 5 min
Surface soil, grassland	17.9	Not disrupted (fixed by roots)	9.7% within 1 h
Surface loose soil, poplar forest	1.2	Not disrupted (fixed by roots)	15.2% within 1 h
Surface dense soil, poplar forest	0.8	Not disrupted (fixed by roots)	9.4% within 1 h

EC, Expansion coefficients.
Source: Jiang Deqi and Zhu Xianmo (1963).

A very dry soil mass may become thoroughly disrupted within 5 min. However, the soil mass will retain its original state without disruption in static water when the water content reaches about 20% (Zhu Xianmo, 1984c). The greater the soil expansion coefficient, the more liable the soil disruption (Wang Yongyan and Zhang Zhonghu, 1980). The antiscourability decreases with the reduction of plant root quantities and the weakening of soil consistency. Land use systems have a great effect on the antiscourability: the strongest effect was found on forest land, followed by grasslands, and the weakest on farmlands. It becomes drastically weakened as soon as forest and grasslands are cultivated. The reverse is true when farmlands are left fallow. In forest, grass, and freshly cultivated lands, the antiscourability is stronger in surface layers and weaker in deep layers. The antiscourability is more important as regards soil erosion than is the antiwashability, since the intensity of the former is remarkably correlated with magnitude of soil erosion.

Soil erosion by water in the Loess Plateau has caused a lot of problems both on-site and off-site (Zhu Xianmo, 1958). On-site, a considerable amount of precipitation is likely to be lost by runoff. This loss has two serious consequences. First, crop plants are deprived of water, which might otherwise have entered the soil; and second, the runoff water carries with it some of the valuable topsoil. As a direct result, this eats into the eroded land, exposing its subsoil, and finally turns a fertile soil into a poor one. Because of this, fields in eroded areas are called "three-thing-losing fields" as they lose water, soil, and fertilizers. Liu Bingzheng et al. (1990b) demonstrated that in the three-thing-losing fields, organic matter, total N, and available nutrients were extremely low in the top layer, and there were fewer fine particles than without erosion. For this reason, the soil fertility is low and the texture coarse. Based on their investigations in three counties located on Weibei dryland, the authors estimated that on average, crop production would be decreased by 18.3% if erosion on gently sloping land with a very low erosion module was increased by 1000 Mg/km^2. As water erosion goes on from year to year in the plateau, large tracts of loess flat highland are lost, and correspondingly, large tracts of hilly land are formed (Plate 3), which are distributed widely in western Shanxi and northern Shaanxi.

Two typical examples can explain the seriousness of the situation. Dongzhiyuan, a large tract of loess area in Gansu Province, was once named Pengyuan in the Tang Dynasty (618 to 907 A.D.). It ws 32 km wide from east to west. Now, the widest part spans only 18 km and the narrowest less than 0.5 km. Over 1000 years, the area lost two-thirds of its land due to serious gully erosion along the Malian and Pu rivers on both sides of the highland (Shi Ninghai, 1976a). The high terrace located in Fengxiang, Qishan, and Fufeng counties along the north bank of the Wei River was the famous Zhou Plain. It was a large intact flat upland when the Zhou nationality settled down at the beginning of the Western Zhou

Plate 3. Seriously eroded farm lands, Wuqi, Shaanxi. Source: Zhu Xianmo (1986a).

Dynasty (about 1100 to 771 A.D.). However, it has been cut into several pieces by some tributaries of the Wei River (Shi Ninghai, 1976b). It is clear that soil erosion is the principal cause of devastation of soil and water resources (Hou Guangjiong et al., 1987).

Throughout the centuries, repeated destruction of the plateau's vegetation by human activities coupled with heavy summer rain has also aggravated the loss of water and soil, resulting in serious problems at distant sites. The Huanghe or Yellow River became choked, causing frequent dyke breaches and bringing disasters to the people along its middle and lower reaches. It is recorded in history that during the past 2000 years, the Yellow River flooded 1593 times, generally twice in 3 years, and great changes of the river's course have occurred 26 times (Xin Shuzhi and Jiang Deqi, 1982). Serious soil erosion has brought about the situation where the Huanghe now transports an average annual sediment load of 1.6 billion tons. Of these, about 400 million tons are deposited in the riverbed of the lower reaches, and the rest, or about 1.2 billion tons empties into the sea. The deposited sediment has made the riverbed rise about 10 cm each year, and therefore the river bank has to be raised every 10 years (Guo Tingfou and Gao Bowen, 1982), which is very costly. Because of this, the riverbed at low reaches is 3 to 5 m above the ground, and the bed in Kaifeng City, Henan Province, is 8 m higher than the city, forming a well-known above-ground or suspended river (Xin Shuzhi and Jiang Deqi, 1982).

Along with the sediments, much fertile soil is lost (Tang Keli et al., 1989). The sediment mainly comes from sheet erosion on the slope surface and gully erosion in the watershed. These amounts of sediment not only constitute a direct threat to the riverbed in the lower reaches, but also give evidence of soil degradation in the Loess Plateau.

Some results have shown that the composition of sediments and soils is very similar. The particle sizes of the latter become gradually finer from the north to the south. Apart from the sediments of the Huangpuchuan Stream and the Kuyehe River, which contain a considerable amount of coarse sand, the mechanical composition of the sediments in the other tributaries of the Huanghe is similar to that of loess in the eroded areas, with coarse silts (0.05 to 0.01 mm) and fine sands (0.25 to 0.05) predominating. This shows that sediments in most tributaries of the Huanghe originate from silty and fine sandy loess. In general, the content of the coarse silts (less than 0.05 mm) in the river sediment is slightly lower than that of soils in the eroded areas, so that coarse grains will deposit while fine grains will be relatively enriched in the transportation of river sediments.

With the sediments, large quantities of nutrients have been lost. A comparison of the nutrient content between eroded soils and sediments shows that the contents of total N, available P, and available K appear to have regional characteristics. They tend to become increasingly greater from north to south in accordance with the geographical distribution of each tributary, roughly agreeing with the geographical regularity of the mechanical composition of loess. The content of total N in sediments in most tributaries is close to or higher than the lowest value of those in soils. Some of them are close to the average value, but those in the Weihe River are higher than the value of the upper limit. The content of available P in sediments is obviously enriched, and most of it is higher than in the soils, even reaching two to three times as high. In general, the content of available K is close to the average value of soils with some individual samples slightly lower or higher than the soils. Although the absolute content of fine particles and nutrients in sediments is roughly equal to that in the soils from the eroded regions, the quality of the sediments have changed greatly because of the increase in the amounts of various nutrients they contain. For instance, the enrichment ratios of fine particles in the Yanhe River and the Wei River are close to 1, but the enrichment ratios of available P in the two rivers are 4.47 and 3.4 respectively. Where the content of fine particles is the same, the enrichment of available P in sediment is three to four times that of the soil. If the loss of available P is estimated in terms of sediment, its value will be greater than that measured in soils.

The people in the Loess Plateau have long waged the hard struggle against soil erosion and have accumulated much experience. Legend has it that as early as more than 4000 years ago, Houji, who taught people to cultivate crops as a state agricultural minister-like official at that time, began to use Quantian, a strip-furrow method of cultivation. According to textual research, more than 3000 years ago, leveling land and controlling water were recorded in an ancient book *Yugong* (Xin Shuzhi, 1964). During the last phase of Warring States (475 to 221 B.C.), Quantian was developed into low and high Qi cultivation. This was to divide a slopy land into several parallel strips (Qi) with ridges and to separate the strips into pairs. Each pair was composed of an upper and a lower strip. On drylands, crops were grown on lower strips, while on wetlands, they were grown on upper strips (Lu Buwei, Warring States). Terraces appeared in an embryonic form during the Western Han Dynasty (206 to 24 B.C.), and Ou Tian (see below) was also recorded at that time. Later on, terraces were widely adopted in the loess regions. The people in the Hongzhao Village, Shanxi Province made terraces long ago. In the meantime, warping lands using flooding or river water containing sediments was also employed. Constructed during the Warring States, the Zhengke Canal drawing muddy water from Jin River has turned the Guanzhong Plain into a vast fertile land. A good example of warping land by diversion of floodwater is the Zhaolaoyu, Fupin County, Shaanxi Province, where the people have used this technique of constructing and improving their lands since the Qin Dynasty (221 to 207 B.C.), and their lands were changed into fertile, high-yielding ones long ago (Water Conservancy Committee of Yellow River, 1954).

Construction of silt arresters to harvest floodwater and the induced sediments is another way to control erosion and construct lands and is widely adopted in the loess regions. The people in Liujiashan, Lishan County, Shanxi Province, have constructed silt arresters which date back to the Ming Dynasty (1368 to 1644 A.D.), and many still in existence have a history of more than 100 years (Water Conservancy Committee of Yellow River, 1954). Of those, one 20 m high, with 5 ha fertile land is more than 200 years old.

The importance of afforestation and planting grasses in controlling water erosion and protecting lands has long been recognized. Enclosing hillsides, mountains, and slopes wasted because of livestock grazing and fuel gathering in order to facilitate afforestation and natural grass growth have been documented in many ancient books such as *Guanzi* (Guan Zi, Zhou Dynasty), *Yizhoushu* (Zhou Dynasty), and *Shiji* (Shi Maqian, Han Dynasty). The first emperor, Qinshihuang, promulgated a decree for the protection of vegetation of the Tai Mountain, and he established an office in charge of the work. In the Southern Song Dynasty (1127 to 1279 A.D.), Wei Xian brilliantly expounded on the function of forests in reducing water velocity and stabilizing and conserving soil. Mei Boyan (1786 to

1856 A.D., Qing Dynasty) further discussed the function in another way. Both concluded that clearing mountain lands for cultivation would lead to serious damage to both the mountain regions and the fertile lands at their feet.

The people in the loess regions also recognized the importance of afforestation and planting grasses, and put these measures into practice. The development of secondary forest to conserve water sources in Ziwu Mountain, Qaoshan Mountain, and Huanglong Mountain in the hinterland of the loess plateau has had a long history of more than 100 years, even more than 1000 years (Zhu Xianmo, 1986a). Great attention has been given to the disasters of the Yellow River overflowing. As early as in the Western Han Dynasty, Jia Rong pointed out that muddy water and sediment deposition were the major factors causing the river to flood. Having recognized the problem, Zhou Yong (Ming Dynasty) offered a good suggestion. He said that having everybody managing arable lands would be equal to having everybody manage the river. This idea was further developed by Xu Zhenming (later Ming Dynasty), a well-known water conservancy specialist. He pointed out that the first step to manage the Yellow River was to manage the water source, i.e., the upper and middle reaches of the river. This implied controlling the hills and slopes of these reaches located in the loess regions where water erosion was serious. Xu Guangqi (Ming Dynasty), a great agronomist, accepted and developed these ideas. He summarized the people's experiences and suggested different ways for management of the lands and for utilization of the water sources. His main idea for managing drylands included use of water from pools, springs, and running water from mountains for irrigation, building ponds and reservoirs for harvesting water, and constructing level terraces, as well as leveling lands, for controlling water water erosion and conserving water and soil. His suggestions were of great value and are still in use.

Based on both past experiences and advanced techniques, the people in the loess regions have adoped various patterns to control water erosion and to construct their land according to the natural conditions. They have made full use of the characteristics of loess such as its thick layer, loose structure, high water-storage capacity, and permeability to construct bench terraces, soil-saving dams, various kinds of small check dams, reservoirs, strip terraces (on gentle slopes), ponds, water cellars, and gully head protection for detaining rainfall and preventing runoff, thus allowing rainfall to be infiltrated into the soil to achieve full utilization of the precious water resources. On the slopes, tillage practices such as furrow-ridge tillage, pit and trench tillage, and contour strip cultivation of grain and forage crops have been used for soil conservation. On steep concave slopes, the land is prepared for reforestation by measures such as fishscale pits, level ditches, and level narrow terraces. In some areas where there are vast lands and sparse population, sowing seeds by airplane has been vigorously used to speed up the establishment of grass and tree covers. In addition, construction of shelter belts and fields enclosed by windbreaks in the wind-drift area are all effective measures. Recently, scientists have used advanced techniques to monitor and predict erosion and to study and evaluate the different measures in the conservation of water and soil and in the increase of crop yields, as well as in the improvement of the ecological environment. These measures include the use of remote sensing (Li Rui et al., 1990; Wang Changyao et al., 1990; Wang Yuru, 1990), false color photos (Liu Goubin et al., 1990), microcomputers (Liu Zhi and Jiang Zhongshan, 1990a, 1990b; Li Bicheng et al., 1990a; Li Bicheng et al., 1990b), statistics (Jiang Zhongshan and Liu Zhi, 1990b), linear planning (Lu Zhiwei et al., 1990), and the dynamic gray model (Zhang Hanxiong, 1989). It is expected that the severe erosion in the Loess Plateau will be controlled in the future by using comprehensive measures.

1. Engineering Methods

a. Slope Erosion Control

Leveling land using engineering methods is an effective way of increasing water percolation into the soil and thus decreasing, even eliminating, water erosion. Runoff is prevented, allowing water or rain to gradually penetrate into the soil for storage. This is particularly effective for controlling erosion on slopes and in gullies, and therefore building terraces is widely adopted for managing sloping lands. Level contour terraces also provide a base for irrigation and use of machinery. If the topsoil is kept and organic fertilizer added, the terrace lands can produce high yields already in the year of building.

The construction of level terraces on sloping lands has been successful in the Loess Plateau. The Gaoxigou production brigade in Mizhi County, Shaanxi Province increased grain yield to 4.5 Mg/ha since 1971 when all the lands were terraced. Experiments from Dinxi and Yanan regions showed that only 0.75 Mg/ha of wheat yield was obtained on sloping land compared with 2.25 Mg/ha on level terraces. Maize yield on level terraces was 10 times higher than that on sloping land under the same conditions (Ding Yonghuang, 1983). Du Bangqing (1987) investigated the effect of terraced lands in Daping Village, Dingxi County, Gansu Province, and revealed that crop yields on the terraced land were 1.6 to 2.0-fold the yields on sloping lands.

Many forms of terraces have been created by farmers. One is the level terrace. It is used to make a slope into several level benches, and each bench is used for planting crops, mainly wheat, millet, and sorghum (Fig. 8a; Plate 4). This kind of terrace is generally built on land with a slope of 5 to 10 degrees, but in the loess hill-gully regions, construction

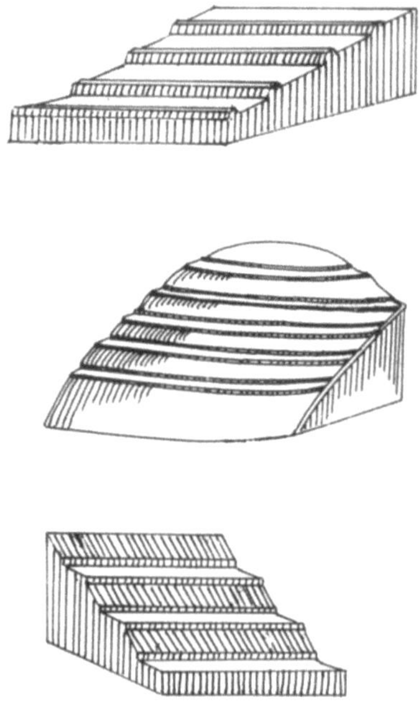

a

b

c

Figure 8. Different forms of terraces. a. Level terraces. b. Sloping terraces. c. Level terraces alternation with slope. Sources: Xin Shuzhi and Jiang Deqi (1982).

Plate 4. Level terraces in Gaoxigou, Mizhi, Shaanxi. Source: Shaanxi Water and Soil Conservation Bureau (1971, 1972).

on slopes of more than 10 degrees is common. Recent research results have shown that narrow level terraces cause more water loss by evaporation both from the land surface and terrace banks. Because of this, the water content in the edge soil is 20% to 35% lower than it is inside (Yang Wenzhi et al., 1990), and crops do not grow well on the outside edge of a terrace; this is what farmers mean by a "fallow edge."

In general, the influence of the terrace bank on soil moisture extends to about 2 m for an earth bank without vegetation and to more than 2 m for one with vegetation, if less than 2% of the difference of soil moisture between outside and inside is used as the criterion (Yang Wenzhi et al., 1990). In fact, the ratio of "fallow edge" width to terrace width depends on terrace width, being decreased with increase of the terrace width. Investigations show that when the width of the terrace is 8 or 14 m, the ratio is relatively low and stable; otherwise it increases. From this point of view, a width of about 8 or 14 m is considered suitable for construction of level terraces in north Shaanxi (Ansai Experimental Area of Soil and Water Conservation, Northwestern Institute of Soil and Water Conservation and the Ansai County Goverment, Ansai Comprehensive Research Station of Soil and Water Conservation, Academia Sinica, 1990). Another type is a sloping terrace with ridges placed at a certain distance along the contour on a slope of 10 to 15 degrees (Fig. 8b) in order to divide the slope into different narrow belts. This kind of terrace can increase water storage and reduce erosion, but it is not as good as the former.

Another terrace worthy of mention is a level terrace alternating with slope, consisting of a terrace ridge, level terrace area, and natural slope (Fig. 8c). This type of terrace is also used on land with a slope of 10 to 15 degrees (Cui Yunpeng, 1990). The ridge and level terrace area can impound runoff water and soil from the slope when runoff is produced by heavy rain, and the slope is left in its natural way to store precipitation and produce runoff water for spreading onto the terrace. The slope is managed in a conventional dryland cropping sequence, but crops are grown annually on the level terrace. The width of the terrace and slope and the layout of the system are variable according to land use and local conditions, but a 1:1 to 1:2 terrace-to-slope width is used for cropping in north Shaanxi, and 1:3 to 1:5 for planting trees.

b. Gully Bottom Control

Controlling gully development should begin with management of uplands and slopes around gullies by letting rain infiltrate into soil in situ and preventing water coming down to the gullies. This can cut down the water source for gully development, and it is therefore the first, important step. Protection of gully head should be followed using engineering methods such as making ridges around the gully head and small reservoirs at a

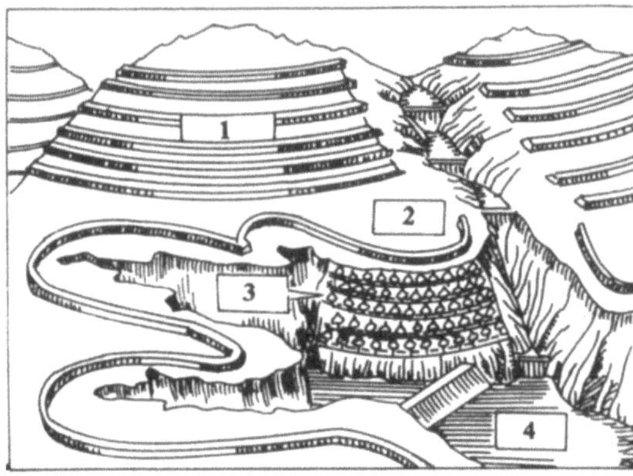

Figure 9. Four defence measures for control of gully erosion: 1, catchment area; 2, protection of gully head and edges; 3, protection of slopes inside gullies by afforestation or grass planting; and 4, control of gully bottom. Sources: Xin Shuzhi and Jiang Deqi (1982).

certain distance from the gully (Cui Yunpeng, 1990), or biotic methods such as planting trees and grasses. The purpose of adopting these measures is to block runoff to the head and/or store water. Slopes formed inside gullies also need controlling to reduce runoff to the bottom. This can be done by planting trees and grasses in fishscale pits, level furrows, level narrow belts, and the level terrace alteration with slopes made using engineering methods. Controlling the gully bottom is the last, most important step (Fig. 9). The principle of the engineering method is to block runoff and the induced sediments by constructing dams, but for differet purposes there are different dams.

Check Dams. The main purpose for building check dams lies in consolidating the eroded base and, therefore, preventing further erosion. This kind of dam can be made by soil (Fig. 10a), stones (Fig. 10b), and fresh willow branches or dead willow stakes (Fig. 10c). The soil and stone dams are made in a general way, but the willow dam is made differently. The latter is generally constructed by sticking three to four rows of fresh willow branches or dead willow stakes into the ground. The distance between the rows is 0.5 to 1.0 m. Bundles of small willow branches are placed between rows for the dead stake, and small branches are used for weaving the live branches in rows. The Yangjiagou River basin reduced gully bottom erosion by 81.3% on average over 20 years after building willow dams (Tu Xingwen, 1987).

The willow dams can let water pass through, and the stone dams can also be used as overflow dams, so there is no need to build up spillways. The soil dams, however, are generally low, about 5 m in height, and their

Figure 10. Check dams. a. Soil check dam. b. Stone check dam. c. Willow check dam. Sources: Xin Shuzhi Jiang Deqi (1982).

size is small. As a result, the second, even third dam is constructed at the same time along a gully to effectively control gully bottom erosion.

Detention Dams or Silt Arresters. Detention dams or silt arresters have traditionally been used for flooding sediment in the gully bottom to make land. Such land is called dam land. The dam may be low or high, the duration may be long or short, and the field made may be large or small, but generally it requires 2 to 3 years for land to be formed and used for cultivation. The advantages of the system are that gully erosion is controlled and gullies are changed into fertile land. Damming gullies in Jiuyuangou, Suide County, Shaanxi Province, reduced soil loss by 45.7% and increased crop yield to 3.75 Mg/ha on average and to 3.5 Mg/ha in a drought year, which was two to ten times higher than that obtained on sloping land. Dam land is used mainly in two ways: one is using the upper dam lands for storing flooding sediment and the lower ones for cropping (Fig. 11a), and the other is cropping and storing rotation, managing one dam land for storing sediment and other lands for cropping in the first year, and then rotating in succeeding years (Fig. 11b).

Small Reservoirs and Other Water Conservancy Projects. Small reservoirs are widely constructed in gully bottoms in the loess regions. These

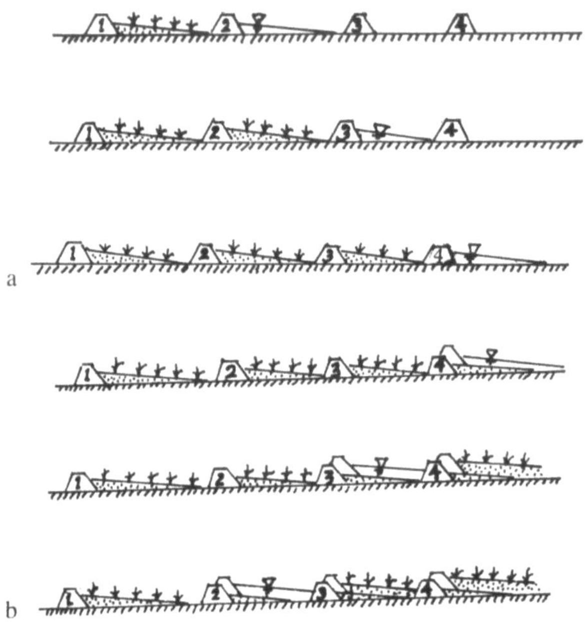

Figure 11. Dam lands and their rotation. a. Diagram of dam lands, showing the upper dam land or lands for storing flooding sediment and the lower one(s) for cropping. 1, 2, 3, and 4 are dams and what remains are dam lands. b. Diagram of dam lands, showing rotation of dam land for storing sediment with dam lands for cropping year after year. Source: Nanjing Institute of Soil Science, Academia Sinica (1978).

reservoirs not only protect water conservancy projects, for example, silt arresters in low reaches, but also store water from runoff for other uses such as irrigation. In the Weibei Highland, small reservoirs play an important role in increasing and stabilizing crop yields (Fig. 12). According to the experiences obtained in controlling gully bottom erosion in the Loess Plateau, a small reservoir and some detention dams and a group of check dams are needed for a watershed. The reservoir should be set up in the lower or middle reach of the main gully, detention dams in the middle or upper reach, or at the junction of the main gully and a small gully, and the check dams in the small gullies. In this way, they form an engineering system for protection and water storage (Cui Yunpeng, 1990). Silt deposition is a big problem for building reservoirs. Wu Faqi and Liu Bingzheng (1990) reported that Chunhua County established 20 reservoirs with an effective volume of 17,240,000 m^3 in 1979, but the volume of 18,610,000 m^3 from 14 reservoirs was deposited in 1980, the lost volume being 25.9% for the 14 reservoirs. Of those reservoirs, 9 of them further lost 10.5% of their volume between 1980 and 1984. It is clear from the

Figure 12. Reservoir. 1, Soil dam; 2, spillway; 3, sluiceway or outlet. Sources: Xin Shuzhi and Jiang Deqi (1982).

results that engineering measures should go hand in hand with other measures, especially biological ones in order to control erosion comprehensively; otherwise the results of engineering measures are limited.

In addition to small reservoirs, other small water conservancy projects, such as ponds and water cellars, are also employed. Located in or near villages, the ponds are mainly used to supply drinking water for livestock and for human needs, the number and size being dependent on the village population, generally with one to three in each village. The water cellars, or "dry" wells, traditional devices for the storage of drinking water for humans, are widely constructed in the areas along the middle reach of Yellow River, where the ground water is too deep or salty. As it is implied by its local name, the dry well is usually built underground on threshing places, roadsides, farmyards, and low sides of grassy slopes where runoff passes and the earth stratum is solid with a catchment area of 2000 to 2500 m^2. Although the shape and volume (20 to 50 m^3) may be different, a small mouth is a specific feature for such cellars (Fig. 13), which can reduce evaporation and keep water from contamination. In the past and also at present in some places, nearly each family has one or two such cellars. These cellars are also constructed on roadsides around arable fields for irrigation at critical times, particularly during a dry spell of seeding.

2. Biological Measures

a. Function of Biological Measures

Biological measures are also very important for controlling soil erosion and other interests. In the Loess Plateau, more than half of the land is occupied by waste mountains and slopes, and therefore great attention has been given to ptotecting it against erosion. Even for flat lands, boilogical control is necessary. The main biological way to manage these lands is to plant trees, shrubs, and grasses. This includes afforestation for

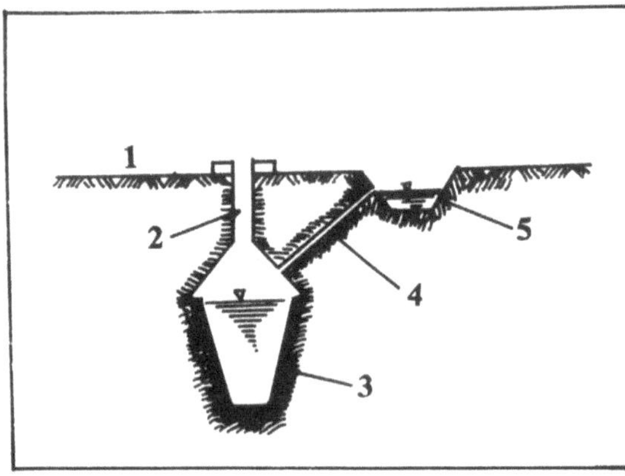

Figure 13. Water cellar. 1, Ground; 2, opening for drawing water; 3, layer for protection of infiltration; 4, water inlet; and 5, pond for collecting sediments. Sources: Xin Shuzhi and Jiang Deqi (1982).

protection of grassland, planting shrubs, grasses (Zhang Xianmin, 1987), and trees for protecting arable fields, hills, gullies, valleys, and river banks (Yang Songwang, 1987; Gao Jishan, 1987). Studies in Huanglong County showed that runoff from forest areas was reduced by 78%, and the sediment with it was reduced by 94% (Institute of Forestry, Academy of Shaanxi Agriculture and Forestry Sciences, 1972). Chen Jimin (1987) reported that Ziwuling forest vegetation intercepted almost all the rainfall, conserved 23.2 Mg water per ha to 2 cm depth, and reduced runoff by 50%.

Grass and Forage Species. The major grasses and forages being used in the Loess Plateau are alfalfa, sainfoin, *Coronilla varia*, and *Astragalus adsurgens*. They are good forage crops and have a definite ability to conserve water and soil. Alfalfa is a traditional forage crop and is regarded as forage king in the Loess Plateau, as it is high in crude protein, Ca, and carotene in its stems and leaves, being a sapid and favorable fodder for domestic animals; furthermore, its roots are long and strong, capable of protecting the soil against water erosion (Zhang Yongping and Wu Zhanqiang, 1990; Zhang Yongping et al., 1990; Wu Zhanqiang and Zhang Yongping, 1990).

Astragalus adsurgens, a species of leguminous family, is distributed in many regions of China, and adapts to different conditions with drought and cold resistance. As it grows fast with luxuriant stems, foliage, and roots, it is considered a pioneer grass species that makes waste mountains green when planted. It can fix sands, conserve soil, protect slope, and control water and soil loss, in addition to providing nutritious forage, fertilizer, and fuel. This grass requires accumulated temperature ($\geq$10° C) of more than 3000° C for flowering and seeding, which is difficult to meet in north China, and this has limited its distribution in the past. In order to solve the problem, the Institute of Soil and Water Conservation treated its dry seeds and overwintering buds by radiation for breeding an early-maturing variety. Through breeding and selection, a steady early-maturing variety was obtained. The new variety flowers more than 60 days earlier than the original one, and is featured by high seeding, natural regeneration, and fine characteristics. The new variety has played a role in making grassland a long-term, stable artificial community and in achieving high and stable yields (Ma Xiumei, 1990; Ma Xiumei and Song Lian, 1990; Liu Goubin and Yang Xiaoying, 1990). Experiments show that alfalfa and sainfoin are very suitable for plantation in the Loess Plateau, and have high production potential. *Coronilla varia* is not as good in production and quality, but it adapts to poor conditions very well and is considered a good species for conservation of water and soil (Gou Zhiquan et al., 1990).

Obviously, different vegetation has different results in soil and water conservation. According to Hou Xilu and Cao Qingyu (1990), the grown-up *Caragana korshinskii*, 6- to 15-year-old *Robinia pseudoacocia*, and 2-to 8-year-old *Astragalus adsurgens* contributed the best benefits for sediment reduction and water conservation, followed by 1- to 6-year-old *Robinia pseudoacocia* and 1- to 5-year-old alfalfa. The runoff from wasted natural grassland is more than that from farmland. For the same tree species, the greater the forest cover, the more significant the benefit of sediment reduction. The layer of the weathered material on forest land has an important effect on water conservation and sediment reduction.

b. Main Species of Trees, Shrubs, and Grasses

Most of the trees, shrubs, and grasses mentioned under wind control are also used for conservation of soil and water in different regions in the Loess Plateau. Here we will give more information on some species that are widely used for the purpose.

Tree Species. The main tree species used are locust (*Robinia pseudoacacia*), simon poplar (*Populus simonii*), elm (*Ulmus* spp.), Chinese tamarisk (*Tamarix chinensis*), Chinese pine (*Pinus tabulaeformis*), and *Pinus sylvestris* var. *mongolica*. Of those , locust widely distributes in waste mountains, gullies, and slopes in the Loess Plateu. According to the investigations by Wang Youmin (1990) in Chunhua County, a sub-humid area, the average height and diameter of 10- to 15-year-old trees is 6.5 m and 5.3 cm, respectively. These trees increase their height by 0.5 m and diameter by 0.5 cm and produce timber 2 to 3 m^3/ha each year. Locust grows very well between 1100 and 1400 m above sea level. Above the upper altitude, its growth is limited by temperature; below the low altitude, it is threatened by droughts. This tree growth takes soil water mainly from the 0- to 160-cm layer, and can obtain some difficultly available water from soil (Li Kairong and Wang Youmin, 1990).

In the Loess Plateau, soil water is the main factor affecting locust growth. When it is planted in regions with an average annual precipitation of more than 500 mm and with proper density (less than 2200/ha), it is expected to become timber. However, when it is planted in areas with 400 to 500 mm rainfall and with unsuitable density (more than 2200/ha), it is impossible for locust to become large-sized timber. This is especially true when it is planted in a loess hilly region where runoff is serious; in this case, the soil is becoming dry after its growth for some years because of intensive evapotranspiration and water absorption by developed roots in the deep layer. As the stored water in the deep layer vanishes, the tree depends only on current precipitation and thus grows very slowly (Ma Yuxi et al., 1990; Yang Xinming and Yang Wenzhi, 1988, 1990). This problem should be kept in mind when considering the development of the tree. Recently, *Pinus armandii* and *Pinus bungeana* have been introduced to the Loess Plateau for afforestation (Liu Zengwen and Wang Youmin, 1990; Xie Dezi et al., 1990), and they seem to have potential for development.

Shrub Species. The shrubs widely used in the Loess Plateau for controlling erosion are *Caragana korshinskii*, *Amorpha froticosa*, and *Hippophae rhamnoides* (Wang Youmin et al., 1990a). The former two can be afforested directly by seeding at a depth of 1 to 5 cm from the end of March to July. Their survival rates are more than 80% and 75% respectively. For *Hippophae rhamnoides*, however, direct seeding is not a good method as its roots grow very slow. Even though its germination rates are high, its survival rates are rather low. For this reason, seeding *Hippophae rhamnoides* should be carefully considered. *Caragana* is more important. It is a perennial legume and deciduous bush with bushy branches, stingy stipules, even pinnate leaves in alternation, yellow flowers, and shales in flat shape. It leaves have a medium-level content of nutrients, and it has a high capability of drought resistance with great vitality. The leaves can be eaten by sheep or goats in all four seasons, and the tips of its branches can be grazed as pasture by sheep and goats in winter and early spring; thus it is called a stubborn pasture. For this reason, *Caragana* is regarded as a good shrub species not only for its high ability in conserving soil and water, but also for its good quality as forage, fertilizer, and fuel. In addition to these shrubs, there are many kinds of wild shrub species distributed everywhere. These include *Ziziphus jujuba*, *Elaeagnus pungens*, *Acanthopanax gracilistylus*, *Prunus pacus*, *Rubus parvifilius*, *Vitis piasezki*, *Ampelopsis* sp., and *Rosa xanthia*. These have large roots, thick tillering, and high ground cover rates. Also, they have valuable fruits and flowers.

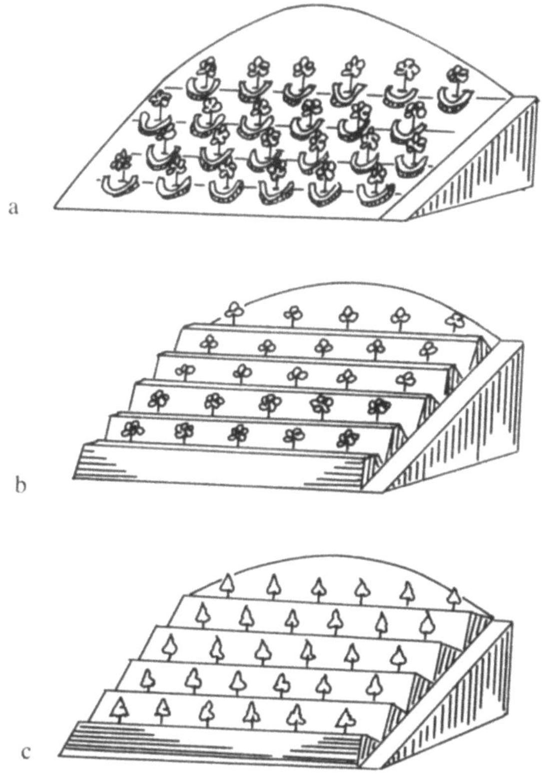

Figure 14. Engineering measures for afforestation. a. Fish scale pits. b. Contour narrow benches (or level ditches). c. Terraces with gentle slope declining inward. Sources: Xin Shuzhi and Jiang Deqi (1982).

c. Soil Preparation for Planting Trees, Shrubs, and Grasses

In order to afforest successfully, engineering measures, such as building fishscale pits (Fig. 14a), contour narrow benches (Fig. 14b), terraces with gentle slope (generally 10 degrees) declining inward (Fig. 14c) (Institute of Forestry, Academy of Shaanxi Agriculture and Forestry Sciences, 1976), and narrow ditches are often used. Wu Yuanji and Li Zhilian (1990) and Jia Houli and Qian Guangshu (1990) reported that planting mulberry (*Morus*, sp.) in level deep ditches on a slope equal to or more than 25 degrees effectively controlled water and soil loss, encouraged mulberry growth, and increased mulberry leaf production by 1026% as compared with traditional pits on the slope. Recent studies (Su Yinquan et al., 1990) showed that planting trees in a V-shaped pit obtained good results in afforestation; this involves digging a pit with two ditches on the upper side in a V-shape. The ditch is 10 cm deep and 10 cm wide, and the soil dug out from the ditch is added on its outside (low side) to make ridges in order to increase the depth of the ditch. The pit is used to plant trees, and the ditch is used to harvest water. The pits with ditches are separated along a contour, and each pit with two ditches occupies a catchment area of $2.4\,m^2$, in general, based on the number of the trees to be planted. This kind of land preparation for afforestation harvests more water from runoff, raises the survival rate of trees, and enables young seedlings to grow much better.

d. Combination of Biological Measures According to Natural Conditions

No matter which biological measures are taken, they should be combined in one system, arranged by productive units according to natural conditions. For example, the Qugu Productive Team in Hequ County, Shanxi Province planted a willow (*Salix* spp.) and locust (*Robinia pseudoacacia*) tree belt along the Yellow River to protect its banks as well as to be a main shelter. A narrow poplar tree belt together with *Salix purpurea* var. *multinervis* and day lily in beach lands were added on both sides of the road as supplement shelters. They also planted *Caragana korshinskii* on the sand tops of hills to fix sands; locust forests on the tops of loess hills to prevent wind and store water; a *Tamarix chinensis* belt on ridges in terraces or slopes; poplars, willows, and *Tamarix chinensis* on the ridges in valleys or tablelands; and fruit trees

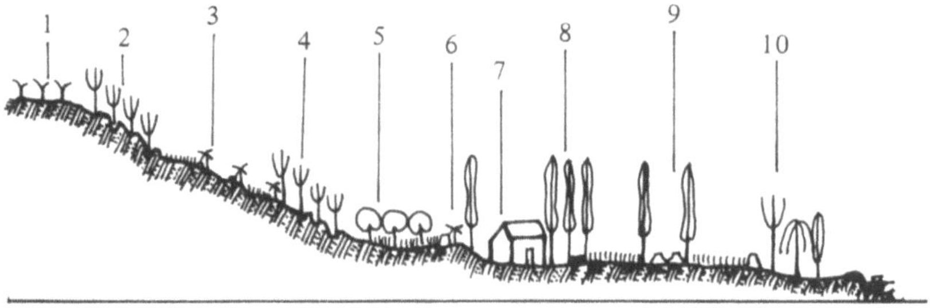

Figure 15. Diagram of biological protection system in Qugu productive team, Hequ County, Shanxi Province: 1, shrub forest for fixing sands and inhibiting winds; 2, forest for protecting slopes; 3, biological belts for protecting ridges of terraces; 4, forest for protecting slopes; 5, belt of interplanting fruit trees with crops in valleys and tablelands; 6, biological forests for protecting field ridges; 7, trees around village and houses; 8, forest belts for protecting roads and fields; 9, forest belts for protecting channels and roads as well as fields; 10, main shelter for protecting banks, beaches, and fields. Source: Nanjing Institute of Soil Science, Academia Sinica (1978).

such as apple (*Malus pumila*), apricot (*Prunus armeniaca—Armeniaca vulgaris*), grape (*Vitis vinifera*), pear (*Pyrus bretschneideri*), and peach (*Prunus persica-Amygdalus persica*) interplanted with crops in dam lands and terraces inside gullies; and locust forests in small gullies (Fig. 15). These measures have constituted a comprehensive protection system for mountains and river sand lands, for fixing sands, inhibiting winds, and keeping production high and stable both in mountains and on beaches (Nanjing Institute of Soil Science, Academia Sinica, 1978).

In loess hilly gully regions, there are obvious differences in soil moisture among hilltops, convex and concave hill slopes, and feet due to redistribution of precipitation by runoff and evaporation. The soil on northern slopes contains more water than that on southern slopes, and that on western slope is in between. Also, the average moisture at a depth of 3 m varies from the hilltop to the foot, the variation being different between the upper and lower layer. For the upper layer, the moisture gradually decreases or remains stable from the hilltop to the foot. However, for the lower layer, it usually decreases downwards. On the whole, the soil moisture increases from the hilltop to the foot (Liu Mei et al., 1990; Chang Caie et al., 1989). Study showed that in the dry season, water content at the middle part of a slope was the highest, storing 388.2 mm at a 2-m depth, which is equivalent to 85.8% of the field capacity. However, on the upper part of a slope, soil contained 323.7 mm water at a 2-m depth, which is equivalent to 49% to 65.2% field capacity (Jiang Dingsheng et al., 1987; Yang Wenzhi et al., 1990). Investigation also showed that the redistribution depended on the slope's degree and precipitation: there would be no redistribution when rainfall was 7, 4.3, 3.2, and 2.8 mm on a slope with 7, 15, 27.5, and 30 degrees respectively (Jiang Dingsheng et al., 1990).

Both engineering and biological measures can weaken the redistribution of precipitation (Liu Mei et al., 1990). In addition to soil water difference at different parts of hills and slopes, sunshine, temperature, and soil fertility are also different. Generally, the period of sunshine is longer at the middle part of a slope, and soil temperature at a 20-cm depth is higher on the middle and lower parts with great variation, which is favorable for fruit trees. Soil organic matter and total N are low at the upper part, but higher at the lower part. Based on this understanding and natural conditions, the Ansai Comprehensive Research Station of Soil and Water Conservation, Academia Sinica, have managed lands both horizontally and vertically to comprehensively control soil erosion and develop agriculture. The horizontal management is to divide a village's land into three circles according to its distance from the village. The first circle is around the village where the land is arranged to construct level terraces for food production as well as ecological and commercial trees, especially fruit trees. The middle circle is a little farther from the village where sloping land is dominant; it is managed by interplanting crops with grass and level ditch tillage on contours, with water and soil conservation as the main purpose. The outside circle is returned to grass and forest land to rehabilitate vegetation and form a green screen for protection of the first two circles.

The vertical arrangement is to manage the liang or mao types of hills or mountains from top to foot using various methods. As mentioned before, the liang-type hill is elongated in shape with an arched top, and the mao type hill has an oval outline or a round top. Both were formed by water erosion, and are surrounded by gullies. Accordingly, five measures have been adopted: (1) putting a "green hat" on the hill's head by planting *Astragalus adsurgens* for a width of about 20 to 30 m on contours at the hilltop; (2) hanging fruits on the slope's "shoulder" by planting fruit trees such as apple, Chinese hawthorne (*Crataegus pinnatifida*) in level terraces on the upper slope; (3) banding ties at the slope's "waist" by constructing level terraces on the middle part of the slope to form high-yielding farmland; (4) putting skirts around the slope's "leg" by afforesting with trees and shrubs to form a mixed forest on the lower part of the slope or at the foot of the hill; (5) putting "boots on the foot" by constructing detention dams and willow dams on the gully bottom to control gully erosion and form dam land (Jiang Dingsheng and Fan Xingke, 1990). Using these comprehensive measures, the water erosion is significantly reduced, agricultural production rapidly raised, and ecological balance obviously improved.

3. Agricultural Measures

On the drylands, any stable form of agriculture relies on both the bulk of the soil being held in place and water from rainfall being conserved in situ at all times. It follows, therefore, that a cultivation system for the drylands must first prevent soil from being washed away by rainfall, which may be badly distributed and fall in intense storms, and provide best conditions to form any shape of basin to hold water from rain or heavy storms until it has had time to run into the soil or to make water flow at a speed too low to cause soil movement. In addition to preventing soil erosion and water loss, cultivation must also maintain the soil in a state suitable for plant growth. In the main, this means that the surface should be permeable to receive rainfall and to allow gases to diffuse freely in and out of the soil. Weed control by cultivation is also important. Although the development of chemical methods may reduce the importance of cultivation simply to kill weeds, it is always cheaper to adopt such methods, which may give other benefits as well. Based on these principles, many agricultural measures have been adopted in the Loess Plateau as well as other places where water stress and soil erosion are the main constraints.

a. Tillage Systems

With regard to conservation of soil and water, the basic principle of tillage is to increase soil-surface roughness. Making low banks of earth between fields and ridges in a field, constructing trenches and digging ditches in fields, or a combination of both of these are the main measures for increasing the roughness. The effect of tillage methods on controlling soil erosion is greatly influenced by slope. Best results were obtained on a slope of less than 5 degrees. Success decreases as slopes increase to 5 to 10 and 10 to 20 degrees, and there is little or no benefit on slopes greater than 20 degrees. On the latter slopes good results can be obtained only when soil is deep and rainfall small.

There are many tillage methods for increasing the soil-surface roughness. Examples include contour plowing, contour planting, digging pits for planting, horizontal plow furrows, digging pits for dibble seeding, digging holes for cropping, cultivation in pits, and cultivation in furrows (Shi Keming, 1986). The major tillage methods are described here.

Contour Tillage. On sloped land, it is important that the cultivation be across the slope rather than in the direction of the slope. This is called contour tillage. Each furrow is thus a line of defense for checking surface runoff, and this method of tillage helps reduce soil erosion. Studies from Jilin Province indicated that the runoff and soil erosion were reduced by 19% and 50%, respectively, and the grain yields increased by 10% after the tillage along slopes had been changed to contour tillage (Xin Shuzhi and Jiang Deqi, 1982).

Ridge-Furrow Tillage. Ridging and furrowing have been shown to foster soil and water conservation, surface permeability, even water distribution and weed control. This system has long been used in the structurally unstable soils of the Loess Plateau. The variations include ridges made along the slope to drain steep hills and those across the slope to conserve soil and water, with cross ties in the furrows to hold surplus water, and with gentle gradients to facilitate water harvest. This is more useful on slopes of 5 to 10 degrees and is used on steep slopes of more than 20 degrees only occasionally. This method allows cultivation to divide a slope into ridges and furrows across the slope and to plant crops in the furrows. Surface runoff is then checked, soil erosion reduced, and applied fertilizer conserved. Furrows with ties are advisable as the ties form basins to hold rainwater from heavy storms, provide an opportunity for more water to soak in, and prevent water flow on either side of the furrows.

The Tianshui District, Gansu Province, has demonstrated that the ridge-furrow system reduced the surface runoff by 77%, and soil erosion by 88% as compared with the fields where this system has not been adopted. This measure can significantly increase crop yields and is therefore widely used. In Yanan District, this approach is widely adopted as a high-yielding method.

The ridges and furrows may be formed in two ways: by plowing in advance of seeding any by intertilling during the plant-growth period for intertilled crops (Xin Shuzhi and Jiang Deqi, 1982; Liang Hongru, 1984). Ren Zhengxi and Li Yingqi (1987) reported that during the crop-growing period, the furrow moisture at 10, 20, 40, and 60 cm depth was 1.2%, 0.9%, 0.4%, and 0.6% higher than in the flat plowing fields respectively, and the spring wheat, potato, flax (*Linum usitatissimum*), and millet yields were increased by 51.4%, 24.7%, 16.3%, and 119.6% respectively. The authors suggested that the furrow: ridge ratio should be 50:50 cm on arid and semiarid land, and 65:50 cm on moist, hilly land. Lu Zongfan et al. (1990a) reported that the ridge-furrow tillage increased crop yield by 28% on flatland, and by 58.5% on 25-degree sloped land, in addition to reducing runoff by 22.8%.

Pit and Furrow Tillage. Pit cultivation and furrow cultivation were used long ago. The former originated from Ou Tian recorded in the ancient book, *Fan Sheng Zhi Shu* (Fanshengzhi's book), which was written as early as 1900 years ago in the Han Dynasty (Shih Shenghan, 1959); the latter came from De Tian (cultivation in furrows), which was widely used in the Ming Dynasty, between 1368 and 1644 A.D. (Wang Yonghou, 1983; Research Group of Ancient Chinese Agriculture, Academy of Agricultural Science and Nanjing Agricultural University, 1986). Both have been tested and imporved and extensively used in arid, semiarid, and subhumid regions.

The procedure of the furrow cultivation is as follows: digging or plowing the land along the contour on sloping land or along a straight line on flatland to make furrows at a certain distance, applying fertilizers into the furrows, deep plowing the furrows, cracking the soil blocks, compacting the loose soil, and smoothing the soil surface, and then seeding in the furrows. The soil removed from the furrows is used to make ridges along the furrows on both sides on flatland or on low sides on sloping land. Sometimes small ridges in the furrows are also made. The ridges and the rest of the land are fallowed. After harvest of the crops, the furrows are moved to where the ridges were, and the original furrows are changed into ridges. In this way, after about 2 to 3 years, depending on the space left for fallow between two furrows, all the soil is used as furrows and therefore improved.

The furrow cultivation method can be simplified in the following way: plowing to make a furrow at the margin of a field along the contours and letting the soil out from the furrow onto the low side, seeding and applying fertilizers into the furrow, plowing to cover the seeds and fertilizers, leaving a space of one plow width (about 20 cm) for fallow, and then repeating the procedure until the whole land is cultivated (Guo

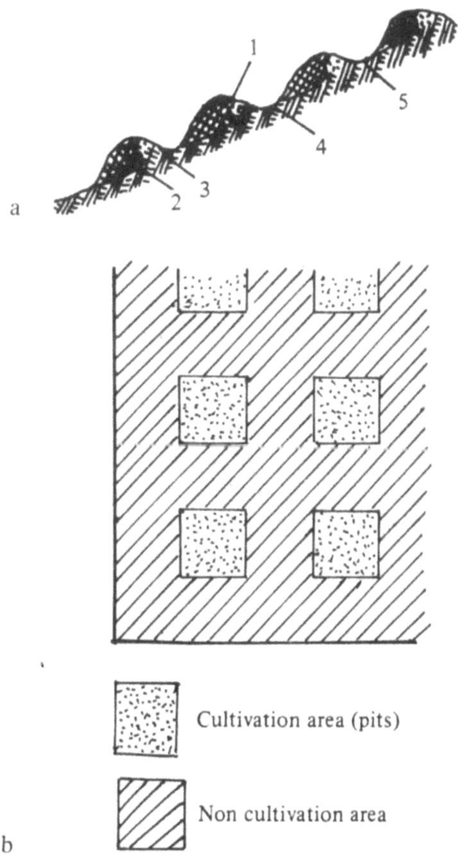

Figure 16. a. Furrow tillage. 1, Seeds and fertilizers; 2, ploughing to make furrow (first plough); 3, ploughing to have soil cover seeds and fertilizers; 4, space of one plough left for fallow; 5, furrow. Source: Nanjing Institute of Soil Science, Academia Sinica (1978). b. Pit tillage , Cultivation area (pits); , non-cultivation area. Source: Research Group of Ancient Chinese Agriculture, Academy of Agriculture Science and Nanjing Agricultutral University (1986).

Tingfu and Gao Bowen, 1982). In this way, the furrows between two ridges are formed (Fig. 16a). The depth of the furrow and the distance between ridges depend on crops.

Recently the Shanxi Water and Soil Conservation Institute created an alternative method for sloping land using the following procedure: plowing to turn over the topsoil on the high side of the contour and plowing once more to turn over the subsoil on the low side to form a ridge, plowing again to loosen the subsoil, followed by fertilization and moving the topsoil into the furrow. In this way the subsoil is used to make ridges and the top-soil remains in the furrows. This method can significantly increase water conservation, improve soil fertility, and increase crop yield (Shi Guangyi, 1982; Zhongyan County Government, 1985; Shi Guangyi and Liu Yongguang, 1981; Hou Zhongtian and Tang Defu, 1984; Wang Tianyi, 1990; Dong Yanqing et al., 1991).

Pit cultivation has the same procedure as the furrow except pits are formed instead of furrows (Fig. 16b). Also, pit cultivation can be simplified by digging pits along the contour, seeding, and applying fertilizer. Both

Table 6. Water content in soil profiles of fields with pit cultivation and traditional cultivation

Pit cultivation		Traditional cultivation	
Depth (cm)	Water (%)	Depth (cm)	Water (%)
0.0–7.1	4.9	0.0– 5.0	5.1
7.2–35.0	13.7	6.0–32.0	19.5
36.0–62.0	26.0	33.0–82.0	31.3
Below 62	18.3	Below 83	18.1

Source: Shaanxi Academy of Agricutural Sciences (1973).

can be used on flatland and sloping land, but they are more suitable for sloping lands, and the pits can be used on slopes of less than 20 degrees.

The furrows and pits concentrate and store water for plant use, and therefore can increase crop production. Experiments from Yanan and Dinxi showed that pit or furrow tillage increased water storage by 10% (Table 6), and crop yields by 20% to 50%, sometimes even doubling them (Shaanxi Academy of Agricultural Science, 1973). The Shanxi Water and Soil Conservation Institute reported that when sloping land was tilled in this way, during 41 days of continuous rain totaling 430.7 mm, the water infiltrated into the soil without erosion. In contrast, the other sloping land with conventional tillage was seriously eroded. The furrow-tilled land conserved 55.6 mm more water and the crop yield doubled over that with conventional tillage (Hou Zhongtian and Tang Defu, 1984).

Level-Ditch Tillage. Plowing along contours to form ditches at a certain distance, generally 5 to 10 m apart, is called level-ditch tillage. This tillage, although very common on fallow land, is often adopted on cropping land as well. The distance between ditches depends on the slope, but is generally 5 to 10 m. Xin Shuzhi and Jiang Deqi (1982) reported that this tillage reduced surface runoff by 70%, and soil erosion by 90% in Tianshui District, Gansu Province. Liang Hongru (1984) concluded that this tillage conserved 62.0% of rainfall and 93.4% of sediments, and increased millet by 15% to 185.5%, wheat by 15% to 186%, and sorghum by 101% in Yanan District. Zheng Fengli et al. (1987) found that in a millet field, level ditch tillage significantly reduced soil rill erosion (Table 7). In north Shaanxi, a very narrow ditch (less than 1 m) is proposed for utilization on slopes of more than 25 degrees (Lu Zongfan et al., 1990a). This kind of level ditch is very similar to furrow-ridge tillage.

b. Cropping Systems

The principle of cropping systems for soil and water conservation is to increase plant cover. Rational close planting, intercropping, and inter-

Table 7. Comparison of rill and total erosion between level ditch and conventional tillages

	Total erosion of soil		Rill erosion of soil	
Tillage	kg/m^2	%	kg/m^2	%
Conventional	0.63	100.0	0.53	100.0
Level ditch	0.35	55.5	0.00	0.0

Notes: Determination was carried out in millet-growing field with a slope of 31 degrees after 56.0 mm precipitation, maximum rainfall intensity being 0.85 mm/min within 30 min.
Source: Zheng Fenli et al. (1987).

planting, or adopting dense belts in a field are widely used for the purpose. An increase in crop canopy can reduce raindrop impact directly on the soil surface and eliminate surface sealing, keeping the soil permeability higher, and thereby decreasing soil erosion. Many ways of increasing crop cover have been practiced in eroded areas by farmers.

Intercropping and Interplanting. As regards soil and water conservation, intercropping and interplanting have two functions, i.e., increasing plant density and thereby plant cover and lasting cover duration. In the loess regions the following intercropping and interplanting systems are largely in use: intercropping millet and soybean (*Glycine max; Glycine hispita*), sorghum and soybean, potato and cow pea (*Vigna unguiculata*), maize and soybean, sorghum and red bean, maize and black soya bean, millet and red bean, maize and potato, and maize and pumpkin (*Cucurbita moschata*); interplanting winter wheat with black soya bean, winter wheat with potato, pea with proso; mixed seeding millet and sorghum as well as mung bean, mung bean and proso as well as radish (*Raphanus sativus*), buckwheat and seedrape, etc. Interplanting tall, low dense crops such as sorghum, maize, and potato with short, dense crops such as oat, millet, proso, and wheat, strip by strip is suitable for gentle slopes with strip width of 3 to 5 m. However, on the steep slopes, interplanting or intercropping different short, dense crops is more useful. Study of the suburban region of Chengde City, Hebei Province, has shown that by intercropping red bean into millet rows on a slope of 22 to 25 degrees, the plant cover degree was doubled, and the cover duration was 30 days longer, since the red bean was still growing after harvest of millet. Because of this, the erosion was reduced by 17% to 47% (Xin Shuzhi and Jiang Deqi, 1982). In the loess hilly gully region, a field trial has been conducted to interplant crops like alfalfa, *Melilotus suaveolens*, and *Astragalus adsurgens* with crops such as millet and potato. Some results show that runoff and sediments were reduced as compared with plating crops. However, this was not better than planting grass.

Interplanting grasses with shrubs is another measure in eroded areas for conservation of soil and water. It is reported that interplanting *Caragana korshinkii* with *Astragalus adsurgens* on a 5-m wide strip for each on a 32-degree slope resulted in significant reduction of runoff and sediments as well as a high production of forage. On a bare slope without vegetation, the annual erosion module was 4013 Mg/km^2, and the runoff module was 25,599 m^3/km^2. However, on the interplanted slope, erosion and runoff modules were 765 Mg/ha and 13,136 m^3/km^2, respectively. That means that this kind of interplanting reduced sediments by 82% and runoff by 51%. In addition, it produced 42.4 Mg/ha fresh forage, much higher than with other interplanting. The other interplanting, such as sainforn + *Caragana*, alfalfa + *Caragana*, and daghestan sweetclover (*Melilotus suaveolens*), also reduced runoff and sediments significantly, but they were not as good. This was also true for planting grass or planting shrub only, as shown by the comparasion made between grass only or shrub planting and interplanting (Lu Zongfan et al., 1990b).

Strip Cropping and Contour Strip Cropping. On long slopes subject to water erosion, the fields may be laid out in narrow strips across the incline, alternating the tilled crops, such as maize and potatoes, with hay and grain. These strips separate the slope into different parts and play the role of a sloping terrace in reducing slope length and erosion. Water cannot achieve high velocities on the narrow strips of cultivated land because the hay and grain definitely check the rate of runoff. Such a layout is called strip cropping and is the basis of much of the erosion control now advocated. When the cross strips are laid out on the contour, the system is called contour strip cropping.

Both strip cropping and contour strip cropping have long been adopted in China for controlling erosion, but they have different names in different locations. Since crop and grass are intercropped in different strips, the term strip alternation of crop and grass is commonly used. The width of the strips depends primarily on the degree of slope and the permeability of the land. Contour strip cropping is often guarded by diversion ditches and waterways between fields. When their grade is high, they should be sown with grass to prevent underwashing. In addition, this system is also used for formation of terraces. Through a few years of cultivation, the elevation of the crop strips gradually decreases, and the elevation of the grass strips increases. As a result, the whole land is changed into gentle terraces, even level terraces.

The grass species used depended mainly on the conditions and degree of slopes. In Huan County, Gansu Province, alfalfa is used with a width of 1 to 2 m, even 3.5 m, and a distance of 10 to 20 m apart on slopes of 10 to 20 degrees. However, in some places of the loess regions, *Tamaxis chinensis* and *Astragarus adsurgens* are preferred (Shaanxi Water and Soil Conservation Bureau and Northwestern Institute of Water and Soil Con-

Table 8. Comparison of total erosion and rill erosion between conventional cropping and strip alternation of crop and grass

Cropping system	Total erosion of soil		Rill erosion of soil	
	kg/m^2	%	kg/m^2	%
Millet	0.63	100.0	0.53	100.0
Millet + *Astragarus adsurgens*	0.42	68.2	0.22	41.1
Millet + *Melilotus*	0.55	86.2	0.43	80.8
Potato	0.79	100.0	0.70	100.0
Potato + *Medicago sativa*	0.41	51.9	0.40	57.4

Notes: Determination was carried out in a millet growing field with a slope of 31 degrees after 56.0 mm precipitation, rainfall intensity being 0.58 mm/min within 30 min. Source: Zheng Fenli et al. (1987).

servation, 1979). Zheng Fenli et al. (1987) reported that the strip alternation of crop and grass could reduce soil erosion to some extent (Table 8).

Dense Planting with Broad Distance Between Rows. Ridge tillage is a basic way of cultivation in northeastern China either on level land or sloping land with a row crop grown on a ridge. Recently, the ridges are being broadened by combining two ridges into one with a width of 100 to 120 cm and a height of 15 to 17 cm. Two rows of the crop are grown on a ridge with high density. The distance between ridges has also been broadened. In addition to other advantages, this system is beneficial for reducing velocity of runoff and sediment deposition, thereby conserving water and soil. It it reported that a productive brigade increased its maize yield by 64% using this cropping system (Xin Shuzhi and Jiang Deqi, 1982).

c. Rotation Systems

Suitable rotation systems can help reduce erosion and have been put into use. Rotating crops with grass is effective in this respect and has been adopted widely. The Tianshui Station of Water and Soil Conservation, Yellow River Committee, reduced surface runoff by 58.0% and 78.2%, soil erosion by 73.8% and 84.8% in the first and second year respectively after changing the lentil (*Lens*)-fallow-winter wheat, millet rotation to lentil + *Melilotus albus–Melilotus albus*-wheat, millet rotation. In the meantime, the soil organic matter was increased by 10.3% to 43.4% and water-stable aggregates by 11.9%. By adding *Melilotus albus* in the rotation, the Wanghebu brigade raised wheat yields from 1.75 Mg/ha to 2.64 Mg/ha in successive 5 years in addition to 2.25 Mg/ha hay and 6.0 Mg/ha straw fuel.

Wang Duqing And Bai Huishan (1987) conducted an experiment in the rotation of crops with alfalfa, *Astragarus adsurgens*, and annual and biennial sweet clover (*Melilotus albus*) in a Suide Soil and Water Conservation Experimental Station of the Yellow River Conservancy Commission during the period 1981 to 1986. The results showed that the effect of alfalfa on soil and water conservation was the highest, followed by *Astragarus adsurgens*, and biennial and annual sweet clover in succession. However, rotating the crops with all of the legumes significantly increase crop yields.

Rotation with grass can be used as an aid to help form terraces on sloping land. In the strip crop-grass cropping system, as mentioned in the previous section, the crop strips gradually change into terraces. If rotation with grass is then carried out, the grass strips, after rotation, are turned into crop. In this way, over several years, these strips are also changed into terraces, and then all the land is terraced.

d. Agroforestry System

In the past, due to population pressure and the problem of food supply, more and more marginal and submarginal lands have been brought under cultivation. Apart from being uneconomical in the long run, especially under dryland conditions, cultivation of such land has led to serious imbalances in the ecosystem. For such drylands, in addition to converting to forest and grass, an efficient land use system rather than arable cropping would be more appropriate. Day by day the demand for food, fodder, and fuel is growing, which could be solved by selecting suitable land use systems.

Alternate land use systems not only help in meeting the diverse demands for land, but also improve the degraded, marginal, and submarginal lands, and therefore minimize productive risk. These systems include ley farming, tree farming, dryland horticulture, and agroforestry. Of those, the so-called tree-cum-crop and tree-cum-forage versions of agroforesty systems could meet the multiple requirements of food, fodder, fuel, fertilizer, etc. This is a kind of system in which food or forage crops are grown in alleys formed by hedge rows of trees, generally fruit trees or shrubs. The essential feature of the system is that hedge rows are pruned before or during cropping to prevent shading and to reduce competition with food crops. This system provides higher total biomass, green fodder during the lean period of fodder availability, and a barrier to runoff water, holds the silt and conserves moisture when planted along the contours, and utilizes off-season precipitation which otherwise would be wasted. Fruit trees such as apple (*Malus pumila*), apricot (*Prunus armeniaca–Armeniaca vulgaris*), walnut (*Juglans*), and kaki persimmon (*Diospyros kaki-D. chinensis*) are widely adopted in the system for more economical benefits.

In loess regions, afforestation is a problem in the areas where soil is drastically eroded. Interplanting crops or forages into the tree rows not only increases plant cover and helps to afforest through management of the crops, but also increase the incomes of the planters. This can be used either as a long-term system or as a method for afforestation. For the former purpose, the distances between two tree rows are wide, and crops are planted in between them each year. Even when the space for cropping is narrow winter crops are still grown. A local saying goes: "Trees are guests passing by, they don't shade winter crops, but shade summer crops." For the latter purpose, after the tree grows up, cropping or foraging will be stopped. Huan County, Gansu Province, interplanted apricot with millet. After 6 years, the trees were 2.7 m in height and 3.5 m in diameter. In contrast, the apricot trees were 1.9 m in height and 2.5 cm in diameter with no millet interplanted. Yanhewan, Ansai County, Shaanxi Province, interplanted trees with crop for 66 ha, and after 3 to 5 years, a wood patch was formed with crop in it and erosion was significantly reduced (Xin Shuzhi and Jiang Deqi, 1982). Recently the tree-cum-crop system has been widely adopted on the Weibei dryland with *Paulownia elongata* as a major tree species. *Paulownia elongata* grows as fast as *Populus daruensis* with timber production of 12.5 to 25 m^3/year per km (Wang Youmin et al., 1990b). Its largest consumption of water appears in July and August, which matches the rainy season (Ren Yong and Wang Youmin, 1990a). Investigation on a 31-m wide terrace lang with one row of *Paulownia elongata*, which was 14 years old, 9.03 m in height and 18.6 cm in diameter on average, showed that *Paulownia elongata* consumed 19.8% of the total water consumed, while wheat consumed 80.2%. In addition, *Paulownia elongata* took water mainly from layers 60 to 120 and 120 to 250 cm deep, so competition for water between *Paulownia elongata* and crop was not serious (Ren Yong and Wang Youmin, 1990b). Wang Zhonglin and Wang Youmin (1990) investigated the protective results of six kinds of agroforesty and found that the *Paulownia elongata*-cum-crop system was the best in protecting against winds. Also, this system reduced evaporation, increased relative air humidity, and raised crop production (Wang Zhonglin, 1990).

Although each of the methods mentioned above has its own role in controlling soil erosion, best results can be obtained only by combining them all together according to different conditions. Experience and agricultural practices have shown that erosion can be successfully controlled by comprehensive measures, but not by one measure alone.

Jiuyuangou Valley in Suide County, Shaanxi Province is an example. Located in the upper part of the Wuding River, this valley covering about 70.1 km^2 was once the most seriously eroded area of the loess hill-gully regions with sediment of 2000 to 2500 Mg/km^2 per year lost by runoff on the whole valley, and crop yield was 300 to 375 kg/ha. In 1953, the people began to control gullies by building dams, but progress was very slow. Since 1964, comprehensive methods have been used, i.e., combining engineering methods with biological methods in managing gullies as well as slopes. This has speeded up the control program, reducing sediment loss by 80%. Controlling soil and water loss has promoted the development of agriculture, forestry, and animal husbandry. In 1973, dam-fields were built on 86.6 ha and terraces on 866.6 ha, totaling half of the cultivated land, with forest land of 800 ha and grassland of 533 ha (Secretariat of Conference on Agriculture Learning from Dazhai, 1975).

Another example is the Taipinggou Catchment. Located in a tributary of the Laoha River, the upper reaches of the west Liao River in Jiangping County, Liaoning Province, the catchment covers an area of 10.6 km^2. In the past, soil erosion was 4600 Mg/km^2 per year due to nature and human activity (extensive farming, overgrazing, and poor forestry management). Since 1980, the people have converted some land into grass and forests, controlled erosion by engineering, as well as implemented biological measures, and established a new land use system. Within 6 years, the forest and grasslands covered 68% of the total land, and thus erosion was reduced by 80%. As a direct result, the total yield was increased by 18.1% for cereals and 132.6% for legume crops (Du Yibin, 1987).

C. Full Store of Precipitation in Soil

Since a lack of groundwater for irrigation occurs in most of the dryland areas of China, the fundamental measure for crop production is to increase the storage of precipitation in the soil for later use. In addition to some methods mentioned above, many other measures are used not only in arid and semiarid areas, but also in subhumid areas susceptible to drought. These will be discussed below.

1. Water Harvesting

During the torrential rain period, runoff is very common in dryland areas. It is produced from sloping fields, roads, waterways, villages, gullies, and river sides. The floodwater is disastrous not only because of the loss of the precious precipitation, but soil, fertilizer, and organic matter are lost as well. However, this floodwater can be harvested for storage in other fields for later or present use by plants, and the sediments can be deposited for improvement of soil in other fields. This kind of water harvesting may be called runoff agriculture (Sun Jianxuan, 1987), and the lands receiving water and sediment are called flooded lands. This technique has long been adopted in China, particularly in the loess regions.

There are different ways of water harvesting for flooded lands. However, harvesting water from gullies, sloping lands, and riverbeds are the most common approaches. For such water harvesting, especially for flooding large areas, engineering projects, such as antiflooding, channeling water, draining, and field ridges should be made. Use of irrigation channels is the easiest and is widely practiced. Flooding sandy lands located on river sides is more beneficial as these lands need sediment to improve their properties. Flooding land techniques are also utilized on a small scale on lowlands, and the water harvested generally comes from the lands nearby, roads, and threshing grounds, even villages. There is generally no need to have engineering projects for such small land water harvesting. However, small ridges are needed for channeling water into the expected field. The water harvested may be all stored in the flooded land, or may flow or drain away, depending on current land use and amounts of the water. However, the sediment will be all deposited on the flooded lands. Using this technique, the Inner Mongolia Autonomous Region has changed 10,700 ha wasteland into fertile farmland and ameliorated 120 ha saline soil on the south side of the Yellow River within the region; yields have stabilized around 2.25 Mg/ha, which is much higher than on other land nearby (Xin Shuzhi and Jiang Deqi, 1982).

2. Level Field Surrounded by Low Bank

In order to store all precipitation in fields, banked-up level fields have been constructed on plains. This involves leveling a piece of land and building low banks on all the edges around it (Plate 5). As the name

Plate 5. Banked level field, Chengcheng, Shaanxi. Source: Shaanxi Water and Soil Conservation Bureau (1971, 1972).

implies, this feature consists of level and banked land. This land can store all precipitation and is easy for irrigation and utilization of machinery. The width may be broad or narrow, depending on the overall size of the land. In any case, convenience for cultivation should be considered. In 1970, Yongshou County ameliorated 400 ha hilly land into banked-up level land, and wheat yield was increased to 5.35 Mg/ha on average with a maximum of 7.5 Mg/ha, which was two to three times higher than that on the other land (Shanxi Water and Soil Conservation Bureau, 1971, 1972).

3. Deep Plowing

Plowing to a depth of about 20 to 30 cm has been widely used in dryland farming as an effective method for storing rainfall. The agricultural proverb "plowing one extra cun (a Chinese unit for length, equal to 3.33 cm) is as good as applying one cart of manure" fully describes its importance. In accordance with the fallow period cropping system and climatic condition, deep plowing is generally carried out in summer, but not in spring. Summer plowing keeps soil loose during the rainy season (from July to September), enhances water penetration, reduces runoff, and thus stores more water in the soil (Soil and Fertilizer Institute, Gansu Academy of Agricultural Science, 1966), while spring plowing increases water loss from the soil because this season is often dry, and drought often occurs.

Data from Chen Guoliang (1983) for different regions of dryland show that deep plowing in summer increased water by more than 12 mm at a 50-cm depth, and the yields of millet (*Panicum miliaceum*), grain sorghum (*Sorghum bicolor*), and maize (*Zea mays*) were increased by 22.3%, 30.2%, and 17%, respectively as compared with the shallow plowing method (at about a 10- to 12-cm depth). In addition to increasing water storage, summer plowing can also increase nutrients available to soil. Through sunning the upturned soil, the weathering process is accelerated, and some unavailable nutrients released. This is particularly true for nitrogen (Dhar, 1933).

Table 9. Effect of time of plowing on amounts of some available nutrients

Time of plowing (month, day)	Amount of sunshine (h)	Available nutrients at 25 cm depth (mg/kg soil)		
		NO_3-N	NH_4-N	P
July 3	329.1	34.4	5.2	0.9
August 7	85.4	18.8	2.6	0.8

Note: Soil samples were taken on August 19 and amount of sunshine was calculated from plowing till sampling. Source: Li Hongen and Yang Yunlian (1965).

Li Hongen and Yang Yunlian (1965) reported that the contents of available nitrate-N, ammonium-N, and P depended on plowed time: more available forms of these elements were obtained with early plowing (Table 9). Summer plowing should be conducted as early as possible, otherwise the effect will be reduced. If plowing is delayed in autumn, harrowing or compacting to pulverize the soil surface should follow immediately to form a dust mulch, reducing the proportion of large pores to decrease evaporation, because in the fall evaporation exceeds precipitation. Measurements obtained the following spring showed that in land harrowed or compacted during the fall, the soil moisture content was 17.2% and 8% to 12.5% higher, respectively, than without harrowing or compacting (Gansu Agricultural Ministry, 1983).

Traditional deep plowing involves turning soil over using a simple moldboard plow pulled by draft animals or moldboard plows pulled by a tractor. Although use of tractors has become widespread, animal traction is still commonly used by a vast majority of the farmers in the dryland areas. The most common combination is a pair of oxen and a simple moldboard plow. The conventional deep plowing method can bury weed seeds or weeds into deep layers, and make the soil surface clean by turning over stubble and manure, which is good for sowing. It is often used when large quantities of manure are applied and weeds, pests, and diseases are serious. However, conventional plowing has its disadvantages; moisture loss, soil structure damage, slopy land soil erosion, and cost increases are the main ones.

Much research done in recent years has shown that subsoiling by loosening deep layers without turning over the soil has advantages over the conventional system for water storage, moisture conservation, machine fuel use, and crop yields. (Zhao Fengbin, 1985, 1987).

Han Siming et al. (1987) demonstrated that subsoiling increased wheat yields by 8.9% to 10.7% as compared with deep plowing (Table 10). An experiment in Shanxi Province (Li Duren, 1986; Li Duren et al., 1987) revealed that subsoiling decreased the soil bulk density by 0.11 to 0.43 Mg/m^3 in the plowpan and 0.06 to 0.14 in the sublayer. It increased

Table 10. Effect of different tillage practices on wheat yields (Mg/ha)

Lacations/year	Deep plowing	Shallow plowing	Subsoiling	Mulching
Chengcheng				
1984–85	1.84	1.95	2.09	2.27
1985–86	3.53	3.37	3.77	4.16
Qianxian				
1984–85	3.30	3.24	3.59	4.30
1985–86	3.89	3.54	4.17	4.36

Source: Han Siming et al. (1987).

Table 11. Effect of subsoiling on soil water storage and wheat root weight

Soil depth (cm)	Subsoiling	Plowing	Difference
	Bulk density (Mg/m^3)		
0–10	1.10	1.14	−0.04
10–20	1.11	1.18	−0.07
20–30	1.23	1.34	−0.11
30–40	1.34	1.26	−0.06
	Soil water storage (mm)		
0–100	304	209	11.0
0–40	120	117	3.0
40–100	184	176	8.0
	Wheat root weight (Mg/ha)		
0–20	2.33	1.83	0.50
20–40	0.62	0.43	0.19
0–40	2.95	2.51	0.44

Source: Li Duren et al. (1987).

soil porosity by 4.2% to 16.3% in plowpan and 2.2% to 5.3% in subsoil. In the 20- to 40-cm layer, the soil infiltration rate was increased by twice that using the conventional method, and the amount of roots in the loosened layer was increased by 29.9% to 47.4%. Consequently, crop yields were significantly increased (Tables 11, 12). Similar results were obtained by Hou Lingsheng et al. (1987) in Hebei Province. They adopted deep plowing and subsoiling at intervals together with concentrated fertilization and wide row spacing with dense planting for summer maize on infertile rainfed land. This method is characterized by combining traditional plowing with subsoiling to form a deep cultivated layer, making soil part soft and part dense. Subsoiling with deep fertilization and close planting can promote the penetration of plant roots into

Table 12. Effects of subsoiling and plowing on wheat yields (Mg/ha)

Locations	Subsoiling	Plowing	Subsoiling–plowing
Duanjiazhang	2.79	2.35	0.54
Duanjiazhang	4.70	4.44	0.29
Duanjiazhang	3.22	2.65	0.47*
Xijia	3.18	2.81	0.37
Gaotousi	3.25	2.25	0.70
Gaotousi	2.80	2.43	0.37*
Chewangzhangou	2.62	2.03	0.59**

*, $P < 0.05$; **, $P < 0.01$.

Source: Li Duren et al. (1987).

the deep layer and ensure yield increases. As a result, the summer maize was increased by 850 kg/ha, and water storage was increased by 47 and 70 mm in the 100-cm layer as compared with traditional and stubble mulch respectively.

Experiments done in Shandong Province indicated that deep plowing obtained good results when soil moisture was adequate and the soil had a plow pan; otherwise, shallow or subsoiling should be more suitable for conservation of soil water (Xu Maozhen, 1987). In Hailun County, Heilongjiang Province, spacing chiselling to loosen soil instead of deep plowing has been considered a key measure of a new system recommended to farmers (Zuo Shuzhen and Chi Renli, 1987). In the Heilonggang region of Hebei Province, the open upturned soil method of deep plowing has been changed to loosening it by raking not only in summer but also in spring before spring crop planting. Excellent results have been obtained when the loosening is alternated with deep plowing (Li Yu and Li Shanxiang, 1987).

4. Summer Fallowing

Fallowing has been considered a basic measure for restoring soil water and fertility. During the fallow period, no crops are grown, and weeds are controlled by cultivation. The fallow period is different from one place to another, mainly depending on precipitation and soil fertility. If a 1-year period of fallow is practiced, the effect on the subsequent crop production will last for at least 3 years. However, in dryland areas of China, summer fallowing is practiced in the rainy season to store more precipitation for later crop use, at least for ensuring its germination. A summer fallow is generally combined with summer deep plowing for controlling weeds, keeping the soil loose, and increasing soil infiltration, and therefore storing more water.

Li Yushan (1962, 1982) found that during the fallow period of the rainy season, the manural loessial soil stored 156.5 to 225.1 mm water at a 2-m

Table 13. Effect of summer fallowing on crop yields (Mg/ha)

	Year						Total yields
	1st		2nd		3rd		
Cropping systems	Wheat	Corn	Wheat	Corn	Wheat	Corn	in 3 years
One crop/year with summer fallowing	3.00		3.20		2.04		8.24
Two crops/year without summer fallowing	1.45	1.44	1.10	1.45	0.66	0.93	7.03

Source: Li Yushan (1982).

Table 14. Accumulated amounts of NO_3-N in soil during summer fallow period (kg/ha)

Summer fallow fields	Accumulated amounts of soil NO_3-N		
	At beginning of fallow	By end of fallow	Accumulated amount
1	22.3	40.0	17.7
2	10.3	34.7	24.4
3	7.8	26.7	18.9
4	12.3	61.8	49.5

Fields 1–3 after wheat harvest; field 4 after pea harvest.
Source: Peng Lin et al. (1981).

depth, of which, 19.2 to 52.2 mm water infiltrated into the deep layer and was stored, and the rest supplemented all the lost water by evaporation and evapotranspiration. In contrast, cropping greatly consumed both the stored and supplemented water, making soil water-deficient. As a result, the production of wheat (*Triticum aestivum*) in the Weibei Loess Plateau greatly depended on summer fallow. Nearly half of the water consumed by wheat came from water stored during the fallow period. Field trials conducted in the Weibei Loess Plateau have indicated that when the summer fallow was cancelled, and the fallow period was used to seed short-period crops, the total yields of the next three harvests were significantly reduced (Table 13). In addition to increasing water storage, summer fallowing also increases available soil nutrients.

Peng Lin et al. (1981) reported that after summer fallowing, about 17.7 to 49.5 kg/ha NO_3-N was accumulated in the top layer (Table 14), which could meet the need for nitrogen of the early growing stage of wheat. Thus, the accumulated N provided a foundation for high yield. The winter wheat yield after summer fallowing was 57.5% to 124.0% higher than that without fallowing. Other reports from different regions have confirmed the good effect of summer fallowing (Li Yongtang, 1966). Based on agricultural practices and scientific experimental results, Shen Yuqinq (1962) fully expounded the role of summer fallowing and concluded that it was a recommendable tillage system ont he drylands.

5. Improvement of Soil Texture to Increase Water Storage

Adding sandy soil to clay soil or clay soil to sandy soil can improve soil texture and therefore soil and water conservation. Although this method is time-consuming and labor-intensive, it is effective in decreasing runoff in heavy soil and increasing the water-holding capacity in sandy soil. Beiheying Agricultural Brigade Experimental Station, Hebei Province (1973) demonstrated that by mixing 600 Mg of clay soil into a 20-cm layer of a sandy soil per hectare the saturated water content in the soil was increased from 21% to 23%; by mixing 600 Mg of sandy soil into a clay soil, the water content retained on the 10th day after 30 mm of precipitation was 0.17 kg/kg, while it was 0.16 kg/kg in fields without mixing. Using flood warping to improve soil texture is an effective method and perfect results have been obtained by Dangpudi Village, as reported by Si Binglian (1987).

6. Return of Crop Residues to Soil to Improve Soil Structure and Water Regime

Return of crop residues, such as wheat straw and maize stovers, after harvest directly to soil by deep plowing has been employed to improve soil structure and its water regime as well as its fertility. Since 1978, the Wanggongzhuang Village in Shanxi Province has applied more chemical fertilizers to obtain more organic matter and returned the crop residues directly to the soil. With grain yield increasing, the crop residues have been correspondingly increased. For example, 3.3 Mg/ha of crop residues were produced in 1978, but 10.0 Mg in 1983, which is equal to the green manure yield in this region. Incorporation of maize stovers for 7 years increased soil organic matter from 0.71% to 1.76%, deepened the cultivated layer by 10 cm, and therefore raised the soil water storage and alleviated the threat of water stress. In 1986, the grain output in Tunliu County close to the village was reduced by 30% because of serious drought (333.5 mm precipitation from September 1985 to October 1986), but the crop reduction in this village was negligible (Jin Shenlin et al., 1987).

Yang Jingqiu et al. (1987) studied the effect of plant residues on soil fertility in a drab soil located in Taiyuan, Shanxi Province. They used stalks, straws, and root stubble of maize, wheat, sorghum, millet, and soybean, as well as *Crotalaric intermedia*, *Vica* sp., *Sesbanis cannabina*, and *sagebrush*. The results showed that although different materials had different decomposition rates, all these plant residues resulted in the increase of soil humus. After application of these residues, soil humus contents were 0.98% for maize stalks, 1.03% for wheat straws, but 0.7% for *Crotalaric intermedia*, and 0.82% for *Vica*. Generally, root stubble increased humus. Application of these residues significantly improved soil porosity and structure. As a result, the soil's water-holding capacity was much higher than in the fields without application of plant residues.

Good results were also obtained in the Mudanjiang Reclamation area. According to Li Guangyue's report (1987), incorporation of plant residues reduced the soil bulk density and increased soil porosity in addition to improving water-holding capacity. The positive role of turning over straw into soil in building up soil fertility was further confirmed, and the problems were discussed by recent studies (Wu Jingmin et al., 1991; Xu Guojun and Cui Qin, 1991; Wang Yanbing and Xu Jianwen, 1991).

Incorporation of crop residues can also reduce salt accumulation. Wang Yingqin et al. (1988) conducted experiments on a waste saline land in Shandong Province. Their results showed that incorporation of 7.5 Mg/ha maize stovers or wheat straw significantly reduced salt content in the soil profile at 0 to 200 cm. On average, over 6 years of incorporation, maize stovers reduced salt content by 75.5%, and wheat straw by 52.5%. Salt tended to accumulate in the treatment without incorporation. Xie Zaizhi (1985) found that direct application of rape straw significantly improved soil-nutrient supplying capacity and increased crop yields. Similar results were obtained from Heilongjiang Province (Liu Fa et al., 1983; Li Shujiang and Xu Shouming, 1987) and Beijing (Chen Zhimin, 1986). As the C:N ratios in most crop residues are high, it is necessary to apply N fertilizer with the incorporation.

D. Most Effective Use of Stored Water

Great potential still exists for the rational use of stored water for agricultural production in dryland areas. Historical and current studies have shown that in the same rainfall area some farmers but not all, have obtained fairly high crop production, depending on how well they used the stored soil water.

1. Raising Soil Fertility

a. Soil Fertility and Agricultural Production on Drylands

Raising soil fertility is a key measure in improving water-production efficiency. Of course, low precipitation and little available water are major constraints for crop production, but in most semiarid and subhumid regions in China it is not the most limiting factor. The Shaanxi Economic Institute studied the relationship between grain production and precipitation in Mizhi County from 1956 to 1977 and found that there was no good correlation between them (Wang Lixiang, 1983). Shan Lun did

Table 15. Rainfall production potentials and current yields (mg/ha)

Locations	Crops	RPP	CY	Sources
Middle Ningxia	Spring wheat	1.04	0.61	Gao Jixiang, and Xu Qiang (1987)
	Winter wheat	2.95	0.14	
	Millet	3.50	0.35	
	Potato (dry)	1.60	0.40	
South Ningxia	Spring wheat	1.80–3.00	0.69–1.50	Shan Lun et al. (1987)
	Proso		0.52–1.71	
	Millet		0.64–1.51	

RPP, rainfall production potentials; CY, current yields.

similar work for Guyuan County, and obtained a very low correlation ($r = 0.17$) (Wang Lixiang, 1983). Gao Jixiang and Xu Qiang (1987) calculated the rainfall production potential in some locations of Ningxia Hui Autonomous Region (Table 15) and found that there was a large gap between the current yields and the rainfall production potential. This shows that in the case of very low water production efficiency, precipitation was not the main factor limiting the production, at least in some places.

What is the main factor controlling crop production now? After studying the potentials of different factors for Guyuan County, a semiarid area with an annual precipitation of 415 mm, Shan Lun (1983) pointed out that the production potential for this climate was about 5.52 Mg/ha. However, the production potential for this soil was only 0.75 Mg/ha. Wang Lixiang et al. (1987) found that there was a very significant correlation between soil fertility and water use efficiency on cropping dryland. The authors pointed out that in addition to the frequent threat of drought to agricultural production, low soil fertility was the direct cause of low production and overcultivation, making the ecosystem fall into a vicious circle. The practice of Dongguanzhuang Agricultural Brigade in Shanxi Province has shown that each 1 mm precipitation produced 10.2 kg/ha wheat grain in soil with more than 1% organic matter, but only 3.36 kg/ha in soil with less than 1% organic matter (Shanxi Agricultural Ministry, 1983). Li Yushan's investigation (1982) also showed that water production efficiency changed with yields (Table 16).

Niu Pu et al. (1987) adopted cement pools to study the relationship between wheat production and soil fertility. The pools were 1.5 m^2 with depths of 60, 120, 180, 240, and 300 cm. Water added to each pool approximated the water-holding capacity (about 22% water content of dry soil) and fertilizers equivalent to 200 kg N and 44 kg P per ha were added before sowing. The pools were covered with plastic sheets. During the plant-growth period, the pools received no irrigation or rain. The experiment revealed that with fertilization, bumper harvests of about 4.14 to 4.50 Mg/ha were obtained when the soil's depth was more than 180 cm. Results showed that soil fertility limited the crop production and,

Table 16. Water consumption by wheat on drylands

Yield ranges (Mg/ha)	Water consumption (mm)	WCC (water, kg/wheat, kg)	WUE (wheat, kg/m^3)
1.50–2.25	270–370	1330–1660	0.56–0.61
3.00	340–350	1100–1200	0.86–0.88
3.75–4.50	360–390	820–1000	1.04–1.15
6.00	450	750	1.33

WCC, water consumption coefficients; WUE, water use efficiency.
Source: Li Yushan (1982).

therefore, the water production efficiency. Recently research in Inner Mongolia, Hebei, Shaaxi, and Ningxia has confirmed that there is a certain potential for precipitation productivity on the drylands; if the soil fertility is improved, this is expected to raise agricultural production (Jiang Jun and Wang Lixiang, 1990; Yu Ling, 1990; Wang Longchang, 1990; Zheng Jianfei et al., 1990; Li Yushan et al., 1990; Liang Zhenji, 1990; Wang Lixiang et al., 1990a). In fact, most dryland soils are poor in plant nutrients and are not as fertile as those in the irrigated areas. For example, a soil survey indicated that in the Weibei Loess Plateau, half of the cultivated land lacks available N, and one-third is deficient in available phosphorus. Most dryland soils are poor for many reasons: historically and also at present, the population in dryland areas has been low. People rely on cultivation of more land per capita with low production to supply their needs. So far the soil fertility has not been improved, and serious erosion has led to the loss of large amounts of nutrients. Moreover, as the peasants are still poor, the input of nutrients into soil is much less than that in irrigated areas. Use of organic fertilizer has not been increased in these areas for many years, and chemical fertilizers are very limited. Data from investigations in northwestern China show that generally the input of N on the drylands is less than 75 kg/ha, and 22.5 to 35.0 kg/ha is common (Li Shengxiu et al., 1988, 1989). Thus, raising soil fertility means increasing the production efficiency of precipitation. This idea has been proved by many field trials and agricultural proctices (Wang Bangxi et al., 1990; Hua Tianmao et al., 1990; Li Ying and Lu Dianqing, 1991).

b. Measures for Raising Soil Fertility

The production practices in many places have shown that applying of fertilizers, arranging legume crops in the rotation, planting green manure crops, adopting intercropping and interplanting, and fallowing are very effective ways of raising and recovering the soil's fertility and making soil water use more efficient.

Cropping and Rotation Systems. Experience from the Yianan area in Shanxi Province has shown that planting oil-bearing crops and forage grasses in rotation can stabilize the production of crops that follow. Experience from other places also shows that intercropping or interplanting crops with green manure or legumes can improve the soil fertility at the same time as crops are grown. Many rotation systems, such as grain and legumes, grain and green manures, legumes and oil-bearing crops, and grain and oil-bearing crops, have been adopted in the dryland areas.

Experiments conducted by the Shaanxi Institute of Cash Crops in Yongshou County have shown that the yield of wheat grown after seed rape increased by 34.1% to 56.5% over those of continous cropping wheat. Wheat rotation with seed rape is able to maintain a sustained, stable, and high yield (Lu Xiangxian and Li Jianjun, 1987). Using leguminous forage crops, particularly alfalfa in rotation, is a very important way of raising soil fertility and has been widely adopted in many areas of the drylands. Experiments from Dingxi region, Gansu Province, indicated that after growing alfalfa for 3 years, the grain yield in semiarid areas was increased by about 30% for 3 years. Having enlarged to 20% the cultivated land to plant alfalfa, Taozhuang Village, located in a semiarid area, significantly raised soil N content and increased livestock numbers and organic manure amounts, which in turn increased N input to the soil. The N input from farmyard manure went up from 57.7% to 67.5% and that from chemical fertilizer dropped down from 42.3% to 32.5%. As a direct result, the water use efficiency increased from 36.3% to 47.8%, and per capita grain increased by 84.5% (Wang Lixiang et al., 1987). Similar results of rotation with one-third land for alfalfa were also obtained in Urumchi, Xinjiang (Zhao Feng et al., 1987), Qianyang, and Long counties, Shaanxi Province (Zhang Xueshang, 1983). Experiments indicated that alfalfa had great potential for production and improvement of soil fertility when adequate fertilizers were applied (Wang Lixiang et al., 1990b; Deng Xinmin, 1990).

Application of Organic and N Fertilizers. Rational utilization of fertilizers can rapidly raise soil fertility and crop yields, and thus the water use efficiency: "Crops can grow very well even under drought or waterlogged conditions when more manure is applied" and "Fertile soil is able to bear drought". These agricultural proverbs reflect the experience of farmers.

Cheng Suyun et al. (1987) found that crop yields as well as the rainfall use efficiency were increased with increasing fertilizer rates. Without fertilization, the wheat yield was 4.00 Mg/ha, and the precipitation use efficiency was 0.95 kg/m^3. After application of 20.0 and 40.0 Mg of organic fertilizer, the yields were increased to 5.51 and 6.83 Mg/ha, and the water use efficiency was raised to 1.3 and 1.5 kg/m^3, respectively (Table 17). Liang Waylea et al. (1987) reported that fertilization in-

Table 17. Effect of applying organic fertilizers (O.F.) on the increase of precipitation-use-efficiency (PUE) of wheat

Treatments	Average wheat yields of 4 years (Mg/ha)	Average PUE (kg/m^3)
Without fertilization	4.02	0.56
With N (50 kg/ha)	4.47	0.62
With N + 19 Mg O.F.	6.11	0.85
With N + 38 Mg O.F.	6.43	0.89
With N + 72 Mg O.F.	6.62	0.92
With N + 124 Mg O.F.	6.67	0.95
With N + 248 Mg O.F.	6.83	0.96

Source: Cheng Suyun et al. (1987).

Table 18. Effect of applying inorganic fertilizers (F) on four crop yields (Mg/ha) and precipitation-use-efficiency (PUE; kg/m^3)

Fertilizers (kg/ha)	Yields				PUE			
	Wheat	Barley	Qingke barley	Rape	Wheat	Barley	Qingke barley	Rape
Without F	3.78	2.84	2.66	1.83	0.66	0.50	0.47	0.32
112.5 N + 24.5 P	4.22	4.20	3.63	2.26	0.74	0.74	0.64	0.40
225.0 N + 49.0 P	4.89	4.74	4.46	2.38	0.86	0.83	0.78	0.42
337.5 N + 73.5 P	5.60	5.21	4.91	2.43	0.98	0.92	0.86	0.43
450.0 N + 98.0 P	6.09	5.48	5.51	3.67	1.07	0.96	0.97	0.64
562.5 N + 112.5 P	5.67	5.31	5.29	2.42	0.99	0.93	0.93	0.42

Source: Wang Lixiang (1983).

creased the ability of wheat to absorb soil moisture after flowering and, therefore, increased the water use efficiency. Experiments from Wang Lixiang (1983) obtained similar results by application of mineral fertilizers (Table 18). Data from the drylands indicated that each 1 mm precipitation could produce only 2.25 to 3.75 kg wheat/ha without application of fertilizers, but 7.75 to 15.0 kg/ha with application of N and P fertilizers.

Combining the use of organic and chemical fertilizers is the best way to raise soil fertility and provide nutrients to plants (Chen Lizhi et al., 1989). Organic fertilizer can provide organic matter and many nutrients, thereby improving the soil's properties and fertility and making chemical fertilizers more efficient (Zhang Fudao, 1984). Ma Heping (1987) reported that application of pig manure at 15 to 45 Mg/ha significantly increased soil organic matter, water-stable aggregates, and aggregate porosity, but decreased bulk density. This, in turn, increased water-holding capacity and water use efficiency. Investigations in Huangling County, Shaanxi Province have revealed that each Mg of organic fertilizer increased wheat production by 6 to 12 kg and maize by 15 to 29 kg (Li Changwei et al., 1987). Since soil is used as bedding material for livestock manure and night soil in dryland areas, the organic fertilizer is low in organic matter and plant nutrients. Analytical data for 30 samples from Guanzhong Plain, Shaanxi Province are taken as a typical example to show this. On the basis of air dry weight, the animal manure contains 3.12% organic matter, 0.15% total N, and 0.092% total P with a wide range, depending on the amount of soil added (Zhao Boshan, 1985, personal communication). In some places where the bedding soil is less, the organic fertilizer may contain 4% to 5% organic matter, 0.27% to

Table 19. Effects of application of organic fertilizer (Mg/ha) with N fertilizer (kg/ha) on the efficiency of both organic and N fertilizers

Organic fertilizer amounts	N 0.00	67.5	135.0	202.5
	Wheat grain increased in kg/ha by adding 1 kg N			
0.0	8.40	8.95	7.40	8.25
37.5	8.00	11.00	9.25	9.45
75.0	9.25	10.50	7.40	9.05
150.0	8.00	10.35	8.40	8.85
	Wheat grain increased in kg/ha by adding 1 Mg organic fertilizers			
37.5	6.66	6.00	14.44	21.10
75.0	11.70	12.50	14.44	11.66
150.0	8.07	7.89	9.33	9.17
Mean	8.81	8.80	12.74	13.98

Source: Li Shengxiu et al. (1979).

0.35% total nitrogen, 0.07% to 0.183% P, and 0.43% to 0.77% K (Sun Xi, 1957). In addition to providing plant nutrients, organic fertilizers supply organic matter to improve soil structure and promote biological activity. However, the nutrients in organic fertilizers are not present in the proportion required for plant uptake. For instance, N in pig manure is too low, but P is too high for plant requirements. The same is true for other organic fertilizers. For this reason, organic fertilizers should be used with N fertilizers, but separately from P fertilizers (Li Shengxiu and Zhao Boshan, 1991).

Experiments from Li Shengxiu et al. (1978) demonstrated that when the applications of organic fertilizers were combined with N fertilizer, the efficiency of both organic and chemical fertilizers was raised, and the yields were higher than when used separately (Table 19). However, when application of organic fertilizer was combined with P fertilizers, the use efficiency of P in the organic fertilizer was reduced, and the yields were not as high as expected. The reason is simple: the N fertilizer supplements the N in organic fertilizer and makes the P in it more useful. However, the P fertilizer inhibits P uptake from organic fertilizer (Table 20). These conclusions were also drawn by other researchers in different places (Hu Luhao et al., 1989). Experiments showed that application of organic fertilizer for 9 years significantly increased soil phosphorus and potassium (Yao Yuanxi and Yang Yanfan, 1989), which further confirms the fact that P in organic fertilizer is relatively high for plant needs. In Shaanxi Province, 1 Mg organic fertilizer has the same effect as 0.44 kg P (Li Changwei et al., 1987). This can be used to determine whether to apply P.

Table 20. Effect of organic fertilizer on phosphate fertilizer efficiency of wheat

Organic fertilizer amounts (Mg/ha)	P added per ha	
	15.0 kg	30.0 kg
	Wheat grain increased in kg by adding 1 kg P	
0.0	4.20	1.77
37.5	2.30	1.09
75.0	1.64	0.66

Source: Li Shengxiu et al. (1979).

Application of P Fertilizers. In most soils on the drylands, total P content is comparatively high, ranging from 0.15% to 0.2%, but available P is rather low, generally in a range of 5 to 15 ppm P. Since crop yields have been increased, and high N and low P application rates used, the imbalance of available N to P in soil occurs in many locations and deficiency of available P in soil becomes more and more serious. In many places, crops have no response to N fertilizers, not because the soil has sufficient N, but because it is deficient in P. Therefore, increasing the

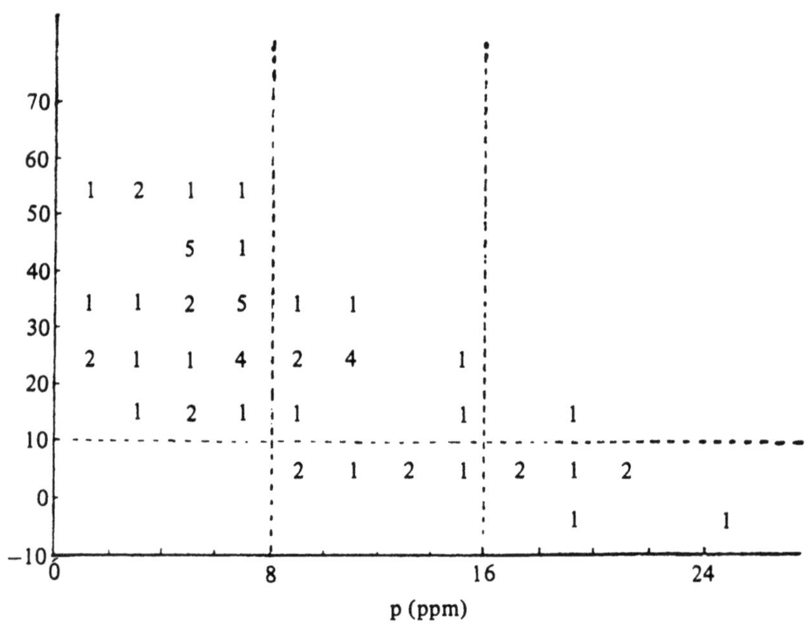

Figure 17. Relationship between available P (ppm) in soil and increased yields of wheat (%). Note: numbers in the figure are field experimental sites arranged by ranks. Sources: Li Shengxiu et al. (1979).

application rates of P cannot only raise crop yields, but also activate N efficiency.

Lu Dianqing and Li Yin (1987) reported that by application of P fertilizer in the loess area, wheat yields were increased by 30% to 50% or even doubled. Cao Xuechang (1987) points out that in the northwest plain of Shandong Province, the available P in soil is more deficient than N, and the benefit of simply applying P fertilizer is greater than that of simply applying N. The same conclusion was also made by Xu Maozhen (1987) in the same province.

Through several years of study, Li Shengxiu et al. (1979) found that crop response to P fertilizer was correlated very well with available soil P (Fig. 17) extracted with 0.05 *M* sodium bicarbonate (Olsen et al., 1954). Based on their experimantal results, the authors suggested P availability indices (Table 21), which have proved to be a useful tool for guiding P fertilizer application in the middle part of Shaanxi Province. It should be pointed out that the deficiency or sufficiency of any nutritive element is changeable, depending on natural conditions and the crop production level. From this point of view, the indices should be worked out from one region to another; recently, different indices have been made for different regions (Shao Yuting and Zhen Yenzhi, 1989; Zhang Guilan et al., 1989).

As mentioned above, the deficiency of N occurs very widely on drylands, which always limits crop response to P fertilizers. In order to make P fertilizers fully play their role, application of P together with N fertilizers is commonly practiced; agricultural practices and research work have demonstrated the advantages of this combination (Lu Dingqing et al., 1989; Wu Jianguo, 1989; Jin Shaoling, 1989).

Every crop has a good response to P fertilizer if the soil is deficient in this element. However, it has been suggested to give priority to leguminous crops as they can supply nitrogen by themselves and thus promote full use of P fertilizer (Li Shengxiu et al., 1988a, 1988b; Li Shengxiu and Zhao Boshan, 1990; Li Shengxiu, 1991). Also, the P fertilizer can

Table 21. Phosphorus availability indices

	Available P (mg/kg)[a] in soil		
	< 8.0	8.0–16.0	> 16.0
Degree of deficiency	Seriously deficient	Deficient	Sufficient
Recommended rates (P, kg/ha)	60	33	None
Application of organic fertilizer	Small amount	None	None
Crops needed to add P fertilizer	Any crop	Winter wheat, barley, peas, and seed rape	None

[a] Determined using Olsen's method (1954).

Source: Li Shengxiu et al. (1978, 1979).

Table 22. Effects of organic fertilizer by base application and by dressing on wheat yields (Mg/ha) and fertilizer efficiency (kg/Mg)

Fertilizer Application ways	Wheat yields	Organic fertilizer efficiency
Ck (28 kg N/ha)	1.24	
Ck + 15 t/ha organic fertilizer (dressing)	1.52	3.80
Ck + 15 t/ha organic fertilizer (basing)	3.26	26.50

Source: Li Shengxiu et al. (1976a).

promote the activity of the N-fixation enzyme nitrogenase (Li Shengxiu et al., 1990b) and raise the N amount. For this reason, applications of P fertilizer to leguminous crops, including green manures, forage, and grain crops, is considered an effective way of improving the soil's fertility and raising crop production; this is called application of P fertilizer to increase N.

Methods for Application of Fertilizers. Applying organic fertilizer as a base by broadcasting and plowing to bury it before seeding is an effective way to raise its efficiency. In this way, the organic fertilizer is available during the whole period of plant growth. It can be applied deeper in the soil where more water is available for soil microorganisms to decompose the material and make the nutrients available for plant uptake. When it is concentrated in the root zone, it also increases the intake and use efficiency of rainfall. Experiments (Li Shengxiu et al., 1976a) have shown that applications of organic fertilizer as a base increased wheat yields by 26.5 kg per 1.0 Mg fertilizer, but only 3.8 kg per 1.0 Mg fertilizer was obtained when applied during the winter (Table 22).

Applying fertilizers as early as possible is another way to raise their efficiency. The fertilizer effect on crop yields depends upon the available water in the soil. It is well known that fertilizers are more efficient when there is sufficient water in the soil. On drylands, available water is a limiting factor not only for crop production but also for fertilizer efficiency. In order to make full use of the precipitation, early application should be encouraged.

Early application has several advantages. Nitrate in fertilizer is not leached by rainfall during the long dry season. Nutrients are also available to plants for a longer period. Sometimes the fertilizer is not effective because the soil is too dry, but it is quickly activated when rainfall occurs. Experimental results (Li Shengxiu et al., 1976b) have shown that early application of all the fertilizers, either organic or inorganic, significantly increased crop yields compared with later application (Table 23).

Placement of fertilizer in different layers has different results. Water stress is the major problem in dryland areas. The top soil is usually dry, but moisture is generally present in deep layers. For instance, in the

Table 23. Effects of fertilizer application time on wheat yields

Fertilizer types	Treatments	Wheat yields (Mg/ha)	Yield increased by fertilization (%)
Organic	Ck (35 kg N/ha)	3.20	
fertilizer	Ck + O.F. (spring)	3.34	4.4
(75 Mg/ha)	Ck + O.F. (winter)	3.59	12.2
	Ck + O.F. (after emergence)	3.69	15.2
	Ck + O.F. (sowing)	3.86	20.6
Phosphate	Ck (135 kg N/ha)	3.17	
fertilizer	Ck + P (spring)	3.13	−0.1
(30 kg P/ha)	Ck + P (winter)	3.67	15.7
	Ck + P (after emergence)	4.05	34.2
	Ck + P (sowing)	4.30	35.6
Nitrogen	Ck (60 kg P/ha)	2.99	
fertilizer	Ck + N (spring)	4.87	62.8
(135 kg N/ha)	Ck + N (winter)	5.69	90.3
	Ck + N (sowing)	5.80	94.0

Source: Li Shengxiu et al. (1976b).

Loess Plateau, the moisture in the top 0 to 10 cm is often lower than wilting point, but the moisture in deep layers is usually above 14% to 16% (by dry weight). For this reason, shallow application of fertilizers cannot be effective because of limited available water, but deep application can. Deep placement of fertilizers is of importance in increasing fertilizer efficiency and crop yields. Deep applications are generally combined with deep plowing, in which fertilizer is mixed with the soil in the deep layers. On arable land, the nutrient content is higher in the top layers than in deep layers, and the deeper the layer, the fewer the nutrients. This is true for any available nutrient, but especially for available phosphorus (Table 24). Crop roots on drylands are concentrated in

Table 24. Available P (mg/kg) in different layers of soils

	Soils		
Depth (cm)	Rusty manured loessial soil[a]	Yellow loessial soil[a]	Eutrophic manured loessial soil[b]
0–10	3.33	3.66	20.4
10–20			15.3
20–30	0.50	0.53	11.2
40–60	0.70	0.16	5.2
~100	0.50	0.18	3.5

Source: [a] Shaanxi Agricultural Survey and Design Academy (1982); [b] Li Shengxiu (1985, unpublished data).

the deeper layers, where the soil moisture is better for root growth and nutrient uptake. For this reason, the nutrient status of deeper layers is more important.

Shallow application of fertilizers concentrates nutrients in the top layer where there are fewer roots. These applications only benefit young plants. When plants grow larger, and roots penetrate into deeper layers, they have difficulty because soluble nutrients from fertilizers are not moved down to deeper layers because of limited rainfall, and P fertilizer can become relatively fixed. Deep applications provide fertilizers in the top layers as well as in the deep layers, meeting the needs of both young and old plants. Furthermore, most of the soils in drylands are calcareous with a high pH. Therefore, loss of N by volatilization of ammonia must be considered. Shallow application encourages such loss while deep application makes it unlikely.

We demonstrated in a laboratory experiment that when N fertilizer (ammonium bicarbonate) was applied to the soil surface at 25° C, the N loss by volatilization reached 79.5% after 5 days. By comparison, the loss was 15.7% when applied at 5 cm, and 5.8% when applied at 10 cm. In addition, deep application places the fertilizers into wet layers, where microorganisms can increase their availability for plant uptake. Li Shengxiu et al. (1976c) showed that deep application of fertilizers definitely increased fertilizer use efficiency and yields compared with shallow application (Table 25). Lu Dianqing and Li Yin (1987) reported that deep

Table 25. Effects of placements of organic, P, and N fertilizers on wheat yields and fertilizer efficiency

Fertilizer types	Fertilizer placements (cm)	Wheat yields (Mg/ha)	Yield increased by fertilizer (%)
Organic (75 Mg/ha)	Ck (35 kg N/ha in rows)	2.76	
	Ck + O.F. (0.0–8.0)	2.97	7.6
	Ck + O.F. (0.0–15.0)	3.03	9.8
	Ck + O.F. (0.0–20.0)	3.56	29.0
P (30 kg P/ha)	Ck (70 kg N/ha)	2.87	
	Ck + P (0.0–8.0)	3.58	24.7
	Ck + P (0.0–15.0)	3.86	34.5
	Ck + P (0.0–20.0)	4.38	52.6
P (60 kg P/ha)	Ck + P (0.0–8.0)	4.10	42.9
	Ck + P (0.0–15.0)	4.60	60.3
	Ck + P (0.0–20.0)	4.85	69.0
N (75 kg N/ha)	Ck (35 kg P)	3.50	
	Ck + N (0.0–8.0)	3.93	12.3
	Ck + N (0.0–15.0)	3.88	10.9
	Ck + N (0.0–20.0)	4.09	16.9

Source: Li Shengxiu et al. (1976c).

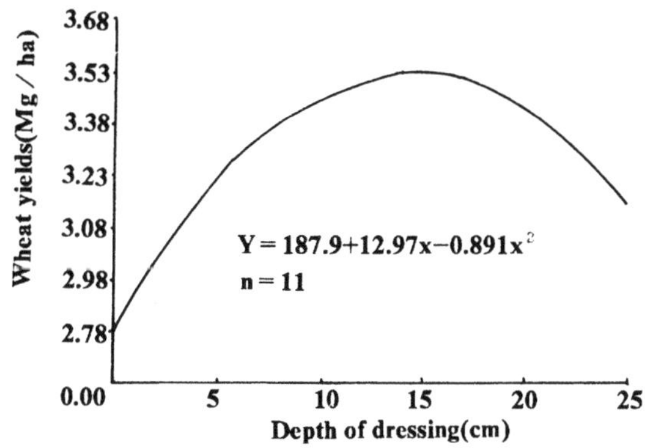

Figure 18. Relationship between depth of N dressing and wheat yields. Sources: Lu Dianqian and Li Yin (1987).

application of the full amounts of N and P fertilizers before sowing wheat had met the crop requirement for the whole growth stage and increased fertilizer use efficiency. The loss of ammonium by volatilization was descreased from 39% to 14% for ammonium bicarbonate, and from 54% to 36% for urea. Consequently, the wheat yield increased by 20%. According to the authors' results, the relationship between wheat yields and depths was as follows:

$$Y = 189.6 + 12.97X - 0.89X^2$$

where Y is wheat yield, and X is applied depth. From this equation and the quadratic curve (Fig. 18), it is found that the suitable depth of applying N to wheat may be 10 to 15 cm.

Balanced Fertilization. The efficiency of fertilizers becomes lower and lower and the loss of nutrients, especially N, is more serious on the drylands with the increase of fertilizer rates. Great attention has been given to the problem, and many measures have been taken to solve it. Of those, balanced fertilization contributes an effective way of applying fertilizers. Balanced fertilization, or prescription fertilization as it is widely referred to in China, is based on three approaches to the application of fertilizers: soil fertility, objective yields, and field trials (Xiang Chongyang, 1989; Liu Gengling, 1989; Lin Bao and Li Jiakang, 1989; Zhang Shixian, 1989). Balanced fertilization has played a great role in increasing fertilizer efficiency and crop production (Jin Shaoling, 1989; Zhan Changgeng, 1989; Zhou Zucheng et al., 1989; Chen Lunshou et al., 1990; Liu Ziyu, 1990; Jiang Bofan, 1990; Li Shangxiu et al., 1990a; Zhang Kuan et al., 1990; Lu Dingqing, 1990; Li Renggang et al., 1990; Zhang Yichun and Guo Zhifeng, 1990; Tian Yuanren and Jiang Xiezheng, 1990). Although much work has been done on this aspect, many problems remain, particularly soil fertility testing. No doubt, there is a long way to go for obtaining solutions to these problems and for efficient utilization of the various approaches.

2. Choice of Suitable Crops

Choice of suitable crops is an economical way to efficiently use limited stored water. The dominant crops on drylands in China are wheat, maize, grain sorghum, millet, proso (*Panicum miliaceum*), potato, and sweet potato. Of those crops, wheat is widely distributed in arid, semiarid, and subhumid areas. With precipitation decreasing, spring-sown maize and sweet potato are reduced, while sorghum, millet, proso, and potato are increased. Where rainfall is less than 400 mm, proso, millet, and sorghum are the dominant spring-sown crops. In the cold and dry areas, highland barley and oats are grown, but they cover a small area.

Shan Lun (1983) studied the relationship between soil water and the adaptability of different crops, including maize, proso, millet, sorghum, and wheat, and showed that sorghum and proso were the most drought-resistant. His research also showed that there were three ways for a crop to adapt to drought: (1) extending its roots to deep layers to obtain more water; (2) effectively regulating evapotranspiration to reduce water use; and (3) maintaining normal activities under dry conditions. Sorghum is effective in all three ways. Zhang Xihai and Shan Lun (1983) found that under controlled conditions, the productivity and drought-resistant ability of proso was higher than that of millet. However, the Dingxi Agricultural Institute (1983) calculated the production stability coefficients (*PSC*) of different crops, based on their yields in Dinxi region for many years, using the empirical formula,

$$PSC = 1/c.v.\%$$

where $c.v.$ is the coefficient of deviation of crop yields, and found that millet (3.67) and wheat (3.44) were more stable than proso (2.49).

Recently Shan Lun et al. (1987) compared the productivity and water use efficiency of wheat, proso, and millet. His experiments showed that there was no significant difference among their yields. The yield variation coefficient of wheat was smaller than that of the other two, and it used more stored water and rainfall: wheat used 57% of stored water prior to sowing and rainfall during the growth period, whereas proso used 52.9% and millet 47.8%. However, proso was the lowest in evaportranspiration and the highest in water use efficiency. The productivity and water use efficiency depended on soil fertility and rainfall. When soil fertility was high, the proso yield was 39.6% and 13.3% higher than that of wheat and millet, but when soil fertility was low, the wheat yield was the highest, being 25.8% and 19% higher than proso and millet respectively. On the whole, wheat is more adaptable to different lands and has most stable yields. Although there is no agreement on the order, all of these are considered to be drought-resistant.

Millet and proso have been planted on drylands for a long time. The discovery of millet and proso seeds in the ancient ruins of the Neolithic Age in Banpo Village, Xian, and Liuzhi Town, Hua County, Shaanxi Province, and Wanyong County, Shanxi Province, have proved that the two crops were planted on the drylands in China as early as 6000 to 7000 years ago. Many poems describing millet and proso in *Shi Jing* (an ancient poem book that appeared 2000 years ago) also indicate that these two were the main crops in the Loess Plateau at least 2000 years ago (Li Fengqi et al., 1983). These two small seed crops have low evapotranpiration coefficients, very well-developed roots, and are able to survive under drought conditions. They can be seeded in spring and summer and adapt to different climates. They occupy a large poportion of dryland areas, and are the main food crops in some regions.

Agricultural practices in many places show that the yield of millet is more stable and higher than other crops on rainfed land (Gansu Academy of Agricultural Science, 1983a). The Jinzhuang Productive Brigade used drought-resistant cultivation methods and produced 7.1 Mg/ha of millet in 1982, when spring and summer drought was serious (Huguan Agricultural Department of Shanxi, 1983), and the millet yield was increased 30% to 50% by seeding in furrows as compared with general seeding (Gansu Academy of Agricultural Science, 1983b). Proso is grown using the same practices as with millet (Wang Yuxi, 1983).

According to ancient records, wheat has been cropped in the Loess Plateau for at least 2000 years. Grain sorghum was introduced there in the Tang Dynasty (618 to 907 A.D.) and sweet potato in the Ming Dynasty (1368 to 1644 A.D.). Proso, millet, grain sorghum, potato, and wheat are still the main crops on drylands. Of these, wheat is particularly worthy of mention. In addition to its drought-enduring ability, it coincides in growth with the dryland climate, making full use of the precipitation. An agricultural proverb, "A good harvest of wheat depends on water stored in the previous year," describes the importance of stored water before or at the time of seeding. When wheat is seeded in autumn, it uses water stored from the previous rainy season, which is generally adequate for winter and early spring use for tillering and rooting. Then, it has a high probability of receiving precipitation during spring for grain production. Therefore, the yield is relatively stable. Over a long period of cultivation, two types of wheat have been selected. Both are drought-resistant, but one is for poor soils and the other for fertile soils. This makes wheat widely adapted to dryland areas (Chen Peiyuan, 1983). Wheat is considered a stable and suitable crop in dryland farming.

3. Sowing Crops Using Appropriate Methods

Using appropriate methods of sowing to ensure germination of a desired density of plants is the prerequisite for efficient use of the limited water. In some dryland areas the shortage of soil moisture often occurs during the sowing period, and special methods have been developed to overcome this problem.

a. Early Sowing

Early sowing is praticed when soil moisture is adequate for germination. Of course, early sowing is less desirable than sowing at the optimum time, but it is better than waiting until the soil moisture is lost and soil is too dry to germinate seeds. "Taking suitable moisture but not waiting for a suitable time for sowing" is an agricultural proverb, warning the farmers not to lose a good opportunity. In dryland areas where two crops a year are grown, an early maturing variety is used for the first crop in order to leave enough moisture in the soil for a second crop. Some crops such as forages that have no definite maturity time are also used for the first crop so they can be harvested early without great effect on their yields.

b. Borrowing Deep Soil Moisture for Sowing

"Borrowing" or seeking moisture in deep soil for sowing is another method widely used on drylands (Yang Sheng and Lian Sheng, 1987). This involves moving the dry surface soil on both sides of the furrows by plowing and sowing seeds into the wet soil of the furrows, covering the seeds with wet soil, and then compacting it if the soil water content is less than 0.18 kg/kg. The depth of soil covering depends on seed sizes: the smaller the seed, the less the soil covering. The furrows for seeding are leveled by cultivation during the plant-growing period. This method was occasionally used when topsoil was too dry to seed, but is now the common practice for producing abundant harvests. Experiments from the Grain Crop Institute, Shaanxi Academy of Agricultural Science (1983) showed that sowing maize using this method resulted in good rooting that made the plants resistant to drought and increased yields as compared with other methods.

c. Moving Subsoil Water Up for Sowing

Much attention has been given to moving water from the subsoil up to the topsoil by compaction with a stone roll before sowing (Li Duren, 1983, 1984). Experience from Shanxi Province shows that when topsoil is dry to 4 to 5 cm and soil water under the dry layer is less than 0.12 kg/kg during the period of sowing millet, cracking the soil blocks, compacting the soil, and then sowing seeds is a successful procedure to ensure a full stand of millet shoots. Compaction increases the soil water in the plow layer by 1% to 3% (Huguan Agricultural Department of Shanxi, 1983). When the topsoil is dry to 2.5 to 3.0 cm and subsoil moisture is not adequate for sowing wheat, farmers in Hebei Province crack the soil blocks, compact it once or twice, and then sow wheat into 3 cm of soil in the morning when the soil surface appears wet. Sowing goes hand in hand with smoothing the soil surface. Compacting and smoothing operations are done again 2 to 3 days after sowing. In this way, the moisture in the deep layers can gradually move up to the topsoil. If topsoil is dry but deep soil contains sufficient available water, sowing wheat 3 cm deep is carried out first. After sowing, compacting and smoothing operations are conducted to move water up to the seeds (Agricutural Ministry of Zhangjiakou Region, Hebei Province, 1983).

d. Other Methods of Sowing

Farmers in dryland areas also use other measures to ensure germination. These include soaking seeds in water for a time to promote germination, sowing millet in winter and waiting for spring rainfall to germinate the seed, and transplanting potato shoots. The principles for sowing are as follows: sow seeds in soil and wait for rain when soil is too dry for seeds to germinate; early sowing when soil moisture is suitable; using different methods to ensure a full stand of plants when soil moisture is deficient. Depending on the actual conditions, these methods can be combined or used separately.

E. Reduction of Soil Water Loss by Evaporation

Although the seasonal distribution of rainfall is different in different regions, many studies have shown that the annual variation of soil moisture has a certain pattern on the drylands (Xi Yaoguo, 1990; Wang Xianming and Danzhengzhu Zasang, 1990; Mei Xurong et al., 1990; Zhao Jubao et al., 1990; Wang Yantian and Xu Zhuling, 1990; Xu Zhuling and Wang Yantian, 1990; Zhang Xiaozhong et al., 1990; Wang Anzhu and He Zhongyang, 1990; Zhong Lianping and Han Shifeng, 1990). Based on the agricultural experiences and soil water dynamics on the drylands, varieties of measures have been taken to reduce water loss from soil, raise water use efficiency and adopt supplementary irrigation to ensure plant growth (Pu Jinchun and Huang Zichen, 1990; Huang Jouchang et al., 1990). Of those, Conserving stored soil water during nongrowing or growing periods is a key to successful management of dryland areas. Serious nonproductive water losses are weed consumption and soil surface evaporation. Since precipitation on drylands is much less than potential evaporation, reduction of soil surface evaporation is very important. The main purpose of a series of cultivation measures used by farmers, in addition to controlling weeds, is to reduce evaporation loss by loosening soil surface to form a dust mulch. Although there are controversial views regarding this practice (Abramova et al., 1956; Greenland, 1978), much work by scientists and farmers in China confirms its positive effect on water conservation (Xi Chengfan, 1961; Agricultural Institute,

Table 26. Effects of compaction of soil on water contents

Locations and treatments	Water content (kg water/kg dry soil) in different layers (cm)			
	0–5	5–10	10–20	20–30
Chaoyang, Liaoning, wheat				
Compaction before sowing	0.057	0.099	0.159	0.191
Without compaction	0.044	0.074	0.152	0.200
	0–20	20–40	40–60	60–80
Suide, Shaanxi, millet				
Rolling after sowing	0.034	0.093	0.102	0.056
Without rolling	0.044	0.046	0.042	0.044
	0–5	5–10		
Henan, millet				
Rolling twice after sowing	0.093	0.147		
Rolling once after sowing	0.088	0.151		
Without rolling	0.059	0.154		
	0–10	10–20	0–10	10–20
	Millet		Maize	
	Water content (kg water/kg dry soil)			
Southeast Shanxi				
Without rolling before sowing				
Sowing	0.099	0.015	0.121	0.153
Final singling of seedlings	0.099	0.158	0.089	0.148
Elongation stage	0.108	0.152	0.056	0.095
With compaction before sowing				
Sowing	0.117	0.155	0.155	0.158
Final singling of seedlings	0.113	0.155	0.105	0.153
Elongation stage	0.106	0.150	0.064	0.104

Source: Chaoyang, Liaoning from Liaoning Institute for Water and Soil conservation (1983); Suide from Suide Experimental Station for Water and Soil Conservation, Yellow River Committee (1983); Henan from Du Xintian and Sun Dunren (1983); South Shanxi from Li Duren (1983).

Chengde region, Hebei Province, 1962; Liu Shuangan, 1984; Zhou Ruihua, 1986).

1. Compaction of Soils with Low Moisture Content

Compacting soil using a stone roller can reduce the proportion of large soil pores, and therefore decrease, even stop, water loss by evaporation. After rolling, the topsoil layer is pulverized and compacted to a high bulk density, so water volatilized from the subsoil can be condensed within the top layer (Table 26). This practice is generally carried out in the dry seasons, i.e., spring and winter (Li Like, 1986; Liu Xueren, 1984), and is used when the topsoil is dry, but the subsoil still has some water. It is harmful to use this practice when sufficient water is present in the soil. Best results are obtained where there is strong wind and poor land preparation for seeding because of large blocks of soil on the surface.

Experiments in Shanxi Province showed that after rolling clay loam soil with 10% to 12% water content in the top layer, the seedlings of wheat, maize, and millet emerged earlier, and yields increased. However, when the soil water content was more than 0.17 kg/kg, rolling either before or after sowing seriously reduced the number of seedlings because of too much compaction. The pressure intensity of rolling varies from 450 to 500 g/cm^2, which produces a bulk density of 1.1 to 1.3 Mg/m^3, suitable for emergence and growth of crops (Li Duren, 1983, 1984). Compacting a loose soil by rolling tends to increase the soil moisture, and therefore it plays a significant role in crop emergence when sowing is done during dry spells.

2. Harrowing or Smoothing Soil with Suitable Moisture Content

Harrowing and smoothing are often used for preparation of a seed bed as well as for water conservation. From the conservation point of view, harrowing is practiced when soil moisture is high and the surface is wet, and smoothing using a special tool (Fig. 19) is conducted either after plowing or after a small rain, or when soil moisture is low and the surface is dry. They may be combined, one followed by the other; or carried out separately, with only smoothing or only harrowing, depending on soil water.

Special attention has been given to harrowing or smoothing soil in early spring when the soil begins to thaw. On the drylands in China, it is dry, cold, and windy in winter, and water stored deep in the soil moves up in the vapor state and condenses in the top layer, resulting in a higher water content in the top layer. When the temperature rises in early spring, the surface soil thaws and is subject to rapid loss by evaporation. This period

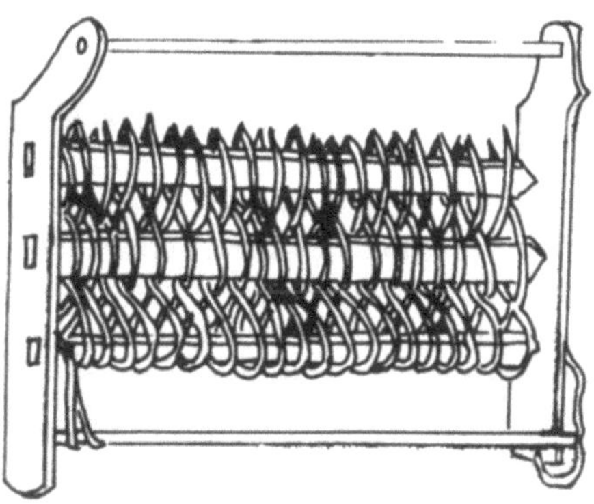

Figure 19. Diagram of "Mo," a special implement made out of bush branches for smoothing and pulverizing surface soil; it is generally drawn by animals, with a person standing on it to weigh it down. Source: Research Group of Ancient Chinese Agriculture, Academy of Agricultural Science and Nanjing Agricultural University (1986).

Table 27. Effects of harrowing or smoothing on soil water content in different layers

Time of measurements (day/month)	Water content (kg water/kg dry soil) in different layers (cm)				
	0–10	10–20	20–30	30–50	50–70
	Without harrowing or smoothing				
5/3	0.100	0.090	0.108	0.100	
9/3	0.095	0.100	0.102	0.098	0.093
4/4	0.101	0.098	0.097	0.095	0.096
23/4	0.073	0.088	0.095	0.093	0.093
24/5	0.044	0.073	0.079	0.084	0.087
Mean	0.083	0.090	0.097	0.094	0.092
	With harrowing or smoothing				
5/3	0.127	0.130	0.142	0.145	
9/3	0.098	0.106	0.115	0.120	0.115
4/4	0.087	0.123	0.124	0.118	0.114
23/4	0.070	0.107	0.115	0.119	0.110
24/5	0.045	0.095	0.097	0.100	0.104
Mean	0.085	0.112	0.119	0.120	0.111

Source: Suide Experimental Station for Water and Soil Conservation, Yellow River Committee (1983).

is short, but water loss is serious. Based on this situation, dryland farmers harrow lands while thawing in order to conserve the condensed water as much as possible for wheat use or for spring sowing.

Experience has shown that harrowing in early spring can increase the soil water by 1% to 3%. The Suide Experimental Station for Soil and Water Conservation, Yellow River Committee (1983) found that harrowing the soil at thawing time conserved 14.6 mm more precipitation in the 0- to 70-cm layer (Table 27).

3. Intertillage after Rain

Intertilling soil, usually by hoeing, after rain during the plant-growth period is commonly undertaken on drylands. The old agricultural proverb, "Hoes have water," gives an idea about the farmers' view of this practice. In the wheat fields intertillage by hoeing in winter can fill the fissures at the surface to reduce water loss by evaporation by 1% or so. In early spring hoeing can break the capillaries to form a dust mulch and conserve water. Hoeing is very important for intertilled crops such as sweet potato, cotton, and maize in order to control weeds and conserve stored water. The depth of hoeing depends on the crop and precipitation with a common pattern of shallow, deep, and *shallow*.

Table 28. Effects of hoeing after rain on conservation of soil water

Treatments	Field experiment: Water loss in 7 days after rain in field free of weeds (mm)	Field experiment: Water loss in 3 days after rain in field with weeds (mm)	Pot experiment: Water loss in 11 days after irrigation (%)
Without hoeing	8.15	9.78	16.00
Hoeing	9.17	2.89	12.35

Source: Wu Shouren et al. (1983).

The first hoeing is shallow, carried out in late spring to conserve soil water while it is still dry; the second is deep, conducted in early summer to loosen the soil and increase infiltration of rainfall that is concentrated in summer. The last hoeing is shallow again, and practiced in early autumn to conserve precipitation occurring late in the season. In addition to conserving water, deep hoeing also promotes root penetration into deep layers, thus raising crop resistance to drought. Wu Shouren et al. (1983) has shown experimentally that hoeing after rain definitely conserved water, and the effect was greater in fields with weeds (Table 28).

4. Covering Soil Surface with Various Materials

Different materials have been used on the drylands to cover the soil surface, which is an effective method for reducing soil water loss by evaporation. The materials used for this purpose depend mainly on local conditions and soil problems so that other benefits in addition to reducing soil water loss may be realized. For instance, in some places where soil is clayed, ash or slag from power stations may be utilized for improvement of soil texture as well as for covering the soil surface. Liu Pengsheng (1987, 1989a, 1989b, 1989c) found that broadcasting ash or slag on clay soil not only improved the soil texture, but also reduced evaporation. The main materials widely used for managing soil surface are plastic film and crop residues as well as gravels in some places.

a. Plastic Film Mulching

Plastic film or sheet mulching is a reputable engineering approach in China that functions in many ways to increase crop production and reduce soil water loss (Liang Yachao et al., 1990; Wang Shusen and Deng Genyun, 1991).

Gao Shiming and Ma Tianen (1987) conducted experiments with a ground plastic sheet mulch of dryland spring wheat at Tangjiapu Village,

Dingxi County. They covered the wheat rows during the entire plant-growth period using plastic sheets with openings for young wheat plants to grow through and for the soil to receive rainfall. Their results revealed that ground plastic sheet mulch effectively reduced soil water evaporation. Before openings were cut, soil water at a depth of 1 m was 47.8 mm higher than that without mulch. After cutting the openings, it was 6.7 to 15.6 mm higher for the remainder of the growing season. Compared with an unmulched field, soil temperature was higher until emergence of the third leaf, resulting in this stage being shortened. However, the soil temperature became lower and stabilized after this stage, which caused the third leaf to stem extension stage and flowering to ripening stage to be lengthened; this is beneficial for wheat enabling more tillers, ears, spikes, and grains to form, as well as an increase in the weight of the grain. As a direct result, the mulched wheat yield was increased by 70.5%. Based on these results, the authors suggested the use of this technique on gentle sloping, terraced, and level rainfed lands receiving about 400 mm precipitation, or fields where irrigation cannot be ensured.

Lu Zongfan et al. (1990c) demonstrated by their experiment done in a hilly gully region of north Shaanxi that maize with plastic film mulching matured 15 days earlier than that without mulching and consumed more water because of high yielding. This technique is more suitable for areas with high elevation, low temperature, high humidity, and frequent freeze injury. It should be noted that with this technique some long-day crops such as spring wheat can be planted in places where the conditions are unstable or the frost-free periods are considered too short for them to grow. For example, in Liaoning Province this technique has allowed big grain groundnut to be grown in Shenyang and Anshan, while in the past it was only grown in Dalian and Jinzhou (Wang Pengkao, 1984).

The use of plastic sheet mulch has been widely practiced in recent years in the northern provinces, not only for vegetables, cotton, and tobacco (Wang Pengkao, 1984), but also for grain crops. This method is excellent as regards conserving moisture, preventing drought, and raising soil temperature, as well as protecting N fertilizer loss against volatility and promoting availability of nutrients (Han Zhongkui, 1990), and thus it has great potential for increasing crop yields.

In 1986, the people in Leide Township, Jinnin County, Gansu Province, adopted this method to grow 10.86 ha of maize on a dry highland, 1700 to 1800 m above sea level, and the average yield per hectare reached 9.225 Mg, which is 4.7 times higher than that without plastic sheet covering. In the same year, Suihua Prefecture, Heilongjiang Province, mulched 210 ha maize with plastic film, and obtained 4.65 Mg/ha grain on average, the yield being increased by 91% as compared with that without plastic sheet mulch (Wang Wunxing, 1987). The maize area with plastic sheet covering was 26,600 ha in Shanxi Province in 1986, and the yields were increased by an average of more than 2.25 Mg/ha. According to incomplete statistics for nine northern provinces, municipalities, and autonomous regions in 1987, the maize area with plastic sheet covering reached 180,000 ha.

It is anticipated that use of the plastic sheet covering technique will continue to grow in northen dry areas. In the course of mulching with plastic films, valuable techniques have been developed. Wang Hongwei and Tian Fuhai (1989) seeded maize 3 to 5 cm away from plastic film early mulched on the ground and found that maize yield was higher than that with a film covering. Without cutting openings in the film, this technique is easy to practice, and the film can be reused. The authors have suggested that the technique should be adopted anywhere except in cold mountainous areas or areas with heavy alkaline or saline soils. Similar results were also obtained by other scientists (Wen Qikai et al., 1991; Wu Xinmin et al., 1990).

Five years of practice by the people in Tugou (a cold region with a long frost period) has shown that mulching with plastic films needs to be early. It is adopted in flat and fertile lands, shady places, for cash or long-term growing crops, and where there is sufficient soil moisture (Zhang Zhanhong, 1989). Application of fertilizer can increase the value of plastic film mulching (Li Zongxin, 1984).

b. Gravel Mulching

In Gansu and Qinghai provinces, since the Ming Dynasty a special material (coarse sand and gravel) has been used as mulch to cover the soil surface. The total area covered by this material amounts to more than 130,000 ha. No tillage has been practiced on this land, and the only agricultural operation is sowing by drilling crop seeds into the soil under the mulch, which is some 5 to 10 cm deep (Plate 6). In this way, besides increasing soil temperature, more rainfall is conserved, evaporation is reduced, water and wind erosion are controlled, and therefore the soil moisture is higher than that without covering (Table 29). Consequently, in the so-called sand land, the crop yield is doubled as compared with the other land (Huang Peiyong, 1983).

c. Crop Residue Mulching with Traditional Tillage

Covering the soil surface with crop residues, especially wheat straw and maize stalks, has been widely tested and is regarded as one of the best ways to store more water and to reduce water loss by evaporation as well as weed consumption (Wu Xinmin et al., 1990). The residue is generally broadcast on the soil surface after intertillage or after preparing the seed bed. It is also used to mulch dryland orchards to reduce water loss (Zhang Yangji et al., 1990; Dai Yutang, 1990). This practice retains the soil moisture for sowing.

Liu Zhongmin and Deng Xiping (1983) found that the germination and maturity of proso and millet were delayed when the seed bed was

a

b

Plate 6. Pebble or sand mulch fields. a. Pebble mulch vegetable field, Lanzhou, Gansu. b. Coarse sand mulch field, Yuzhong, Gansu. Source: Zhu Xianmo (1986a).

Table 29. Comparison of water content between fields with and without gravel mulch

	Water content (kg water/kg dry soil)						
Month	Mar.	Apr.	May	Jun.	Jul.	Aug.	Sept.
With gravel mulch	0.134	0.123	0.133	0.157	0.126	0.138	0.189
Without	0.068	0.062	0.071	0.079	0.088	0.085	0.102

Source: Huang Peiyong (1983).

mulched with wheat straw 1 month before sowing. This was caused by low temperatures and higher moisture. However, the yields were increased. Wu Xinmin et al. (1990) demonstrated that the soil moisture in mulched farmlands was much improved, especially during the seeding stage, while the evapotranspiration in the mulched fields remained similar to that of the controlled (without mulching) farmlands. This indicates that some soil water used by soil evaporation is transpired by crops, and thus water use efficiency is improved.

Straw mulch can reduce salt accumulation on the soil surface and therefore reduce soil salinization. Gong Jiake (1984) mulched an alkaline and saline soil surface with 7.5 Mg wheat chaff per ha and incorporated it with soil at a depth of 15 cm. He found that the incorporation reduced the total amount of salt at 15 cm by 18.5% to 50.4%, which is much higher than achieved by mulching (2.8% to 28.6%). However, when the wheat chaff was increased to 15.0 Mg, the mulching effect on the reduction of salt was significantly increased (24.8% to 40.7%). In any case, the highest wheat yield was obtained by mulching an alkaline soil.

d. Crop Residue Mulching with Minimum or No Tillage

Xu Weikui and Lu Bingzhang (1990) reported that the stubble mulch no-tillage system was very effective in improving newly opened saline soil.

Table 30. Effects of stubble mulch no-tillage and conventional tillage on salt content in soil

	Range of salt content	Salt content (%) at different layers (cm)		
Treatments	(%)	0–5	5–20	20–40
Stubble mulch no-tillage	0.15–0.20	0.039	0.067	0.039
Conventional		0.218	0.206	0.131
Stubble mulch no-tillage	0.20–0.30	0.105	0.131	0.254
Conventional		0.786	0.524	0.314
Stubble mulch no-tillage	0.30–0.40	0.069	0.116	0.380
Conventional		1.066	0.256	0.367

Source: Xu Weikui and Lu Bingzhang (1990).

They conducted an experiment for 5 years and found that the organic matter content in soil was increased from 0.82% to 1.21%, and the salt content at a depth of 1 m was decreased from 0.30% to 0.13% (Table 30). As a result, cotton yields were increased by 10%, 21%, and 26% in 3 successive years as compared with conventional tillage. Through 5 years of studies in the southern Loess Plateau, Han Siming et al. (1988) showed that using wheat straw as a mulch covering the soil surface, after plowing and pulverizing the fields, had an excellent effect on conserving soil water as well as increasing nutrient content and earthworm number, but decreased the soil bulk density and soil temperature. The authors also

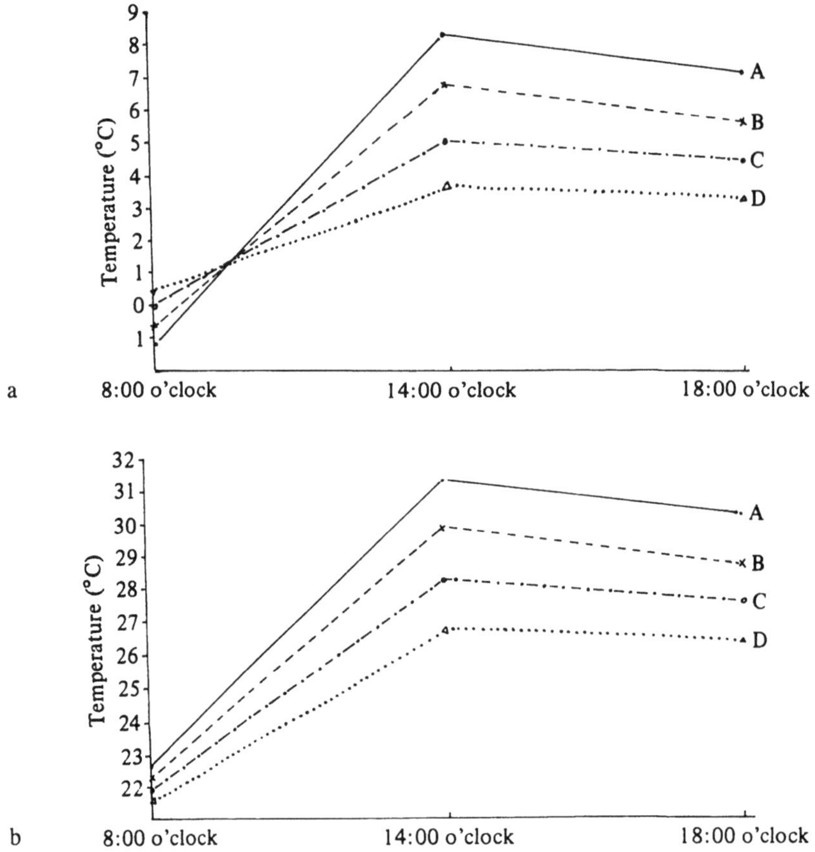

Figure 20. Effects of straw mulching on soil temperature. a. Day temperature variation at 5 cm soil depth in winter. A, Without mulching; B, mulching 3000 kg/ha; C, mulching 4500 kg/ha; D, mulching 6000 kg/ha. b. Day temperature variation at 5 cm soil depth in summer. A, Without mulching; B, mulching 3000 kg/ha; C, mulching 4500 kg/ha; D, mulching 6000 kg/ha. Source: Han Siming et al. (1988).

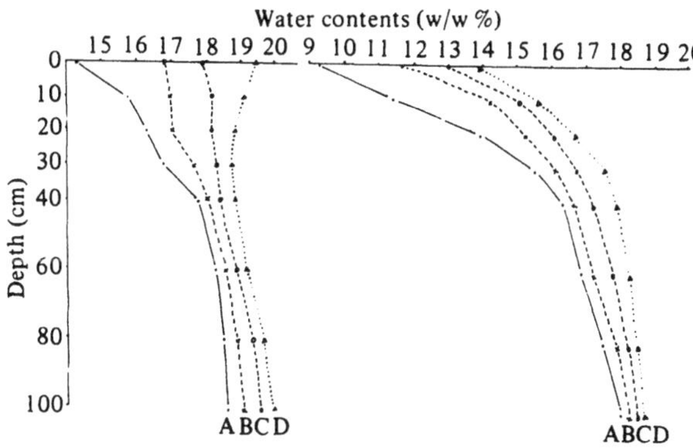

a

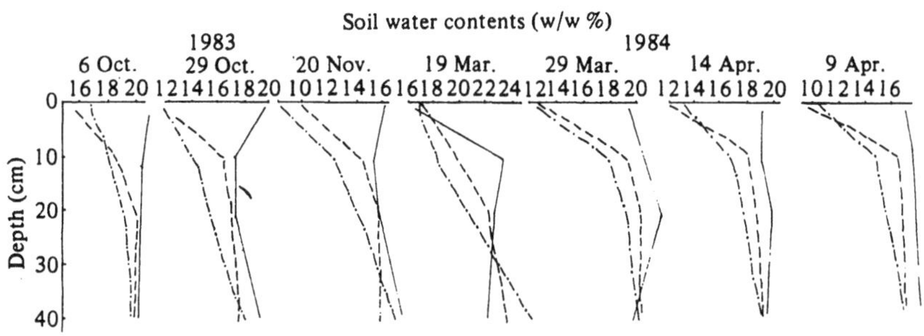

b

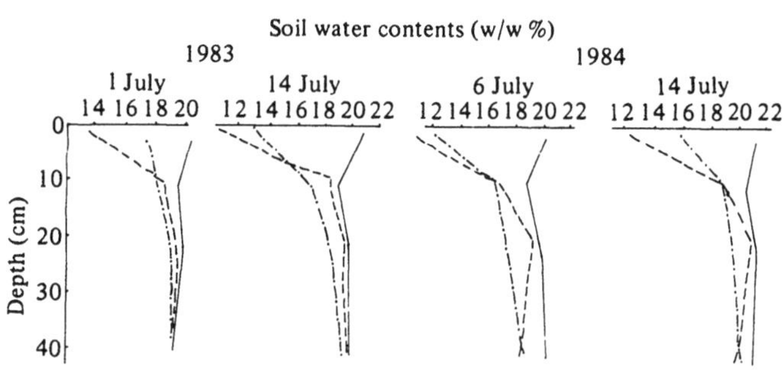

c

Figure 21. Water content distribution in soil profile. a. Water distribution in 100 cm soil profile. A, Without mulching; B, mulching 3000 kg/ha; C, mulching 4500 kg/ha; D, mulching 6000 kg/ha. Source: Han Siming et al. (1988). b. Water distribution in soil profile during early growth stage of maize. ———, No till with mulch; ·–·–·–·, no till without mulch; ----, ploughing tillage. Source: Zhu Wenshan and Cao Mingkui (1988). c. Water distribution in soil profile during early growth stage of wheat. ———, No till with mulch; ·–·–·–·, no till without mulch; -----, ploughing tillage. Source: Zhu Wenshan and Cao Mingkui (1988).

Table 31. Amount of water stored (mm) in different soil layers with different amounts of straw mulch (kg/ha)

Depth (cm)	Without mulch	Mulch 3000	Mulch 4500	Mulch 6000
		Amount of water stored		
0–30	66.1	71.2	74.4	76.2
30–100	191.5	194.5	198.3	201.1
100–200	248.9	252.5	259.9	271.1
0–200	506.5	518.2	532.6	548.4

Note: Two years' average values in two locations. Source: Han Siming et al. (1988).

showed that the bulk density in the top 0 to 10 cm was 1.36 Mg/m^3 without a straw mulch compared with 1.29 and 1.23 in fields with 4.5 and 6.0 Mg/ha straw mulch, respectively. The earthworm number per m^2 in the top 0 to 15 cm was 2 without straw mulch, but 12, 32, and 34 with straw mulch of 3.0, 4.5, and 6.0 Mg/ha, respectively. Organic matter in the top layer is 1.61% with no straw mulch, and 1.67% and 1.76% with 4.5 and 6.0 Mg/ha straw mulch, respectively. The same was true for total N and available N and P. The amounts were increased with increasing rates of straw. The temperature of the top 0 to 5 cm is lower in summer and higher in winter (Fig. 20a,b) with straw mulch. Therefore, the straw mulch was favorable for the conservation of soil moisture and protection from heat in summer and from cold in winter. The straw mulch significantly increased soil infiltration and reduced water loss by evaporation and increased water storage both in summer and winter (Fig. 21, Table 31).

Experiments from different dryland locations have demonstrated that this method can increase wheat production by 13.1% to 23.5% (Table 32). In Tengxian County, Shandong Province, 600 km south of Beijing, similar experiments were conducted by Xu Xinyu et al. (1985) and Dang Tongqi (1986, 1987) from 1975 to 1984. The authors top-dressed 2.25 to 4.50 Mg/ha of wheat trash between growing wheat rows in early April. After wheat was harvested, they seeded soybean or maize directly into

Table 32. Wheat yields (Mg/ha) with different amounts of straw mulch (kg/ha)

Year	Locations	Without mulch	Mulch 3000	Mulch 4500	Mulch 6000
			Wheat yields		
1986	Chengcheng	3.52	3.74	3.98	4.03
	Qinnxian	3.80	4.00	4.25	4.30
	NWAU	3.55	3.70	3.83	4.18
1987	Qianxian	3.63	3.99	3.83	4.38
	NWAU	4.16	4.63	4.71	5.13

NWAU, Northwestern Agricultural University.
Source: Han Siming et al. (1988).

the mulched space between rows of wheat stubble. They mulched the soil surface again with 2.25 to 4.50 Mg/ha of wheat trash when the summer crops grew about 30 cm before the rainy season. Having harvested the summer crop, heavy disking was performed to thoroughly mix the partly decayed wheat trash and the stubble with soil before sowing wheat. Under straw mulch, the soil moisture was conserved, soil structure improved, and organic matter steadily increased. Mulching made the crop residues decompose slowly. As a result, splash erosion and surface runoff were significantly reduced, weed hazard lessened, and labor and fuel saved. In addition, the activities of *Azotoabacter* and cellulose decomposing bacteria were improved, and the release of carbon dioxide from straw under canopy was slowed, which might be useful for crop assimilation. Consequently, the crop yields increased by 16% for soybean and 8% for maize and wheat. In 1984, this practice was demonstrated in fields of 30,000 ha.

A similar experiment was done by Wang Shuangzhuang and Xu Shuzhen (1989) in the east piedmont of the Taihang mountains, Hebei Province, a semiarid region with a serious shortage of water. They reported that mulching chopped maize stover in the space between wheat rows in early winter increased the soil water content, inhibited weed growth, and increased the soil organic matter and humus contents. The temperature of the topsoil was increased in winter and early spring, which is favorable for wheat to survive the winter. To determine the effect of mulch on irrigation, plots with and without mulch were irrigated in a limited manner. The authors found that the soil surface hardening and crusting as well as soil pore sealing were decreased in the plots with mulch as compared with the plots without mulch after irrigation. For example, the soil bulk density was increased by 0.065 to 0.075 Mg/m^3 and porosity was decreased by 2.34% to 2.77% in the plot without mulch after irrigation, while little change occurred in plots with maize stover mulch. The experiment further showed that mulching wheat rows not only increased water for wheat use but also saved water for sowing maize (the next crop), and that in the case of limited irrigation, mulching reduced irrigation water by 22.8% to 27.8%.

Good results of straw mulching were also obtained by Shen Laishui (1989), Gou Yangping and Ji Huan (1989), Dang Tongqi (1987), Zeng Jingqing and Wang Wenshan (1984), and Zhu Wenshan (1987). In addition to crop residue mulching, green manure mulching has recently been reported (Xu Zhuomin, 1986). This involves mulching the top layer after summer plowing and filling rows between wheat plants in winter and spring with green manure or forage. This kind of mulch not only increased water storage, but also provided crop nutrients. As a result, wheat yields were increased by 45% to 125%.

When combined with minmum tillage or no tillage, crop residue mulching has played a major role in conservation of soil moisture. The minimum or no-tillage method has long been used in China, especially in the state farms. For example, the General Bureau of State Farms in Heilongjiang Province employed drilling wheat into the harrowed soybean stubble in the early 1950s. Later on, the Bureau adopted other forms of minimum tillage, such as chiselling once instead of several plowings and directly seeding on the previous crop row margins. Since 1981, minimum tillage (just chiselling and harrowing) has been widely extended in the state farms, 26% to 27% of the total land is now cultivated using this technique.

According to Jin Rongwan's report (1987), minimum tillage has certain advantages in Heilongjiang Province: it keeps the topsoil in its own position without mixing with subsoil and makes the soil less compact; it coordinates water, nutrients, air, and heat; it limits soil erosion and encourages accumulation of organic matter and improves soil structure; and it is easy to manage, saving time and labor. As a result, crop yields were increased in 6 successive years.

Recent studies and agricultural practices in many locations in dryland areas have shown that combining residue mulch with minimum or no tillage has more attractive results. Zhu Wenshan and Cao Mingkui (1988) conducted experiments for 9 successive years by directly sowing the next crop seeds in the fields without plowing after the first crop harvest, and mulching the soil surface with the previous straw crop before or after seeding. The results showed that straw mulch combined with no tillage saved labor and energy and improved soil structure and fertility. By using this technique for 2 to 3 successive years, the water-stable aggregates of soil were increased by 24.2% as compared with using traditional tillage or no tillage without mulch. The total soil porosity remained above 50% with air-filled porosity more than 10%, which is suitable for plant growth. Water conductivity was increased both in saturated and unsaturated states and so was the soil's water-holding capacity. Therefore, water loss was decreased and the soil water content was increased (Fig. 22a, b). The organic matter in the topsoil (0 to 20 cm) was also raised (Table 33), and the mineralizable N was increased by 33.2% and 29.2% as compared with using no tillage (without mulching) and traditional tillage, respectively. As a result, the crop yields were successively increased (Table 34). From these experimental results the authors concluded that no tillage combined

Table 33. Organic matter content (g/kg) at top soil (0–20 cm) with different tillage systems

Years	No-till with straw mulch	No-till without straw mulch	Traditional tillage
1985	18.0	18.2	17.6
1987	19.4	17.8	16.7

Source: Zhu Wenshan and Cao Mingkui (1988).

Table 34. Annual average yields (Mg/ha) of wheat and maize produced in different tillage systems

Tillage methods	1980	1981	1982	Year 1983	1984	1987	Mean
A	8.3	10.8	8.2	11.0	10.0	9.6	9.7
B	8.3	10.9	7.3	10.3	9.2	7.8	8.9
C	9.2	11.6	8.6	11.2	10.2	11.9	10.5

A, traditional tillage; B, no-till without mulch; C, no-till with straw mulch.
Source: Zhu Wenshan and Cao Mingkui (1988).

with straw mulch was an effective way to minimize drought and low fertility problems in the dryland areas of north China.

Experiments conducted by Zheng Fenli et al. (1987) in Shaanxi Province showed that the stubble mulch no-tillage system was very effective in reducing soil erosion (Table 35). Similar results were obtained by other authors in different locations (Wu Shantan, 1988; Yang Chunfeng and Han Siming, 1988; Zhao Fengbin, 1987; Du Xuanwen, 1984; Yang Shusen, 1986; Xu Xinyu et al., 1988). All this shows that this technique has great potential for use on drylands.

IV. Future Strategies for Management of Drylands

Increasing concern about food supply and agricultural product diversity as well as ecology and the environment has led to the establishment of guidelines for the protection, exploitation, utilization, and development of the drylands. To put an end to the backwardness of the dryland areas, the Central Chinese government and the Ministry of Agriculture, Animal Husbandry, and Fisheries of China, have done much work, including working out policies and concrete measures. Based on scientific informa-

Table 35. Effects of stubble mulch and crop residue mulch on soil erosion (artificial rain simulation experimental results)

	Total erosion of soil (kg/m^2)			Rill erosion of soil (kg/m^2)		
Degree of slope	Plowing fallow	Crop residue mulching	Stubble mulch with no-till	Plowing fallow	Crop residue mulching	Stubble mulch with no-till
5	2.90		1.78			
10	3.81	0.00	2.99	1.03	0.00	0.00
20	8.52	0.27	6.51	1.63	0.22	0.70
25	22.11	6.06		12.84	4.76	

Source: Zheng Fenli et al. (1987).

tion, investigation, and experimentation, substantial knowledge has been accumulated for understanding the drylands, and many good proposals have been submitted for future management (Wan Jiangzhong et al., 1987; Xin Naiquan and Zhang Shaoli, 1987).

A. Priorities for Management of the Drylands

As mentioned in the previous section, the ecology of the drylands is fragile. Frequent drought, desertification, and soil erosion have led to ecological imbalances and soil degradation. These problems will be more severe if management is inadequate. The exploitation, utilization, and development of the drylands should involve rehabilitation and protection. Of course, utilization of the drylands is the primary objective, but, rehabilitation and protection cannot be overlooked. They must go hand in hand with exploitation, utilization, and development in order to gain both ecological benefit and economic result.

From the viewpoint of rehabilitation, protection, and development, priority has been and should be given to planting trees and grass, since sparse vegetation is the main cause of the constraints in the dryland areas. Many years' experience in many places show that it is difficult to develop dryland areas only with cropping. Use of the natural resources to plant trees and grasses can improve the ability to fight against natural disasters and also provide the best ecologically sound environment for high and stable dryland farming production and also for the development of other agricultural industries. For this reason, the Chinese government has given great attention to afforestation and the planting of grass in these areas.

In 1978, the State Council decided to establish a great forest protective system in the "three north areas," i.e., the northwest, northeast, and north of China. The areas for constructing such a system include most parts of the northwest and the north, the west part of the northeast, occupying 40.54% of the geographical land of China with an area of 3,892,000 km^2. Since then, numerous investigations have been carried out, and much information relating to this work has been obtained. As a result of several years of study, a detailed plan has been prepared for such construction (Zhu Junfeng, 1985; Mao Zhenlong et al., 1983).

The system includes windbreaks and shelter belts as well as protection of natural vegetation for stabilization of sand dunes in the desert and desertified areas; forests for protection of livestock in the grasslands; forests for conservation of head water in the hilly ranges; forests for conservation of water and soil on the Loess Plateau; tree networks for protection of arable lands in oases and in plain areas as well as on slopy lands; and commerical orchards in the areas close to cities or areas having better conditions for cultivation. When this comprehensive protective system is established, drought and sand drifts as well as soil losses will be alleviated, even prevented. Natural disasters will be controlled, a better economic system and environment will be formed, and the stable and high production of agriculture and animal husbandry will be ensured. In addition, economic benefits such as timber, fuel, and fodder will be obtained. This system will form a "Great Green Wall" in the northwest areas and a protective green screen for the southeast of China. It will greatly improve the agro-eco-environment in these areas and will bring benefit to future generations. Since this decision was made, different disciplines and branches have made contributions.

In 1983, the Ministry of Agriculture, Animal Husbandry, and Fisheries organized a conference on the management of northern dryland areas. In addition to emphasizing the importance of planting trees and grasses, attention was given to extending animal husbandry and adopting comprehensive developments in agriculture, forestry, stock raising, and economic sidelines. The conference has boosted the development of dryland agriculture. Since then, a certain amount of financial support from the state has been granted to these areas for planting trees and grasses and for constructing fields as well as extension activities. In the past few years, crop yields have been increased, much research carried out, and great achievements made in the management of the drylands. The project is presently in full swing.

The dryland areas are the traditional livestock production base of China. Grasslands or pasture cover 30% of the total area. There are still great opportunities and potential for the development of pastures from arable land, alkali-saline land, and nondrift sandy land. Through enclosing pastures and rotation grazing, development of artificial pastures, and rationally arranging the combination of animal breeds, not only can the drylands be protected, but also a commercial breed production base can be set up, and the leading role of animal husbandry in the areas can be brought into full play in the ecology and economic system.

B. Development of Dryland Agriculture in Accordance with Natural Conditions

As a whole, dryland agriculture should be protected by forests, supported by and combined with animal husbandry. Thus, on the drylands, agriculture, forest, and animal husbandry should be developed in a comprehensive way. However, the comprehensive development should be based on the natural conditions.

1. Arid Areas

In the agricultural region, farming should be based on rational use and water conservation, raising water use efficiency, preventing soil desertification and salinization, and establishing the commercial base for producing long-staple cotton, beets, melons, and fruits. In the agricultural-pastoral region, emphasis should be given to both agriculture and pasture. Grassland construction should be strengthened in order to raise the animal carrying capacity and the rate of meat production, as well as to gradually form a commercial base of animal products.

2. Semiarid Areas

In the windy-sandy agricultural regions, growing grass to support animals should be a priority. With the development of animal husbandry, agriculture will be promoted. In these regions, animal husbandry should be of great importance. Combined with forestry, diversified economies should be encouraged. For the agricultural-pastoral regions, unsuitable farmland should be returned to grassland and grazing by turns should be changed to rotation with crops and grass. Along the Great Wall, cropping should be continued. In the hilly-gully regions, attention should be given to proper land use, increasing vegetation cover, controlling floods from the Yellow River, and establishing self-sufficient agriculture.

3. Subhumid Areas

In these areas, cropping should be the main practice. Agriculture, forestry, animal husbandry, sideline production, as well as fisheries should be developed comprehensively. In addition, proper exploration, use, and protection of water resources should be carried out. Dryland farming should go hand in hand with irrigation farming. Exploitation of the natural resources, application of fertilizers, adoption of drought-tolerant crops, and improvement of soil fertility, as well as management of water resources should be encouraged in order to increase crop yield per unit area, establish a big commercial cotton and grain base, and to improve the ecological environment.

C. Increasing Production Stability and Product Diversity

For steady improvement of the dryland economy, it is necessary to increase production stability and product diversity. This can be done in three steps. The first is to improve the basic production conditions of arable lands by engineering constructions in combination with biological measures to intercept and store more natural rainfall. The second step is to develop grain production by adjusting the internal structure of forest and animal husbandry as well as the cropping system so that the industrial crops or cash crops and legume species and forage crops can be grown, through which the ecological and economic environment can be improved, and better ecological and economic cycles gradually formed. Based on the stabilization of agricultural production, the third step is to encourage a diversified economy of agriculture, animal husbandry, forestry, and fruit production, and the development of nonagricultural industries suitable to the local conditions in order to increase farmers' income and bring about a comprehensive development of the rural economy in the dryland areas.

D. Coordination of Urban and Rural Endeavors for Promotion of Specilized Commercial Products in Different Regions

Merely depending on agriculture and animal husbandry, cannot eliminate the poor situation. For further development of the dryland economy, it is necessary to take full advantage of the local natural resources to develop specialized commercial products in different regions. The present situation of either producing primary products or providing raw materials should be changed. The growth of a third industry should be promoted with the development of industrial construction. The regional economies must be expanded in accordance with the local ecological environment to raise economic returns and social benefits from the ecosystem. A number of small new towns will be gradually set up which will provide more opportunities for the growth of commercial economies.

E. Utilization of New Technology

In the vast arid and semiarid areas, rainfed agriculture is dominant. However, there are rich groundwater resources in some regions that are worthy of exploitation and utilization for the development of irrigation agriculture. In the irrigated areas scientific irrigation and drainage systems should be set up, and advanced irrigation techniques adopted so that there can be a rational balance of exploitation and rehabilitation for the limited water resources. For full use of the water and soil resources, arable land and irrigation should be further developed, where possible.

In addition to establishing a stable economic policy to promote the development of commercial economies in rural areas, we should rely, to a large extent, on the progress of science and technology, accelerate the process of implementing agricultural techniques, and raise production levels. To do this, it is important to organize integrated research to tackle the key problems in different regions, actively extend the achievements in scientific research and advanced techniques as productive forces, popularize them with the farmers, and raise scientific and technical levels. As far as the use of advanced technology is concerned, agricultural education or training should be offered at all levels. The involved provinces, municipalities, and autonomous regions should draw up a plan of training by turns and the plan should be implemented in a variety of ways.

Increasing capital and material input is also important for the development of dryland agriculture. The farmers in these areas are still poor, and therefore both material and capital inputs are low. As can be concluded from the agricultureal survey, it is necessary to increase the application of fertilizers, insecticides, plastic membranes, irrigation facilities, and farm implements, all of which, however, need financial support. The Chinese government, is giving special consideration to these areas—which will certainly promote the development of agriculture at a much greater speed.

References

Abramova, M.M., A.F. Bolshakov, N.S. Oreshken, and M.M. Roge. 1956. Evaporation of mobile water from soil. *Soil Sci.* 2: 27–41 (in Russian).

Agricultural Institute, Chengde Region, Heibei Province. 1962. An investigation summary account for soil water state and conservation of moisture and protection of seedlings by tillage methods in Chengde region, Hebei province. *J. Soil Sci.* 5: 21–27.

Agricultural Ministry of Zhangjiakou Region, Hebei province. 1983. Developing agriculture according to dryland farming properties. In: Shan Lun, Niu Pu (eds.) *Selected Works on Arid and Semi-Arid Farming*, vol. 1, pp. 169–175. Xian Agricultural Information Centre Press.

Ansai Experimental Area of Soil and Water Conservation, Northwestern Institute of Soil and Water Conservation and the Ansai County Government, Ansai Comprehensive Research Station of Soil and Water Conservation, Academia Sinica, 1990. The structure model and practice on ecological agriculture with soil and water conservation. In: Lu Zongfan (ed.) *Studies on Ecological Agriculture with Soil and Water Conservation in Loess Hilly Gully Regions*, vol. 1, pp. 13–27. Tianze Eldonejo, Yangling, Shaanxi, China.

Beiheying Agricultural Brigade Experimental Station, Hebei Province. 1973. Abundant harvest by changing production conditions in hilly drylands and by irrigation from pumping water, Proceedings of Fighting Against Drought in the North China Area. pp. 116–121. Agricultural Information Press. Xian.

Brady, N.C. 1974. The nature and properties of soils, 8th edn. pp. 240–251. Macmillan, New York.

Brown, L.R., and E.C. Wolf. 1984. Soil erosion: quiet crisis in the World economy, p. 49. World Watch Paper 60, World Watch Institute. Washington, D.C.

Cao Xuechang. 1987. Manure fertilizer on dryland wheat crop. Proceedings of the International Symposium on Dryland Farming. vol. 2, pp. 402–404. Yangling, Shaanxi, China.

Cao Yinzhen. 1981. The machanism and prediction of gravitational erosion. *Bull. Soil Water Cons.* 4: 19–23.

Central Arid Zone Research Institute, Jodhpur. 1988. *Progress in Arid Zone Research* (1952–1987), pp. 3–15. Jodhpur, India.

Central Meteorological Bureau. 1979. *Climate Atlas of the People's Republic of China* (Zhonghuarenmin gongheguo qihou tuji), pp. 3–4 and 222–223. Map Publishing House. Beijing.

Chang Caie, Jiang Dingsheng, Liu Mei, Huang Guojun, and Fan Xingke. 1989. An approach to the redistribution laws of precipitation on convex and concave hillslopes. *Acta Conservationis Soli et Aquae Sinica* 2: 36–44.

Chen Changyu. 1990. The suitable areas in oasis and cultivated lands and preventive measures against sand disasters in Hexi Corridor. *Agric. Res. Arid Areas* 2: 30–37.
Chen Guoliang. 1983. The agricultural climate resource in loess plateau and its rational use. In: Shan Lun, Niu Pu (eds.) *Selected Works on Arid and Semi-Arid Farming*, vol. 1, pp. 82–89. Xian Agricultural Information Centre Press.
Chen Jimin. 1987. Effect of vegetation cover of trees on Ziwuling in controlling loss of soil and water. *Soil Water Cons. China* 5: 8–10.
Chen Lizhi, Xia Zhenglian, and Wu Shengjun. 1989. Application of organic and inorganic fertilizers in China. Proceedings of the International Symposium on Balanced Fertilization. pp. 380–389. Agricultural Press, Beijing.
Chen Lunshou, Mao Daru, and Zhang Chendong. 1990. Several technical problems with respect to recommended (formulated) fertilization. *Soils* 22: 169–172.
Chen Peiyuan. 1983. Two ecological types of drought-resistant wheat varieties and their physical properties. In: Shan Lun, Niu Pu (eds.) *Selected Works on Arid and Semi-Arid Farming*, vol. 3, pp. 17–21. Xian Agricultural Information Centre Press.
Chen Yongzong. 1988. *Current Erosion and Control of Loess Plateau* (Huangtu gaoyuan xiandai qinshu yu zhili), pp. 5–6, 31–53, 115–137, 183–192. Science press, Beijing.
Chen Wenrui, and Zhang Jixian. 1962. Moisture status in shifting sand dunes and sand stabilization through afforestation. *Sand Control Res.* 4: 246–254.
Chen Zhimin. 1986. The effect of some crop stalks used as fertilizer on crop yield increase and soil improvement. *Soil Fertilizer.* 4: 25–29.
Chen Longheng. 1963. Development of Sand Dune soil in Baijia Beach Region, Hedong, Ningxiao. *J. Soil Sci.* 3: 26–32.
Chen Longhen, and Chen Wenrui, 1965. Soil formation on the natural fixed sand in southwestern Tenggeili desert. *Sand Control Res.* 7: 98–108.
Cheng Suyun, Qiao Cunfang, Han Guangmin, and Guo Xiaodong. 1987. The effect of fertilizer application upon the improvement of soil moisture productive efficiency in wheat on the drylands. *Agric. Res. Arid Areas* 2: 58–65.
Ci Longjun. 1963. Water Conservancy Study. *Sand Control Res.* No. 5.
Contributing Committee of Natural Geography of China, Academia Sinica. 1981. China's natural geography (Zhongguo ziran dili), vol. 1, pp. 187–412; vol. 2. pp. 21–31, 56–92. Science Press. Beijing.
Contributing Group of China's Comprehensive Agricultural Regionalization, National Agricultural Regionalization Committee. 1981. *China's Comprehensive Agricultural Regionalization* (Zhongguo zonghe nongye quhua), pp. 56–68, 227–237, and 294–316. Agricultural Publishing House. Beijing.
Cui Yunpeng. 1990. Suggestions for arrangements of engineering project system of comprehensive conservation of water and soil in small watershed of gully areas on Weibei highland. In: Wang Youmin (ed.) *Studies on Comprehensively Controlling and Managing Gully Areas in the Loess Plateau and Their Results*, pp. 210–214. China's Forestry Publishing House, Beijing.
Cui Wencai. 1975. Construction of high and stable yield farmlands. *Xinjiang Agricultural Science and Technology* No. 4.
Cui Wencai. 1976. The properties of Xingjiang hardened and impervious soil (desert soil) and its amelioration and utilization. *Soils* 1: 20–25.

Dai Yutang. 1990. Effect of straw mulching orchards on hilly sloping lands. *Agri. Res. Arid Areas* 1: 43–48.
Dang Tongqi. 1986. Straw mulch tillage, yield increase and soil fertility improvement. *Commun. Agric. Techn.* 11: 26–27.
Dang Tongqi. 1987. The effect of crop stalk mulching on the increase of crop yields and improvement of soil fertility. *Soil Fertilizer* 3: 40–42.
Deng Xinmin. 1990. A study on the effects of N and P fertilizers upon yields of alfalfa production. *Agric. Res. Arid Areas* supplement, 65–70.
Desert Research Laboratory, Lanzhou Institute of Glacier and Frozen Earth, Academia Sinica. 1976. China's Desert and its Control. *Scientia Sinica* No. 4–5.
Desert Research Laboratory, Lanzhou Institute of Glacier and Frozen Earth, Academia Sinica. 1977. *Control and Management of Sand Disasters of Arable Lands* (Nongtian shahaide fangzhi), pp. 19–62, 85–95, and 103–119. Science Press.
Dhar, N.A. 1933. Photonitrification in soil. *Soil Sci.* No. 35.
Ding Yonghuang. 1983. Some problems on agricultural production in arid and semi-arid areas in Shaanxi, Gansu and Ningxia. In: Shan Lun, Niu Pu (eds.) *Selected Works on Arid and Semi-Arid Farming*, vol. 1, pp. 72–81. Xian Agricultural Information Centre Press.
Dingxi Agricultural Institute. 1983. Investigation on the problem of dryland rotation in Dingxi region. In: Shan Lun, Niu Pu (eds.) *Selected Works on Arid and Semi-Arid Farming*, vol. 2, pp. 287–289. Xian Agricultural Information Centre Press.
Dong Yanqing, Zhang Jitao, Zhangsun Zhengpeng, and Li Junhou. 1991. Agronomical efficiency of furrow seeding in marginal farmland and study on coupling implements. *Agric. Res. Arid Areas* 2: 32–39.
Du Bangqing. 1987. Investigation on benefits of terraced fields in Daping village of Dingxi county. *Soil and Water Conservation in China* 3: 32–33.
Du Xintian, and Sun Dunren. 1983. Studies on the way to increase crop yields on the drylands of Henan province. In: Shan Lun, Niu Pu (eds.) *Selected Works on Arid and Semi-Arid Farming*, vol. 1. pp. 176–181. Xian Agricultural Information Centre Press.
Du Xuanwen. 1984. The preliminary research on minimum-tillage on drylands. *Shanxi Agricultural Sciences* 3: 2.
Du Yibin. 1987. An approach of controlling soil and water erosion in Taipinggou catchment. Proceedings of the international symposium on dryland farming, vol. 1, pp. 227–282. Yangling, Shaanxi, China.
Economic Geography Research Group, Geography Institute, Academia Sinica. 1985. *An Outline of Agricultural Geography of China* (Zhongguo nongnie dili zonglun), pp. 337–421. Science Press. Beijing.
Forest Institute of Inner Mongolia Autonomous Region. 1972. *Techniques for Afforestation to Stabilize Sands*. Inner Mongolia People's Publishing House.
Gansu Academy of Agricultural Science. 1983a. Millet, a crop suitable for planting in dryland areas in middle Gansu. In: Shan Lun, Niu Pu (eds.) *Selected Works on Arid and Semi-Arid Farming*, vol. 3, pp. 194–195. Xian Agricultural Information Centre Press.
Gansu Academy of Agricultural Science. 1983b. The effects of seeding millet in furrows and the reason for yield increase. In: Shan Lun, Niu Pu (eds.) *Selected*

Works on Arid and Semi-Arid Farming, vol. 3, pp. 200–202. Xian Agricultural Information Centre Press.
Gansu Agricultural Ministry. 1983. Taking correct ideas, regulating agricultural structure and adopting science to develop dryland farming. In: Shan Lun, Niu Pu (eds.) *Selected Works on Arid and Semi-Arid Farming*, vol. 1, pp. 100–107. Xian Agricultural Information Centre Press.
Gao Jishan. 1987. Investigation of loss of soil and water in Zang Autonomous Prefecture of Gannan. *Soil and Water Conservation in China* 6: 27–29.
Gao Jixiang, and Xu Qiang. 1987. The analysis and forecast of planting production in dry and wind erosion area of Yan, Tong, Xiang mountain region. Proceedings of the international symposium on dryland farming, vol. 2, pp. 231–234. Yangling, Shaanxi, China.
Gao Shiming, and Ma Tianen. 1987. Effects of plastic ground covering during the whole growing period in dryland spring wheat. Proceedings of the international symposium on dryland farming, vol. 2, pp. 269–371. Yangling, Shaanxi, China.
Geng Kuanhong. 1961. The preliminary result of stabilization of sand by sand obstacles made from clay soil in Minle. *Geography* 5: 200–205.
Gong Jiake. 1984. Studies of the effect of wheat chaff mulching on protection of seedlings and yield increasing in heavy saline soils. *Soil Fertilizer* 4: 7–10.
Gong Zitong, Lei Wenjing, and Chen Hongzho. 1988. Aridsols of China. *Arid Zone Res.* 2: 1–9.
Gou Yangping, and Ji Huan. 1989. Crop stalk mulch can increase wheat yield. *Gansu Agricultural Science and Technology* 1: 22.
Gou Zhiquan, Ma Tianchang, and Pei Shuying, 1990. An experimental report on cultivation of three forage crops. In: Wang Youmin (ed.) *Studies on Comprehensively Controlling and Managing Gully Areas in the Loess Plateau and Their Results*, pp. 198–200. China's Forestry Publishing House. Beijing.
Grain Crop Institute, Shaanxi Academy of Agricultural Sciences. 1983. Studies on methods for sowing maize in order to resist drought in Weibei dryland. In: Shan Lun, Niu Pu (eds.) *Selected Works on Arid and Semi-Arid Farming*, vol. 3, pp. 299–303. Xian Agricultural Information Centre Press.
Greenland, D.J. 1978. In Plenary papers, vol. 2. 11th International Congress of Soil Science. Edmonton, Canada.
Guangzhou Earthquate Brigade, Institute of National Earthquate Bureau. 1977. *An Outline of Global Structure of China*, pp. 3–11, 14–22, 33–37. Science Press.
Guan Zi. Zhou Dynasty. Diyuanpian (A chapter dealing with soil distribution and its propeties, interpretated by Xia Weiying), pp. 1–107. Printed in 1958. China's Publishing House.
Guo Tingfu, and Gao Bowen. 1982. *Water and Soil Conservation* (Shitu baochi), pp. 23–96. Agricultural Press, Beijing.
Han Siming, Yang Chunfeng, Shi Juntong, and Pang Huancheng. 1988. Stubble mulching in dryland on the loess plateau in China. *Agric. Res. Arid Areas* 3: 1–12.
Han Siming, Shi Juntong, Yang Chunfeng, and Li Feixiong. 1987. Research on drought-resistant tillage on dryland of Weibei Highland in Shaanxi Province. Proceedings of the international symposium on dryland farming, vol. 1, pp. 149–163. Yangling, Shaanxi, China.

Han Zhongkui. 1990. Cultivation methods of mulched corn. In: Zhu Xiangsan (ed.) *Studies on Exploitation and Harness of the Loess Plateau*, vol. 2, pp. 161–167. Shaanxi Science and Technology Publishing House.

Hartge, K.H. 1986. Demands on soils increasing in diversity and intensity. Transactions of the XIII Congress of International Society of Soil Science, vol. 1, pp. 1–12. Hamburg, Federal Republic of Germany.

Hou Guangjiong, Zhu Lianqing, and Li Lianjie. 1935. The report on soil survey in Ding County, Hebei Province. *Soil Bull.* 13: 1–57.

Hou Guangjiong, Li Xiaopin, Liu Zhongchen, Hu Bangyou, Hu Huaizhou, Deng Xuepin, Zhang Xulin, and Li Zhenjiang. 1987. Soil erosion and sediment deposition is the principal origin of the soil and water resources to be suffered from devastation. *Bull. Soil Water Cons.* 6: 1–4.

Hou Lingsheng, Li Xunmin, and Liu Weijin. 1987. A complete set of cultural techniques based on plowing and subsoiling at intervals for maize on infertile rainfed land. Proceedings of the international symposium on dryland farming, vol. 2, pp. 364–365. Yangling, Shaanxi, China.

Hou Xilu, and Cao Qingyu, 1990. Study on the benefits of plants to reduce sediment in the loess rolling region of north Shaaxi. *Bull. Soil Water Cons.* 2: 33–40.

Hou Zhongtian, and Tang Defu. 1984. *Dryland Farming* (Handi nongye), pp. 27–49, 126–136, and 227–259. Heilongjiang Science and Technology Publishing House.

Huang Jouchang, Wang Bangxi, Wang Hui, Zhang Xiaoping, and Zhao Daqian. 1990. Effects of supplementary water supply on wheat yield and water use efficiency in arid areas. *Agric. Res. Arid Areas* 4: 105–109.

Hua Tianmao, Li Changwei, Li Huitao, Zou Jianbin, and Hu Tianmin. 1990. Research on the rational application of fertilizers to winter wheat on rainfed farmlands. *Agric. Res. Arid Areas* supplement, 49–56.

Huguan Agricultural Department of Shanxi. 1983. Successive abundant millet harvests obtained in Jinzhuang brigade by adopting organic dryland farming. In: Shan Lun, Niu Pu (eds.) *Selected Works on Arid and Semi-Arid Farming*, vol. 3, pp. 175–178. Xian Agricultural Information Centre Press.

Huang Peiyong. 1983. Gravel mulch lands in Gansu province. In: Shan Lun, Niu Pu (eds.) *Selected Works on Arid and Semi-Arid Farming*, vol. 2, pp. 125–129. Xian Agricultural Information Centre Press.

Hu Luhao, Jing Qingyuan, Yang Gang, and Hu Guancai. 1989. The effects of continuous application of organic and inorganic fertilizers on the increase of wheat and millet yields. Proceedings of the International Symposium on Balanced Fertilization, pp. 397–409. Agricultural Press, Beijing.

Institute of Forestry, Academy of Shaanxi Forestry Sciences. 1972. *Forest and Agriculture* (Senlin yu nongnie), pp. 18–25. China's Forestry Publishing House. Beijing.

Institute of Forestry, Academy of Shaanxi Agriculture and Forestry Sciences. 1976. *Narrow Benches with Gentle Slope Declining Inward*. China's Forestry Sciences.

Jiang Bofan. 1990. Chemical behavior of soil phosphorus and testing of soil available phosphorus. *Soils* 22: 181–183.

Jiang Deqi, and Zhu Xianmo. 1963. Water and soil conservation. *A Collection of Works of Agricultural Soil of China*, pp. 384–440. Shanghai Science and Technology Press.

Jiang Dingsheng, Liu Meimei, and Huang Guojun. 1987. An initial approach to the redistribution laws of precipitation on convex and concave hillslopes. *Bull. Soil Water Conservation* 7: 45–50.

Jiang Dingsheng, and Fan Xingke. 1990. On the benefits and optimal disposition for measures of soil and water conservation in hilly and gully region in the Loess Plateau. In: Lu Zongfan (ed.) *Studies on Ecological Agriculture with Soil and Water Conservation in Loess Hilly Gully Regions*, vol. 1, pp. 88–101. Tianze Eldonejo, Yangling, Shaanxi, China.

Jiang Dingsheng, Jia Zhiwei, and Liu Zhi. 1990. Study on relation between precipitation and landform and soil erosion in slopeland. In: Lu Zongfan (ed.) *Studies on Ecological Agriculture with Soil and Water Conservation in Loess Hilly Gully Regions*, vol. 1, pp. 77–87. Tianze Eldonejo, Yangling, Shaanxi, China.

Jiang Jun, and Wang Lixiang. 1990. Potential of rainfall productivity for major food crops in different rainfall regimes in South Ningxia. *Agric. Res. Arid Areas* 4: 55–61.

Jiang Zhongshan, and Liu Zhi. 1989. Effect of rainfall factors and slope on splash erosion. *Acta Conservationis Soli et Aquae Sinica* 2: 29–35.

Jiang Zhongshan, and Liu Zhi. 1990a. Effect of natural raindrop impact on loess crust, infiltration and erosion. In: Lu Zongfan (ed.) *Studies on Ecological Agriculture with Soil and Water Conservation in Loess Hilly Gully Regions*, vol. 2, pp. 115–121. Tianze Eldonejo, Yangling, Shaanxi, China.

Jiang Zhongshan, and Liu Zhi. 1990b. The statistics and analysis in quantitative soil erosion study. In: Lu Zongfan (ed.) *Studies on Ecological Agriculture with Soil and Water Conservation in Loess Hilly Gully Regions*, vol. 2, pp. 150–154. Tianze Eldonejo, Yangling, Shaanxi, China.

Jiang Yuan. 1980. *Amelioration of Aeolian Soil* (Fengshatude gailiang), pp. 25–27. Agricultural Press. Beijing.

Jia Houli, and Qian Guangshu. 1990. Study on mulberry as water conservation agent at reclaimed arable land at Shaanbei. In Zhu Xiangsan (ed.) *Studies on Exploitation and Harness of the Loess Plateau*, vol. 2, pp. 261–268. Shaanxi Science and Technology Publishing House.

Jia Rong. Han Dynasty. Three tactics for control of rivers. In: *Hanshu* (A historical book of the Han Dynasty written by Ban Gu in Han Dynasty), vol. 29, records of gullies and rivers. New copy printed by China's Publishing House in 1962. p. 1692.

Jia Zhiwei, Jiang Zhongshan, and Liu Zhi. 1990. Study on relationship between rainfall characteristics and the loss of soil and water. In: Lu Zongfan (ed.) *Studies on Ecological Agriculture with Soil and Water Conservation in Loess Hilly Gully Regions*, vol. 2, pp. 107–114. Tianze Eldonejo, Yangling, Shaanxi, China.

Jin Rongwan. 1987. The study on basic theory and application techniques of zero and minimum-tillage. Proceedings of the international symposium on dryland farming, vol. 2, pp. 381–384. Yangling, Shaanxi, China.

Jin Shaoling. 1989. Balanced fertilization of spring wheat in the Hexi Corridor. Proceedings of the International Symposium on Balanced Fertilization, pp. 179–184. Agricultural Press, Beijing.

Jin Shenlin, Wang Shanrong, Zhang Xiwen, and Song Dianzhen. 1987. A study on the path of raising grain yield stability in the subhumid region apt to drought. Proceedings of the international symposium on dryland farming, vol. 2, pp. 377–380. Yangling, Shaanxi, China.

Li Bicheng, Li Rui, Ma Xiaoyun, and Wang Yuru. 1990a. Establishment and application of an information system of soil and water conservation in a small catchment. In: Lu Zongfan (ed.) *Studies on Ecological Agriculture with Soil and Water Conservation in Loess Hilly Gully Regions*, vol. 2, pp. 214–220. Tianze Eldonejo, Yangling, Shaanxi, China.

Li Bicheng, Lu Zhiwel, Ma Xiaoyun, Wang Jijun, and Wang Peisen. 1990b. The Strategic plan and forecasting on the social and economical development of Ansai county. *Bull. Soil Water Cons.* 2: 1–7.

Li Changwei, Zhao Boshan, Hua Tianmao, and Li Huitao. 1987. Some matter of fertilization in Weibai dryland region. Proceedings of the international symposium on dryland farming, vol. 2, pp. 408–411. Yangling, Shaanxi, China.

Li Duren. 1983. Studies on the yield-increasing techniques of conserving and pumping soil water by compacting soil before sowing. In: Shan Lun, Niu Pu (eds.) *Selected Works on Arid and Semi-Arid Farming*, vol. 2, pp. 223–229. Xian Agricultural Information Centre Press.

Li Duren. 1984. The effect of soil compaction before sowing on crop yield increase. *Shanxi Agricultural Science* 1: 2.

Li Duren. 1986. Studies of the effect of wheat field subsoiling on crop yield increase and soil moisture conservation. *Soil Fertilizer* 3: 1–5.

Li Duren, Gao Xuke, Wang Deshui, Wand Xiaobin, and Zeng Lingwen. 1987. A study on the effect of subsoiling on soil water storage and rainfed wheat yields in southeastern Shanxi. Proceedings of the international symposium on dryland farming, vol. 2, pp. 342–346. Yangling, Shaanxi, China.

Li Fengqi, Zhang Bo, and Fan Zhimin. 1983. Preliminary investigation on the drought-resistant farming experience in ancient times. In: Shan Lun, Niu Pu (eds.) *Selected Works on Arid and Semi-Arid Farming*, vol. 2, pp. 18–33. Xian Agricultural Information Centre Press.

Li Guangyue. 1987. A study of stalks returning to fields and suitable tillage operations. Proceedings of the international symposium on dryland farming, vol. 2, pp. 390–391. Yangling, Shaanxi, China.

Li Guihua. 1963. Studies on the amelioration of white hardened and impervious soil (desert soil) by summer plowing. *A Collection of Works of Xinjiang Soil Science Society* Xinjing People's Publishing House.

Li Guihua. 1965. Summer plowing, an important measure for resistance of drought and increase of yields. *Xinjiang Agricultural Sciences* 7: 266.

Li Hongen, and Yang Yunlian. 1965. Effect of summer plowing on the winter wheat. *Acta Pedologica Sinica* 13: 404–410.

Li Kairong, and Wang Youmin. 1990. Water regime on artificial locust forest land in the Loess Plateau. In: Wang Youmin (ed.) *Studies on Comprehensively Controlling and Managing Gully Areas in the Loess Plateau and Their Results*, pp. 56–64. China's Forestry Publishing House. Beijing.

Li Like. 1986. Studies of the effect of compacting wheat fields in east loess plateau. *Shaanxi Agricultural Sciences* 6: 37–38.

Li Maosong. 1987. Regionalized zones and their evaluation of the rainfed agricultural in North China. Proceedings of the international symposium on dryland farming, vol. 1, pp. 68–78. Yangling, Shaanxi, China.

Li Mingang, Gao Shangwu, Zhang Hanhao, Zhang Jingye, Hu Keqin, Han Jiong, Zhao Tianxi, and Wang Kangfu. 1963. Experiment summary of stabilization of sands by afforestation. *Sand Control Res.* 5: 19–36.

Li Minggang. 1980. The principles and measures of shifting-sand fixation on both sides of railways. In: Lanzhou Desert Institute, Academia Sinica (ed.) *Research on Stabilization of Shifting Sands on the Desert Slope Head of Tenggeili Desert Area* (Tenggeili shamo shapotou diqu liusha zhili yangjiu), pp. 27–49. Ningxia People's Publishing House.

Ling Yuquan. 1980. The protective efficiency of grass checkboard-barriers. In: Lanzhou Desert Institute, Academia Sinica (ed.) *Research on Stabilization of Shifting Sands on the Desert Slope Head of Tenggeili Desert Area* (Tenggeili shamo shapotou diqu liusha zhili yangjiu), pp. 49–59. Ningxia People's Publishing House.

Li Rengang, Wang Kewu, Fan Jingzhong, Sun Zhuxuan, Li Bingshen. 1990. Preliminary studies on fertilization pattern for winter wheat. *Soils* 22: 210–211.

Li Rui, B.A. Harrison, and D.L.B. Jupp. 1990. Using remote sensing to map landform and cover in the Loess Plateau, China. In: Lu Zongfan (ed.) *Studies on Ecological Agriculture with Soil and Water Conservation in Loess Hilly Gully Regions*, vol. 2, pp. 221–226. Tianze Eldonejo, Yangling, Shaanxi, China.

Li Shengxiu, Zhao Boshan, and Li Changwei. 1976a. The effects of organic fertilizer by base application and by dressing on wheat yields and fertilizer efficiency. In: Shaanxi Agricultural Ministry (ed.) *Selected Works of the Shaaxi Agricultural Science and Technology*, pp. 83–85. Shaanxi People's Publishing House.

Li Shengxiu, Zhao Boshan, and Li Changwei. 1976b. The effects of fertilizer application time on wheat yields. In: Shaanxi Agricultural Ministry (ed.) *Selected Works of the Shaaxi Agricultural Science and Technology*, pp. 78–82. Shaanxi People's Publishing House.

Li Shengxiu, Zhao Boshan, and Li Changwei. 1976c. Increasing fertilizer efficiency by deep application. In: Shaanxi Agricultural Ministry (ed.) *Selected Works of the Shaaxi Agricultural Science and Technology*, pp. 73–76. Shaanxi People's Publishing House.

Li Shengxiu, Zhao Boshan, and Li Changwei. 1978. Rational application of phosphate fertilizer to wheat, part I. *Acta Collegii Septentrionali Occidentali Agriculturae* 1: 77–85.

Li Shengxiu, Zhao Boshan, and Li Changwei. 1979. Rational application of phosphate fertilizer to wheat, part II. *Acta Collegii Septentrionali Occidentali Agriculturae* 1: 55–99.

Li Shengxiu, Zhao Boshan, Li Changwei, and Li Shiqing. 1988a. Some factors influencing the effect of phosphatic fertilizer on pea. *Agric. Res. Arid Areas* 1: 56–63.

Li Shengxiu, Li Changwei, and Zhao Boshan. 1988b. The effect of N fertilizer on the efficiency of P fertilizer in several legume and non-legume plants. *J. Soil Sci.* 17: 268–271.

Li Shengxiu, Xiao Junzhang, and Cheng Suyun. 1988. Soil water management on drylands in China. In: Unger, P.W., Sneed, T.V., Jordan, W.R., Jensen, R. (eds.) *Challenges in Dryland Agriculture—A Global Perspective* (Proceedings of the International Conference on Dryland farming. August 15–19, 1988. Amarillo/Bushland, Texas U.S.A.), pp. 201–204. Texas Agricultural Experiment Station.

Li Shengxiu, Xiao Junzhang, and Cheng Suyun. 1989. On Dryland water management in China. *Agric. Res. Arid Areas* 1: 1–10.

Li Shengxiu, and Zhao Boshan. 1990. The effect of soil nitrogen supplying capacity on phosphate fertilizer efficiency for some legume and non-legume crops. *Soil Fertilizers* 4: 19–23.
Li Shengxiu, Fu Huifang, Yuan Hulin, and Xiao Junzhong. 1990a. Comparison of several methods reflecting the nitrogen supplying power in rainfed land soils. *Soils* 22: 190–193.
Li Shengxiu, Zhao Boshan, and Li Changwei. 1990b. The effect of N, P and organic manure application on the N-fixing performance of leguminous crops. *J. Soil Sci.* 21:13–15.
Li Shengxiu. 1991. Some factors affecting pea response to N fertilizers. *Agric. Res. Arid Areas* 1: 1–10.
Li Shujiang, and Xu Shouming. 1987. The function of crop straw and stubble in supplement of soil organic matter. *Soil Fertilizer* 4: 14–16.
Li Shengxiu, and Zhao Boshan. 1991. Proposals for rational fertilization of arid soil in China. *Chinese J. Soil Sci.* 22 (4): 145–148.
Li Ying, and Lu Dianqing. 1991. Effect of fertilization on wheat yield and its quality in wheat-producing region on Weibei rainfed highland. *Agric. Res. Arid Areas* 2: 1–9.
Li Yongtang. 1966. Some drought-resistant methods for soil preparation in Inner Mongolia Automonous Region. *Tillage Fertilizer* 2: 9–11.
Li Yu, and Li Shanxing. 1987. On the technical ways to carry out water-saving farming and dry farming from drought and shotage of water in Heilongang region of Hebei Province. Proceedings of the international symposium on dryland farming, vol. 1, pp. 114–123. Yangling, Shaanxi, China.
Li Yuanfang, Hu Jisheng, Ning Guozhan, Wu Yuying, Huang Yanling, and Yu Daiguan. 1984. Studies of the effect of wrapping legume forage seeds in a granular form with *Rhizobium*-containing material for airplane sowing. *Soil Fertilizer* 2: 27–28.
Li Yushan. 1962. Water regime in Loessial soil and crop growth. *Acta Pedologica Sinica* 10: 289–304.
Li Yushan. 1982. The relationship between soil water dynamics and wheat production. *Shaanxi Agricultural Science* 2: 5–8.
Li Yushan. 1983. Evaluation of agricultural field moisture on Weibei dryland and the way to raise water production efficiency of dryland farming. In: Shan Lun, Niu Pu (eds.) *Selected Works on Arid and Semi-Arid Farming*, vol. 2, pp. 70–77. Xian Agricultural Information Centre Press.
Li Yushan, Zhang Xiaozhong, and Guo Minhang. 1990. The field experiments on the field-response of crops to water and fertilizer in the south of the Loess Plateau. *Acta Pedologica Sinica* 27: 1–7.
Li Zongxin. 1984. The effect of plastic film mulch in hilly area and some notes. *Commun. Agric. Sci. Techn.* 1: 12–13.
Liang Hongru. 1984. Water and soil conserving tillage and measures for conversion of soil and water loss on slopy lands. *Soil Fertilizer* 1: 11–19.
Liang Waylea, Liu Hanli, and Hu Hengjui. 1987. The effect of nitrogen and phosphorus fertilization on water use efficiency of dryland wheat. Proceedings of the international symposium on dryland farming, vol. 2, pp. 412–414. Yangling, Shaanxi, China.
Liang Yachao, Yu Guixia, Yang Dianrong, and Liu Qingjiang. 1990. A study of the machanism of maize cultivated with plastic film cover for water storage and moisture conservation. *Agric. Res. Arid Areas* 1: 27–32.

Liang Zhenji. 1990. An analysis of soil moisture productive potential in winter wheat on Weibei Loess Plateau. *Agric. Res. Arid Areas* 2: 12–17.

Liao Ciyuan, Jiang Jin, and Qian Taitao. 1980. The studies on the selection of sand binder species and their characterics. In: Lanzhou Desert Institute, Academia Sinica (ed.) *Research on Stabilization of Shifting Sands on the Desert Slope Head of Tenggeili Desert Area* (Tenggeili shamo shapotou diqu liusha zhili yangjiu), pp. 60–120. Ningxia People's Publishing House.

Liaoning Institute for Water and Soil Conservation. 1983. The tillage techniques for preservation of soil water against drought and protecting plants. In: Shan Lun, Niu Pu (eds.) *Selected Works on Arid and Semi-Arid Farming*, vol. 2, pp. 257–264. Xian Agricultural Information Centre Press.

Lin Bao, and Li Jiakang. 1989. The development of chemical fertilizers in the past fifty years and balanced fertilization. Proceedings of the International Symposium on Balanced Fertilization, pp. 43–51. Agricultural Press, Beijing.

Lin Yuquan, and Zou Bengong. 1984. The function of railings in preventing sand accumulation in their frontlines. *J. Desert Res.* 3.

Liu Bingzheng, Zei Mingzhu, and Wu Faqi. 1990a. The preliminary study on gully erosion on Weibei Highland. In: Wang Youmin (ed.) *Studies on Comprehensively Controlling and Managing Gully Areas in the Loess Plateau and Their Results*, pp. 215–223. China's Forestry Publishing House. Beijing.

Liu Bingzheng, Wu Faqi, and Chen Jiming. 1990b. A preliminary study on the reduction of soil fertility and productivity by water erosion on the Weibei highland. In: Wang Youmin (ed.) *Studies on Comprehensively Controlling and Managing Gully Areas in the Loess Plateau and Their Results*, pp. 236–247. China's Forestry Publishing House, Beijing.

Liu Bingzheng, and Chen Jiming. 1990. Erosion by rainstorm and water and soil conservation in Nihe gully. In: Wang Youmin (ed.) *Studies on Comprehensively Controlling and Managing Gully Areas in the Loess Plateau and Their Results*, pp. 224–229. China's Forestry Publishing House, Beijing.

Liu Dongsheng. 1964. Loess soils in the middle part of Yellow River Valley (Huanghe zhongyou huangtu), pp. 181–185. Science Press, Beijing.

Liu Fa, Wang Keyu, and Liu Yinghua. 1983. Effect of direct application of wheat straw on the improvement of soil and increase of yield in high-latitude frost region. *Soils* 15: 175–178.

Liu Gengling. 1989. Food production and balanced fertilization in China. Proceedings of the International Symposium on Balanced Fertilization, pp. 16–21. Agricultural Preess, Beijing.

Liu Goubin, Song Lian, Wang Zhangyao, and Guan Yanning. 1990. The classification of grassland in Ansai County by using false color photo and study on the growth dynamics of locusts in the Loess Plateau. In: Lu Zongfan (ed.) *Studies on Ecological Agriculture with Soil and Water Conservation in Loess Hilly Gully Regions*, vol. 2, pp. 355–363. Tianze Eldonejo, Yangling, Shaanxi, China.

Liu Goubin, and Yang Xiaoying. 1990. A study on laws of formation of biomass and nutritive substance for *Astragalus*. In: Lu Zongfan (ed.) *Studies on Ecological Agriculture with Soil and Water Conservation in Loess Hilly Gully Region*, vol. 2, pp. 328–338. Tianze Eldonejo, Yangling, Shaanxi, China.

Liu Xunhao, and Han Xianglin. 1987. China farming system regionalization (Zhongguo Gengzuo zhidu quhao), pp. 1–31, 25–49, 94–96. Beijing Agricultural University Publishing House.

Liu Mei, Jiang Dingsheng, Huang Guojun, Fan Xingke, and Chang Caie. 1990. An analysis to the differences of soil water in varied places of hillslopes. *Bull. Soil Water Cons.* 2: 16–20.
Liu Pengsheng. 1963. Manural soils in Guanzhong. I. The formation of manural soil and its fertility development, pp. 46–55. *Selected Works of the Shaanxi Soil Science Society.*
Liu Pengsheng. 1987. *Selected Papers on the Agricultural Use of Slag from Power Stations*, pp. 75–86.Northwestern Agricultural University Press.
Liu Pengsheng. 1989a. Slag from power stations is a good soil amendment with multiple effects. In: Liu Pengsheng (ed.) *Collection Papers Studying Soil*, pp. 148–178. Tianze Eldonjo, Yangling, Shaanxi, China.
Liu Pengsheng. 1989b. Experimental report of the effect of slag from power stations on cotton growth and yields. In: Liu Pengsheng (ed.) *Collection of Papers Studying Soil*, pp. 179–193. Tianze Eldonjo, Yangling, Shaanxi, China.
Liu Pengsheng. 1989c. The effect of slag from power stations on soil amelioration. In: Liu Pengsheng (ed.) *Collection of Papers Studying Soil*, pp. 194–205. Tianze Eldonjo, Yangling, Shaanxi, China.
Liu Shu, and Shi Qinghui. 1980. The studies on the density and arrangements of sand binders. In: Lanzhou Desert Institute, Academia Sinica (ed.) *Research on Stabilization of Shifting Sands on the Desert Slope Head of Tenggeili Desert Area* (Tenggeili shamo shapotou diqu liusha zhili yangjiu), pp. 121–153. Ningxia People's Publishing House.
Liu Shuangan. 1984. Attention of harrowing, smoothing, rolling and hoeing in dryland farming. *Commun. Agric. Sci. Techn.* 1: 9.
Liu Xinmin, Wu Zuoqi, Wang Honglou, and Ren Aicheng. 1982. Studies on controlling desertification of north part of Linzei oasis, Gansu province. *J. Desert Res.* 3: 9–15.
Liu Xueren. 1984. The effects of harrowing, smoothing and compacting of wheat fields in winter and spring on moisture conservation. *Shanxi Agricultural Sciences* 1: 37.
Liu Yingxin, and Yang Xilin. 1980. Studies on elevating survival rate of sand-fixing plants and accelerating their growth. In: Lanzhou Desert Institute, Academia Sinica (ed.) *Research on Stabilization of Shifting Sands on the Desert Slope Head of Tenggeili Desert Area* (Tenggeili shamo shapotou diqu liusha zhili yangjiu). pp. 179–204. Ningxia People's Publishing House.
Liu Yingxin. 1988. Phyto-reclamation of sand dunes in Northwest, North, and Northeast of China. *J. Desert Res.* 4: 11–18.
Liu Yumin. 1962. The development of sandy soils in Caidam Basin. *J. Soil Sci.* 4: 45–48.
Liu Zengxiang. 1965. Nitrogen deficiency in Xinjiang agriculture methods of solution. *J. Soil Sci.* 1: 24–28.
Liu Zengwen, and Wang Youmin. 1990. Water consumption by transpiration and water characteristics on artificial forest land of *Pinus tabulaeformis*. In: Wang Youmin (ed.) *Studies on Comprehensively Controlling and Managing Gully Areas in the Loess Plateau and Their Results*, pp. 89–96. China's Forestry Publishing House. Beijing.
Liu Zhi, and Jiang Zhongshan. 1990a. Application of microcomputer to data processing and management of sediment-runoff in the SWC field. In: Lu Zongfan (ed.) *Studies on Ecological Agriculture with Soil and Water*

Conservation in Loess Hilly Gully Region, vol. 2, pp. 145–149. Tianze Eldonejo, Yangling, Shaanxi, China.

Liu Zhi, and Jiang Zhongshan. 1990b. Microcomputer mapping of digital topographic elements. In: Lu Zongfan (ed.) *Studies on Ecological Agriculture with Soil and Water Conservation in Loess Hilly Gully Region*, vol. 2, pp. 275–281. Tianze Eldonejo, Yangling, Shaanxi, China.

Liu Zhi, and Jiang Zhongshan. 1988. Study on natural raindrop impacting effects on loess crusting of upland. *Bull. Soil Water Cons.* 1: 62–64.

Liu Zhongmin, and Deng Xiping. 1983. A preliminary report on the effect of millet yield increase by mulching tillage in semi-arid and temperate-cool regions. In: Shan Lun, Niu Pu (eds.) *Selected Works on Arid and Semi-Arid Farming*, vol. 3, pp. 206–209. Xian Agricultural Information Centre Press.

Liu Ziyu. 1990. Plant nutrition diagnosis in retrospect and in prospect. *Soils* 22: 173–176.

Lu Buwei. Warring States. Lu's Spring and Autumn, vol. 154, p. 500. Shanghai Ancient Literature Publishing House. Printed in 1987.

Lu Dianqing, and Li Yin. 1987. Enhancement of the rate of nitrogen utilization for increasing wheat production on Loess Plateau. Proceedings of the international symposium on dryland farming, vol. 1, pp. 236–248. Yangling, Shaanxi, China.

Lu Dingqing, Li Ying, and Tan Wenlan. 1989. Wheat response to N and P balanced fertilization and the factors affecting the response in Shaanxi loess region. Proceedings of the International Symposium on Balanced Fertilization, pp. 317–324. Agricultural Press, Beijing.

Lu Dingqing. 1990. Study on a technical system of recommended fertilization and its use in northwestern China. *Soils* 22: 201–204.

Lu Zhiwei, Wang Jijun, Wang Peisen, and Kang Zhangpeng. 1990. Study on the planning and forecasting of input and output in the small watershed of Zifanggou in Ansai County. In: Lu Zongfan (ed.) *Studies on Ecological Agriculture with Soil and Water Conservation in Loess Hilly Gully Region*, vol. 2, pp. 236–241. Tianze Eldonejo, Yangling, Shaanxi, China.

Lu Zongfan, Su Min, and Zhang Xingchang. 1990a. Study on yield increasing system with soil and water conservation in hilly gullied region in the Loess Plateau. In: Lu Zongfan (ed.) *Studies on Ecological Agriculture with Soil and Water Conservation in Loess Hilly Gully Region*, vol, 1, pp. 100–105. Tianze Eldonejo, Yangling, Shaanxi, China.

Lu Zongfan, Su Ming, Li Gouxia, Zhang Junfeng, Li Ming, and Feng Fu. 1990b. Study on grass and bush and tillage measures to conserve soil and water in loess region. In: Lu Zongfan (ed.) *Studies on Ecological Agriculture with Soil and Water Conservation in Loess Hilly Gully Region*, vol. 2, pp. 11–12. Tianze Eldonejo, Yangling, Shaanxi, China.

Lu Zongfan, Su Ming, Li Gouxia, Zhang Junfeng, Li Ming, and Feng Yu. 1990c. An experiment report on corn with ground plastics film in the hilly gully region, North of Shaanxi. In: Lu Zongfan (ed.) *Studies on Ecological Agriculture with Soil and Water Conservation in Loess Hilly Gully Region*, vol. 2, pp. 29–33. Tianze Eldonejo, Yangling, Shaanxi, China.

Luo Rongxi, Ren Aihu, Fu Cunren, and Zhang Chunzuo. 1987. Wind erosion characteristics in Wumeng dry steppe soil, Inner Mongolia, and its control. *Soil and Water Conservation in China* 7: 11–12.

Lu Xiangxian, and Li Jianjun. 1987. A study of optimal structure modeled by enhancing agricultural after effects and increase of yield and income from crop plantation on the Weibei rainfed highland. Proceedings of the international symposium on dryland farming, vol. 2, pp. 331–335. Yangling, Shaanxi, China.

Ma Heping. 1987. Studies on the relationship between fertilized black soil and its physical properties. Proceedings of the international symposium on dryland farming, vol. 1, pp. 249–257. Yangling, Shaanxi, China.

Ma Xiumei, and Song Lian. 1990. The cultivation and extention of fine herbage early-maturing *Astragalus adsurgens* Pall bred by radiation. *Acta Conservationis Soli et Aquare Sinica* 1: 56–60.

Ma Xiumei. 1990. The experimental report on breeding of early-maturing varieties of species *Astragalus adsurgens*. In: Lu Zongfan (ed.) *Studies on Ecological Agriculture with Soil and Water Conservation in Loess Hilly Gully Regions*, vol. 2, pp. 296–322. Tianze Eldonejo, Yangling, Shaanxi, China.

Ma Yongzhi. 1938. Soils in Northwest Gansu. *Soil Bull.* 19: 1–79.

Ma Yongzhi. 1944. Soil Georgraphy in Northwest China. *Geography* 4.

Ma Yongzhi. 1945. Soil Geography in the middle part of Xinjiang. *Quarterly Soil Sci.* 3–4.

Ma Yongzhi, and Wen Qixiao. 1951. Soils in North Shaanxi and their utilization. *Soil Bull.* 26: 73–98.

Ma Yongzhi. 1957. Geographical distribution patterns of soils in China. *Acta Pedologica Sinica* 5: 1–18.

Ma Yuxi, Yang Wengzhi, Han Shifeng, and Yang Xingming. 1990. On the growth dynamics of locust in the Loess Plateau. *Acta Conservationis Soli et Aquae Sinica* 2: 26–32.

Maowusu Survey Team, Geology Department, Beijing University. 1976. The experiences and problems of afforestation using the method of "tragging sands in front of trees and blocking sands behind trees." In: Lanzhou Glacier, Frozen Earth and Desert Institute, Academia Sinica (ed.) *Desert Control* (Shamode Zhili), pp. 72–80. Science Press. Beijing.

Mao Zhenglong, Cai Yuwen, Shi Zhaoqin, and Sun Meiying. 1983. Soils and landscape in the shelter belt constructing areas of the "Three Norths" (Sanbei fanghudai jianshede turanghedimao), pp. 3–15. Office of Agricultural Regionization, Shelter Belt Constructing Areas of the Three Norths. Forestry Press, Beijing.

Mei Boyan (1786–1856). Qing Dynasty. Discussion of people's matter at the study. In: *A Collection of Ancient Literature* (continued edition. Edited by Wang Xianqian in Qing Dynasty), vol. 27, p. 2. Changsha Wang's Woodcut Edition printed in 1882.

Mei Xurong, Zhan Jubao, and Lu Xuedu. 1990. Field moisture dynamics of winter wheat in rainfed farmland in Tunliu experimental areas. *Agric. Res. Arid Areas* 4: 1–6.

Nanjing Institute of Soil Science, Academia Sinica. 1978. *Soils of China*, pp. 145–194, 573–594, and 655–690. Science Press. Beijing.

Niu Pu, Yang Gaihe, and Shen Yunxia. 1987. Winter wheat cultivation under the condition of protection from rain. Proceedings of the international symposium on dryland farming, vol. 2, pp. 351–352. Yangling, Shaanxi, China.

Olsen, S.R., C.V. Cole, F.S. Watanabe, and L.A. Dean. 1954. Estimation of available phosphorus in soils by extraction with sodium bicarbonate. U.S. Dep. of Agric. Circ. 939.

Pan Borong. 1987. The past and future of introduction of sand-holding species in China. *J. Desert Res.* 1: 1–8.

Peng Lin, Peng Xianglin, and Lou Zhenfan. 1981. The seasonal variation of soil nitrate and the effect of summer fallow on the fertility of manured Loessial soil. *Acta Pedologica Sinica* 18: 212–222.

Pu Jinchun, and Huang Zichen. 1990. Experiment of optimum irrigation norm for spring wheat in the arid of Gansu Province. *Agric. Res. Arid Areas* 1: 49–58.

Qi Jiangzhong. 1988. Study on improving afforestation quality in Maowusu sand land. *J. Desert Res.* 3: 53–62.

Ren Yong, and Wang Youmin, 1990a. Study on water physiology of *Paulownia elongata*. In: Wang Youmin (ed.) *Studies on Comprehensively Controlling and Managing Gully Areas in the Loess Plateau and Their Results*, pp. 65–74. China's Forestry Publishing House. Beijing.

Ren Yong, and Wang Youmin. 1990b. Water relationship and water characteristics in the system of crop-cum-*Paulownia elongata*. In: Wang Youmin (ed.) *Studies on Comprehensively Controlling and Managing Gully Areas in the Loess Plateau and Their Results*, pp. 75–88. China's Forestry Publishing House. Beijing.

Ren Zhengxi, and Li Yingqi. 1987. On ridge and furrow cultivation at rainfed hilly in Ningxia. Proceedings of the international symposium on dryland farming, vol. 2, pp. 392–393. Yangling, Shaanxi, China.

Research Group of Ancient Chinese Agriculture, Academy of Agricultural Science and Nanjing Agricultural University. 1986. Dryland Farming in North China (Beifang handi nongnie), pp. 9–30. Chinese Agricultural Science and Technology Press, Beijing.

Sand Control Working Team, Academia Sinica. 1958. Soils in West of Inner Mongolia and Hexi Corridor. *Comprehensive Investigation and Research Reports for Desert Regions*, vol. 1. Science Press. Beijing.

Secretariat of Conference on Agriculture Learning from Dazhai. 1975. *A Collection of Typical Experiences on Agricultural Learning from Dazhai*. Xian.

Shaanxi Academy of Agricultural Science. 1973. The investigation on the tillage methods against drought in Yanan region. Proceedings of fighting against drought in North China Areas, pp. 139–148. Xian Agricultural Information Press.

Shaanxi Agricultural Survey and Design Academy. 1982. *Shaanxi Agricultural Soils*, pp. 71–73. Shaanxi Science and Technology Publishing House.

Shaanxi Water and Soil Conservation Bureau. 1971, 1972. *A Handbook for Soil and Water Conservation*, pp. 1–123, 137–158. Shaaxi People's Publishing House.

Shaanxi Water and Soil Conservation Bureau and Northwestern Institute of Water and Soil Conservation. 1979. *Forest and Grass Measures for Conservation of Water and Soil* (Shuitu baochi lincao cuoshi). pp. 68–134 and 291–492. Agricultural Press, Beijing.

Shaanxi Water and Soil Conservation Bureau. 1974. *Construction of Lands by Letting in Water to Flush Sands* (Yinshui lasha zaotian). Water Conservancy and Electricity Power Publishing House.

Shan Lun. 1983. Suggestions for improving grain production on drylands in the Loess Plateau. In: Shan Lun, Niu Pu (eds.) *Selected Works on Arid and Semi-Arid Farming*, vol. 3, pp. 1–4. Xian Agricultural Information Centre Press.

Shan Lun, Sun Jibin, Liu Zhongmin, and Du Souyu. 1987. Studies on productivity and water use for major crops in the mountain region, Southern Ningxia. Proceedings of the international symposium on dryland farming, vol. 1, pp. 104–113. Yangling, Shaanxi, China.

Shanxi Agricultural Ministry. 1983. The general condition of Shanxi dryland farming and advice on the development in future. In: Shan Lun, Niu Pu (eds.) *Selected Works on Arid and Semi-Arid Farming*, vol. 1, pp. 137–143. Xian Agricultural Information Centre Press.

Shao Yuting, and Zhen Yenzhi. 1989. Studies on the indices of phosphorus in Oasis soil of the Hexi Corridor and phosphate application rates. Proceedings of the International Symposium on Balanced Fertilization, pp. 76–82. Agricultural Press, Beijing.

Shao Liyie, and Chen Guishang. 1987. Study on the prevention of wind erosion of soil from dry cropland. *Soil and Water Conservation in China* 1: 46–48.

Shen Yuqing. 1962. On summer fallow. *J. China's Agriculture* 7: 1–5.

Shen Laishui. 1989. The effect of straw mulch on crop yields under dryland farming condition. *Shanxi Agricultural Sciences* 6: 2.

Shi Guangyi and Liu Yongguang. 1981. Contour furrow tillage—a new way for high yields in hilly areas. *Shanxi Agricultural Sciences* 1: 17.

Shi Guangyi. 1982. Water storing and fertilizer concentrating tillage for dryland farming. *Shanxi Agricultural Sciences* 9: 7.

Shi Keming. 1986. Studies on the tillage of water and soil conservation. *Water and Soil Conservation in China* 5: 3–5.

Shi Maqian. Han Dynasty. Qin Shihuang's Biography. In: Shiji (A historical record book written by Shi Maqian in Han Dynasty), vol. 2, pp. 1–115. New copy printed by Ancient Chinese Literature Publishing House in 1954.

Shi Ninghai. 1976a. Yellow River between Shaanxi and Shanxi. *Shaanxi Normal University* 2: 96–107.

Shi Ninghai. 1976b. The change of Zhouyuan. J. *Shaanxi Normal University* 3: 111–129, 132.

Shih Shenghan. 1959. *On "Fan Sheng-Chih Shu," An agriculturist book of China written in the first century* B.C., pp. 30–31. Science Press, Peking, China.

Si Binglian. 1987. Constructing agroecologically is a strategic measure to develop agriculture in arid areas—an investigation about the constructing experience of ecological agriculture of Dangpudi village. Proceedings of the international symposium on dryland farming, vol. 1, pp. 124–133. Yangling, Shaanxi, China.

Soil and Fertilizer Institute, Gansu Academy of Agricultural Science. 1963. The soil crusting problem in Shulehe River basin. *Tillage Fertilizer* 5: 35–38.

Soil and Fertilizer Institute, Gansu Academy of Agricultural Science. 1966. The role of summer plowing and sunning the upturned soil in agricultural production. *Tillage Fertilizer* 3: 48–51.

Soil Classification Research Group, Institute of Soil Science, Academia Sinica, Nanjing. 1985. Chinese Soil Taxonomic classification (1st project). *Soils* 17: 290–318.

Soil Survey Staff, USDA, 1975. *Soil. Taxonomy*. Handbook No. 436. Washington, D.C.

Song Wanlin, and Chen Ailian. 1988. Comments on sowing herbage with airplanes in Changji. *Arid Zone Res.* 4: 56–59.

Su Yinquan, Zhang Kanjan, Xie Dezi, Qiu Junkui. 1990. The effect of lands prepared by V-shape on soil water and seedling growth. In: Wang Youmin (ed.) *Studies on Comprehensively Controlling and Managing Gully Areas in the Loess Plateau and Their Results*, pp. 129–133. China's Forestry Publishing House. Beijing.

Suide Experimental Station for Water and Soil Conservation, Yellow River Committee. 1983. The cultivation technique for high and stable millet production in hill-gully loess dryland of North Shanxi. In: Shan Lun, Niu Pu (eds.) *Selected Works on Arid and Semi-Arid Farming*, vol. 3, pp. 179–185. Xian Agricultural Information Centre Press.

Sun Jianxuan. 1987. Discussion on runoff agriculture. *Shanxi Science and Technology of Water and Soil Conservation* (Special issue), p. 39.

Sun Xi. 1957. *Agrochemistry*, p. 231. Educational Publishing House. Beijing.

Tang Keli, Zhang Zhongzi, and Kong Xiaoling. 1989. A study on soil loss and soil degradation on the Loess Plateau. In: Brush, L.M. et al. (eds.) *Taming the Yellow River: Silt and Floods*, pp. 299–312. Kluwer Academic Publishers, Leyden.

Thorp, James and Hou Guangjiong. 1935. *Soils in the North and Northwest of China*. Soil Bull. 15: 1–34.

Tian Jiying, and Huang Yiduan. 1964. Investigation on the physical properties of soil in relation to the index of soil resistance to erosion in the region of Ziwuling, Gansu. *Acta Pedologica Sinica* 12: 286–296.

Tian Yuanren, and Jiang Xizheng. 1990. Assessing the rate of split application of fertilizer nitrogen based on soil testing. *Soils* 22: 208–200.

Tu Xingwen. 1987. Stabilization of gullies by plant engineering—willow dam. *Soil and Water Conservation in China* 4: 20–23.

Wan Jianzhong, Jia Yongying, and Luo Zhicheng. 1987. Allocation and exploitation of the arid and semi-arid areas in China. Proceedings of the international symposium on dryland farming, vol. 1, pp. 1–12. Yangling, Shaanxi, China.

Wang Anzhu, and He Zhongyang. 1990. Soil moisture behaviors in farmlands and its productive efficiency on Heyang Rainfed Highlnd. *Agric. Res. Arid Areas* 4: 37–43.

Wang Bangxi, Huang Jiuchang, Zhao Daqian, Wang Hui, and Zhang Xiaoping. 1990. Effects of fertilization on yields and water use efficiency in spring wheat in rainfed farmland. *Agric. Res. Arid Areas* 4: 98–104.

Wang Changyao, Guan Yanning, Zhen Tanhe, Liu Goubin, Song Lian, Seyingbatu. 1990. A report on grassland resources survey by remote sensing in Ansai County. In: Lu Zongfan (ed.) *Studies on Ecological Agriculture with Soil and Water Conservation in Loess Hilly Gully Regions*, vol. 2, pp. 347–354. Tianze Eldonejo, Yangling, Shaanxi, China.

Wang Duqing, and Bai Huishan. 1987. Experimentation of rotation of crops with several kinds of herbage of the legume family on cropland on slopes in North Shaanxi. *Soil and Water Conservation in China* 10: 26–31.

Wang Fukong. 1980. The effect of compactness of sand layers on the growth of sand binders. In: Lanzhou Desert Institute, Academia Sinica (ed.) *Research on Stabilization of Shifting Sands on the Desert Slope Head of Tenggeili Desert Area* (Tenggeili shamo shapotou diqu liusha zhili yangjiu), pp. 154–178. Ningxia People's Publishing House.

Wang Hongwei, and Tian Fuhai. 1989. The techniques of seeding maize on plastic film sides. *Shanxi Agricultural Science* 1: 16–18.
Wang Jizhi. 1986. Main soils in Ningxia Autonomous Region and Alashanyuoqi. In: Soil Science Society of China (ed.) *Current Progress in Soil Research in the People's Republic of China*, pp. 562–573. Jiangsu Science and Technology Publishing House.
Wang Kangfu, and Dong Shudi. 1988. Jingtong railway in sandy area of Naiman Banner. *J. Desert Res.* 2: 65–73.
Wang Lixian. 1981. Identification and classification of soil erosion intensity in middle reaches of the Huanghe River. Trans. Scientific Symp. on Comprehensive Control of Water and Soil Erosion in the Loess Plateau, p. 176.
Wang Lixiang. 1983. Studies on the precipitation production potentials in Northwestern Loess Plateau and its exploitation. In: Shan Lun and Niu Pu (eds.) *Selected Works on Arid and Semi-Arid Farming*, vol. 2, pp. 54–61. Xian Agricultural Information Centre Press.
Wang Lixiang, Wang Liufang, and Fan Fangqiang. 1987. Reform of agrostructure and its economical benefits in the semi-arid rainfed agricultural zone in Guyuan, Ningxia. Proceedings of the international symposium on dryland farming, vol. 2, pp. 313–315. Yangling, Shaanxi, China.
Wang Lixiang, Wang Lioufang, Wang Longchang, Yang Zhengli, Fan Fangqiang, and Ren Shichun. 1990b. Studies on the productive potential of alfalfa. *Agric. Res. Arid Areas* 4: 86–91.
Wang Lixiang, Wang Liufang, Jiang Jun, Wang Longchang, Fan Fangqiang, Zhang Xitian, and Ren Shichun. 1990a. Experimental summary of rainfall productive potential on Weibei Rainfed Highland. *Agric. Res. Arid Areas* supplement, 57–64.
Wang Longchang. 1990. A systematic analysis of food self-sufficient ability in Qingshui River watershed of Guyuan profecture. *Agric. Res. Arid Areas* 4: 69–78.
Wang Pengkao. 1984. The use of plastic film mulch in Liaoning province. *Commun. Agric. Sci. Techn.* 10: 36.
Wang Qian. 1983. The distribution and Climatic characteristic of arid and semi-arid areas in China. *Agric. Res. Arid Areas* (Trial issue), pp. 11–24.
Wang Shaoming. 1987. Wind erosion and its control in Qinghai Plateau. Proceedings of the international symposium on dryland farming, vol. 2, pp. 415–418. Yangling, Shaanxi, China.
Wang Shuangzhuang, and Xu Shuzhen. 1989. Effect of straw mulch in wheat fields and preliminary study of water economization. *Agric. Res. Arid Areas* 2: 7–15.
Wang Shusen, and Deng Genyun. 1991. A study on the machanism of soil temperature increasing under plastic mulch. Scientia Agricultura Sinica 24 (3): 74–78.
Wang Tianyi. 1990. Studies on yield increase laws of wheat cultivated by level furrows in mountain land. *Agri. Res. Arid Areas* 1: 33–42.
Wang Xianming, and Danzhengzhu Zasang. 1990. The changing laws of soil moisture and rainfed agriculture in agricultural Areas of river vallays in Tibet. *Agric. Res. Arid Areas* 2: 1–11.
Wang Yanbing, and Xu Jianwen, 1991. Experiments on increasing soil fertility and ameliorating by turning over stwaw into soil. *Chinese J. Soil Sci.* 22 (4): 145–148.

Wang Yantian and Xu Zhuling. 1990. Soil water behaviors in farmlands in the semiarid areas liable to drought in Wuchuan of Inner Mongolia. *Agric. Res. Arid Areas* 4: 14–19.

Wang Yingqin, Gao Junling, Ma Weiping, and Xie Senxiang. 1988. The role of wheat straw and maize stalks in ameliorating saline wasteland. *J. Soil Sci.* 19: 274–275.

Wang Youghou. 1983. The technique achievements on the dryland farming in ancient times of China. In: Shan Lun, Niu Pu (eds.) *Selected Works on Arid and Semi-Arid Farming*, vol. 2, pp. 13–17. Xian Agricultural Information Centre Press.

Wang Yongyan, and Zhang Zhonghu (chief eds.). 1980. *Loess in China*. Part III: Constituent Materials and Micro-textures of Loess in China. Shaanxi People's Art Publishing House (a photography book with no page numbers).

Wang Yunxing. 1987. Broad prospects of dryland agriculture by covering with plastic film. *Soil and Water Conservation in China* 12: 17–19.

Wang Youmin. 1990. An investigation on the locust growth in Chunhua county. In: Wang Youmin (ed.) *Studies on Comprehensively Contrölling and Managing Gully Areass in the Loess Plateau and Their Resultes*, pp. 34–43. China's Forestry Publishing House. Beijing.

Wang Youmin, Li Kairong, and Zhu Qingke. 1990a. The preliminary report on the experimental results by seeding three shrubs directly. In: Wang Youmin (ed.) *Studies on Comprehensively Controlling and Managing Gully Areas in the Loess Plateau and Their Results*, pp. 123–128. China's Forestry Publishing House. Beijing.

Wang Youmin, Feng Ting, and Guo Peicai. 1990b. Investigations on the growth of *Populus daruensis* and *Paulownia elongata* in Weibei dryland. In: Wang Youmin (ed.) *Studies on Comprehensively Controlling and Managing Gully Areas in the Loess Plateau and Their Results*, pp. 44–49. China's Forestry Publishing House, Beijing.

Wang Yuru, 1990. GIS and comprehensive analysis evaluation system based on microcomputers. In: Lu Zongfan (ed.) *Studies on Ecological Agriculture with Soil and Water Conservation in Loess Hilly Gully Regions*, vol. 2, pp. 257–267. Tianze Eldonejo, Yangling, Shaanxi, China.

Wang Yuxi. 1983. The position of proso in dryland farming of South Ningxia. In: Shan Lun, Niu Pu (eds.) *Selected Works on Arid and Semi-Arid Farming*, vol. 3, pp. 211–214. Xian Agricultural Information Centre Press.

Wang Wanzhong. 1983. The relationship between rainfall character and soil loss in the loess region. *Bull. Soil Water Cons.* 4: 7–13.

Wang Zhonglin, and Wang Youmin. 1990. The preliminary report on the protective results of structure of forest mixing with crops. In: Wang Youmin (ed.) *Studies on Comprehensively Controlling and Managing Gully Areas in the Loess Plateau and Their Results*, pp. 103–107. China's Forestry Publishing House. Beijing.

Wang Zhonglin, 1990. The protective results of interplanting *Paulownia elongata* with crops on Weibei dryland. In: Wang Youmin (ed.) *Studies on Comprehensively Controlling and Managing Gully Areas in the Loess Plateau and Their Results*, pp. 108–115. China's Forestry Publishing House. Beijing.

Water Conservancy Committee of Yellow River. 1954. The experience of construction of dams in Liujiashan, Lishan, Shaanxi. *New Yellow River* 12.

Wei Xian. Song Dynasty. The outline of water conservancy of Simingta Mountain. In "Division of History and Geography, Siming Series of books," vol 1. pp. 1–21 and vol 2, pp. 1–19. Guangling Ancient Book Publishing House. Reprinted in 1981 according to Zhang Yueyuan's woodcut edition printed in 1932–1948.

Wen Zhenwang, Wang Anqiu, Su Bingxun, Liu Wensheng, Ruan Chengxun, Wang Longxiu. 1957. Outline of soil geography in Rehe Province. *Soil Bull.* 30: 1–33.

Wen Qikai, Liu Yulan, Yang Guangbi, Lai Zongsheng, and Liu Naian. 1991. Studies on ecological conditions of soil under cultivation of corn outside plastic film cover. *Chinese J. Soil Sci.* 22: 12–15.

Wen Zhenwang. 1959. Geographical regionalization of soils in Inner Mongolia. *Soil Bull.* 34.

Working Committee of Natural Regionalization, Academia Sinica. 1959. China's Comprehensive Natural Regionalization, (Zhongguo zonghe ziran quhua), pp. 1–197, and 260–287. Science Press, Beijing.

Wu Faqi, and Liu Bingzheng. 1990. Water and soil loss around reservoirs in Chunhua county. In: Wang Youmin (ed.) *Studies on Comprehensively Controlling and Managing Gully Areas in the Loess Plateau and Their Results*, pp. 230–235. China's Forestry Publishing House. Beijing.

Wu Jiangou. 1989. The effects of combination of N, P and B application on increase of wheat production. Proceedings of the International Symposium on Balanced fertilization, pp. 357–360. Agricultural Press, Beijing.

Wu Jingmin, Xu Wenyuan, Dong Baishu, and Wang Zhenzhong. 1991. The effect of turning straw over into soil as fertilizer and its role in building up soil fertility. *Chinese J. Soil Sci.* 22: 211–215.

Wu Xinmin, Xue Junhong, and Guan Shucai. 1990. Discussion on soil water conditions and crop increase production in mulched farmlands. *Agric. Res. Arid Areas* 4: 92–97.

Wu Shantan. 1988. Studies of no-tillage in maize field. *Tillage and cultivation* 1: 50–52.

Wu Shouren, Wei Yunxiang, and Cheng Suyun. 1983. Research on hoeing for preservation of soil moisture and on soil moisture loss by evaporation in the rainfed highland in Wugong, Shaanxi. *Agric. Res. Arid Areas* 1: 61–76.

Wu Xinmin, Zhou Bai, Lu Longquan, and Du Raodong. 1990. A discussion on agricultural productivity in Wuchuan county of Inner Mongolia. *Agric. Res. Arid Areas* 4: 92–97.

Wu Yuanju, and Li Zhilian. 1990. Planting mulberry at leveled deep ditches. In: Zhu Xiangsan (ed.) *Studies on Exploitation and Harness of the Loess Plateau*, vol. 2., pp. 172–174. Shaanxi Science and Technology Publishing House.

Wu Zhanqiang, and Zhang Yongping, 1990. Sowing patterns of alfalfa at slope land. In: Zhu Xiangsan (ed.) *Studies on Exploitation and Harness of the Loess Plateau*, vol. 2, pp. 193–194. Shaanxi Science and Technology Publishing House.

Xiang Chongyang. 1989. Opening address on the International Symposium on Balanced Fertilization, Novermber 7–11, Beijing. Proceedings of the International Symposium on Balanced Fertilization, pp. 2–3. Agricultural Press, Beijing.

Xi Chengfan. 1961. Soil moisture and conservation of water for resisting drought. *Soils* 4: 27–29

Xi Yaoguo. 1990. The soil moisture evaporation and water balance on Weibei Loess Plateau. *Agric. Res. Arid Areas* 1: 70–75.

Xie Dezi, Ma Kanxin, Su Yinquan, and Qiu Junkui. 1990. The effect of seed size of *Pinus armandii* and *Pinus bybgeana* on its vitality and seedling growth. In: Wang Youmin (ed.) *Studies on Comprehensively Controlling and Managing Gully Areas in the Loess Plateau and Their Results*, pp. 116–122. China's Forestry Publishing House, Beijing.

Xie Zaizhi. 1985. Effect of direct application of rape straw on the promotion of soil fertility. *Soils* 17: 88–89, 96.

Xinjiang Comprehensive Survey Team and Soil Institute of Academia Sinica (ed.). 1965. Xinjiang Soil Geography (Xinjiang turang dili), pp. 52–74. Science Press, Beijing.

Xinjiang Forest Institute. 1976. Studies on the effect of tree network on prevention of winds in Turfan County. In: Lanzhou Glacier, Frozen Earth and Desert Institute, AcademiaSinica (ed.) *Desert Control* (Shamode zhili), pp. 81–87. Science Press, Beijing.

Xin Naiquan, and Zhang Shaoli. 1987. Strategy and policy for dryland farming development in the northern part of China. Proceedings of the international symposium on dryland farming, vol. 1, pp. 40–55. Yangling, Shaanxi, China.

Xin Shuzhi, 1964. *An Interpretation to Yugong*. (An ancient book dealing with agriculture, written in Zhou Dynasty), pp. 1–8. Agricultural Press. Beijing.

Xin Shuzhi, and Jiang Deqi. 1982. *An Introduction to China's Water and Soil Conservation* (Zhongguo shuitu baochi gailum), pp. 12–38, 101–102, 115–145, 187–285. Agricultural Press, Beijing.

Xing Xuexiang, Tang Kunlun, Huang Zhongcai, Liu Shenghua, Zhu Jicheng, Li Jifeng, and Zhao Peixiong. 1987. Report on airplane seeding grass in Yulin sandy loess region. *China's Grassland* 2: 16–23.

Xiong Yi, and Xi Chengfan. 1965. *Soils in the North China Plain* (Huabei pingyuan turang), pp. 105–167. Science Press, Beijing.

Xu Chengxiang. 1987. Studies on the effect of *Linum usitatissimum* stalk sand obstacles on stabilization of sand. *Soil and Water Conservation in China* 12: 24–26.

Xu Guangqi. Ming Dynasty. A memorial to the throne for utilization of water on drylands. In: *Nongzhengquanshu* (The complete book of agriculture and family economics, written by Xu Guangqi in Ming Dynasty). A new copy printed by Shanghai Ancient Book Publishing House, 1979, pp. 283–290.

Xu Guojun, and Cui Qin. 1991. The role of turning over straw into soil and some problems in agricultural practice. *Soil Fertilizers* 4: 30–32.

Xu Maozhen. 1987. Studies on the techniques of increasing production of cereal crops on dryland. Proceedings of the international symposium on dryland farming, vol. 2, pp. 366–368. Yangling, Shaanxi, China.

Xu Weikui, and Lu Bingzhang. 1990. Effect of no-tillage mulch farming on newly reclaimed saline wastelans. *Soils* 22: 17–19, and 35.

Xu Xinyu, Zhang Yumei, Xiang Hua, Hu Jisheng, Zhou Xianming, Yu Fuxiang, and Wang Fupin. 1988. The effect of covering field with crop residues plus minimum tillage on increasing soil fertility and yield. *J. Soil Sci.* 19: 86–88, 77.

Xu Xinyu, Zhang Yumei, Xiang Hua, Li Rensheng, and Hu Jisheng. 1985. Studies on the practical significance of plant residue mulch and its effect on soil organisms. *Scientia Agricultura Sinica* 5: 42–48.

Xu Zhenming. Ming Dynasty. Notes of talking with guests in Lushui. In: Wang Yunwu (ed.) *Collection of a Series of Books*. Shanghai Ancient Book Publishing House. Printed in 1937, book no. 3020, pp. 1–12.

Xu Zhuling, and Wang Yantian. 1990. A study of field water balance of dryland faming experimental areas in Wuchuan of Inner Mongolia. *Agric. Res. Arid Areas* 4: 20–26.

Xu Zhuomin. 1986. Green manure mulch tillage in Loess Plateau. *Agric. Res. Arid Areas* 2: 1–16; *Shanxi Agric. Sci.* 3: 1.

Yang Chunfeng, and Han Siming. 1988. Drought-resistant tillage method on Weibei drylands. *Tillage Cultivation* 4: 11.

Yang Jingqiu, Liu Jincheng, and Zhang Xiaoming. 1987. Effect of plant residues on the fertility of the soil in semi-arid regions. Proceedings of the international symposium on dryland farming, vol. 2, pp. 419–422. Yangling, Shaanxi, China.

Yang Sheng, and Lian Sheng. 1987. The result of comprehensive management of Shanwanggou small river basin. *Soil and Water Conservation in China* 12: 47–52.

Yang Shusen. 1986. No-till of summer fallow fields on Weibei drylands. *Shaanxi Agricultural Sciences* 4: 8–9.

Yang Songwang. 1987. Studies of forest protecting system for water and soil conservation in gully area of the Loess Plateau, South Shanxi. *Shanxi Science and Technology of Water and Soil Conservation* 3: 11–13.

Yang Xinming, and Yang Wenzhi. 1988. Study on the soil water balance of artificial forest land in arid region. *Bull. Soil Water Cons.* 3: 32–38.

Yang Xinming, and Yang Wenzhi, 1990. A prime report on soil moisture in Caragana woodland. In: Lu Zongfan (ed.) *Studies on Ecological Agriculture with Soil and Water Conservation in Loess Hilly Gully Regions*, vol. 2, pp. 133–136. Tianze Eldonejo, Yangling, Shaanxi, China.

Yang Wenbing. 1988. The evaporation velocities of some trees in arid zone and their relationships with environment. *Arid Zone Res.* 4: 47–55.

Yang Wenzhi, Yang Xinmin, and Ma Yuxi. 1990. Study on soil water resource and its utilization in loess hilly region. In: Lu Zongfan (ed.) *Studies on Econlogical Agriculture with Soil and Water Conservation in Loess Hilly Gully Regions*, vol. 1, pp. 56–75. Tianze Eldonejo, Yangling, Shaanxi, China.

Yang Xiaoyin. 1987. Great effect of planting trees and grasses in water and soil loss area of North Shaanxi. *Soil and Water Conservation in China* 4: 28–30.

Yao Yuanxi, and Yang Yanfan. 1989. Studies on the effects of application of organic and inorganic fertilizers on regulating soil P and K balance. Proceedings of the International Symposium on Balanced Fertilization, pp. 403–409. Agricultural Press, Beijing.

Yizhoushu. Zhou Dynasty (Yizhoushu meaning a missing historical book of Zhou Dynasty, written in Zhou Dynasty; the Yizhoushu cited here was found in a tomb, and interpreted by Kong Yao in Jin Dynasty). In: Wang Yunwu (ed.) *Collection of a Series of Books*. Shanghai Ancient Book Publishing House. Printed in 1937, book no. 3693, pp. 102–109.

Yu Ling, 1990. The productive potential structure and developing approaches to productive potential of field crops on Bashang rainfed highland in Hebei province. *Agric. Res. Arid Areas* 4: 62–68.

Yulin Prefecture of Shaanxi Province and Shaanxi Water and Soil Conservation Bureau. 1975. *Sand Control in North Shaanxi* (Shaanbei zhisha), pp. 29–125. Shaanxi People's Publishing House.

Yulin Prefecture Office for Directing Airplane Seeding. 1977. Advances of afforestation by airplane seeding and sand stabilization. *China's Forestry Sciences* 2.

Yu Zhifu, and Zhang Renming. 1984. High economic benefit of arable land protective tree network. *Communication on Agricultural Science and Technology* 2: 27.

Zeng Baoling, and Chen Zuoming. 1959. The experiences of Manasi People's Commune on amelioration of white hardened and impervious soil (desert soil). *A Collection of Achievements in Agricultural Scientific Research in Xinjiang*, vol. 3.

Zeng Baoling, and Xu Zhikun. 1963. Discussion on raising and consolidating Xinjiang soil fertility. *Collected Papers of Xinjiang Soil Science Society*.

Zeng Jingqing, and Wang Wenshan. 1984. The role of crop stubble in increasing soil fertility. *J. Soil Sci.* 15: 63–64.

Zhang Guilan, Zhang Ziwu, and Bao Dejun. 1989. Studies on phosphorus supplying capacity in wheat fields and application rates. Proceedings of the International Symposium on Balanced Fertilization, pp. 83–88. Agricultural Press, Beijing.

Zhang Fudao. 1984. Organic-inorganic fertilizer conbination is the developing direction of mordern technique for applying fertilizers. *Soil and Fertilizer* 1: 16–19.

Zhang Hanxiong. 1989. Predicting whole benefit of soil and water conservation with dynamic grey model group. *Acta Conservationis Soli et Aquae Sinica* 4: 40–48.

Zhang Kangjian, Sun Changzhong, and Zhao Zhong. 1987. A new method for evaluating site quality of afforesting areas in the loess plateau—quantification analysis of soil moisture content. Proceedings of the international symposium on dryland farming, vol. 1, pp. 268–276. Yangling, Shaanxi, China.

Zhang Kuan, Wang Xiufang, Wu Wei, Hu Heyun, and Wang Xiaolin. 1990. The application of fertilizer effect function to the formulated fertilization. *Soils* 22: 198–200.

Zhang Liyuan, Zhang Ximin, and Wu Cheng. 1988. Preliminary study on the growing characteristics and the output of the *Haloxylon persicum* twigs in Xinjiang Mosouwan. *Arid Zone Res.* 4: 30–35.

Zhang Shixian. 1989. China's agriculture development and adoption of balanced fertilization in agriculture. Proceedings of the International Symposium on Balanced Fertilization, pp. 10–15. Agricultural Press, Beijing.

Zhang Xianmin. 1987. Preliminary report on adoption of grass of different species for growing mixedly over large areas. *Soil and Water Conservation in China* 8: 14–16.

Zhang Xiaozhong, Han Shifeng, and Li Yushan. 1990. The laws of soil water behaviors on Weibei Rainfed Highland. *Agric. Res. Arid Areas* 4: 27–36.

Zhang Xihai, and Shan Lun. 1983. The preliminary study of millet and proso on the drought-resistance. In: Shan Lun, Niu Pu (eds.) *Selected Works on Arid and Semi-Arid Farming*, vol. 2, pp. 202–205. Xian Agricultural Information Centre Press.

Zhang Xueshang. 1983. Thr preliminary understanding of the biological role of raising soil fertility on Weibei dryland. In: Shan Lun, Niu Pu (eds.) *Selected Works on Arid and Semi-Arid Farming*, vol. 2, pp. 115–120. Xian Agricultural Information Centre Press.

Zhang Yangji, Li Jiarui, and Fan Xiaoyi. 1990. A comparative study of mulching effects upon dryland orchards. *Agric. Res. Arid Areas* supplement, 91–95.

Zhang Yichun, and Guo Zhifeng. 1990. Nutrient uptake by cotton plants and their recommended fertilization. *Soils* 22: 212–217.

Zhang Yongping, Wu Zhanqiang, Si Xiewu, and Li Zhuqin. 1990. Introduction of alfalfa cultivars. In: Zhu Xiangsan (ed.) *Studies on Exploitation and Harness of the Loess Plateau*, vol. 2, pp. 188–192. Shaanxi Science and Technology Publishing House.

Zhang Yongping, and Wu Zhanqiang. 1990. A study on alfalfa production at Mizhi County. In: Zhu Xiangsan (ed.) *Studies on Exploitation and Harness of the Loess Plateau*, vol. 2, pp. 259–260. Shaanxi Science and Technology Publishing House.

Zhang Zhanhong. 1989. Ten situations for mulching and not-mulching in use of plastic film cultivation. *Shansi Agricultural Sciences* 7: 31–33.

Zhan Changgeng. 1989. The effects of application of K with N and P fertilizers on production increase for some crops. Proceedings of the International Symposium on Balanced Fertilization, pp. 191–199. Agricultural Press, Beijing.

Zhao Feng, Xia Ailin, and Zhou Bin. 1987. Research on grassland agriculture in six plots in Ninjiang. Proceedings of the international symposium on dryland farming, vol. 2, p. 389. Yangling. Shaanxi, China.

Zhao Fengbin. 1985. Deep subsoiling of wheat fields on drylands of southeast Shanxi. *Shanxi Agricultural Sciences* 1: 2.

Zhao Fengbin. 1987. Studies on soil tillage methods on southeast Shanxi drylands. *Shanxi Agricultural Sciences* 5: 30.

Zhao Jubao, Mei Xurong, and Lu Xuedu. 1990. Evaluation of the characteristics of water supplies and demands in winter wheat fields and study on the economizing irrigation in the southeast profecture. *Agric. Res. Arid Areas* 4: 7–13.

Zhao Songqiao. 1985. Human impacts on China's arid lands: desertification or de-desertification. In: Whitehead, E.E., Hutchinson, C.F., Timmermann, B.N., Varady, R.G. (eds.) *Arid Lands, Today and Tomorrow*. Proceedings of an International Research and Development, pp. 1127–1138. Westview Press. Boulder, Colorado, and Belhaven Press, London.

Zheng Fenli, Tang Keli, and Zhou Peihua. 1987. Approach to the genesis and development of rill erosion on slope land and the way to control. *Acta Conservationis Soli et Aquae Sinica* 1: 36–48.

Zheng Guoying, and Liu Cuirong. 1991. Utilization of water resources and oasis ecological agriculture in arid areas. *Agric. Res. Arid Areas* 2: 79–84.

Zheng Jiangfei, Zhou Bai, Lu Longquan, and Du Raodong. 1990. A discussion on agricultural productivity in Wuchuan county of Inner Mongolia. *Agric. Res. Arid Areas* 4: 79–84.

Zhong Lianping, and Han Shifeng. 1990. Statistical analysis of soil moisture regime on rainfed farmlands in Chengcheng County of Shaansi Province. *Agric. Res. Arid Areas* 4: 44–53.

Zhongyan County Government. 1985. High yielding contour furrow tillage—a way for farmers to be rich. *Shanxi Agricultural Sciences* 2: 6.

Zhou Peizi, Hou Caixia, and Cheng Shimin. 1988. Physiological and biochemical specificities of Haloxylon ammodendron responding to water stress. *Arid Zone Res.* 1: 1–8.

Zhou Ruihua. 1986. Studies on tillage techniques for moisture conservation in east loess plateau. *Shaanxi Agricultural Science* 6: 31–33.

Zhou Tingru. 1965. Discussion on some matters of palageogaraphy in Xinjiang global history of recent times. *J. Natural Science of Universities* 3.

Zhou Yong. Ming Dynasty. A memorial to the throne on the matter of river control. In: *A compilation of Confucian Classics and Current Works of Ming Dynasty* (Edited by Chen Zhilong in Ming Dynasty). China's Publishing House printed 1962, new vol. 2 (original vol. 146), pp. 1458–1459.

Zhou Zucheng, Sun Weichang, Wang Linbao, Wang Hongyu, Jin Zhengyu, and Wang Xidong. 1989. Comparison of two methods for prescription fertilization. Proceedings of the International Symposium on Balanced Fertilization, pp. 311–316. Agricultural Press, Beijing.

Zhu Junfeng (Chief ed.). 1985. *Natural Resources and Comprehensive Agricultural Regionalization in the Shelter Belt Constructing Areas of the* "Three Norths" (Sanbei fanghulin diqu ziren ziyuan yu zonghe nongnie quhua), pp. 3–17, 797–900, 1133–1174. China's Forestry Publishing House, Beijing.

Zhu Haizhi. 1964. Microstructure in the loess and optics orientation of the clays in covered soils. *Quaternaria Sinica* 1: 72–76.

Zhu Lianqing, Ma Yongzhi, and Li Qingkui. 1941. An Introduction to China Soils. *Quarterly Soil Science* 1.

Zhu Maoshun. 1965. *Xinjiang Agriculture*. Science Press, Beijing.

Zhu Wenshan, and Cao Mingkui. 1988. Effects of saving water and bettering land fertility by stubble mulching no-till and its future utilization. *Agric. Res. Arid Areas* 4: 12–17.

Zhu Wenshan. 1987. An agricultural technique for moisture conservation—mulching with no-tillage. *Hebei Agricultural Science and Technology* 9: 17.

Zhu Xianmo. 1956. Classification of soil erosion in the loess region. *Acta Pedologica Sinica* 4: 99–115.

Zhu Xianmo. 1957. Soil distribution patterns in loess areas. *Acta Scientia Sinica* 15: 477–478.

Zhu Xianmo. 1958. Investigation report on soil erosion in the centre of Gansu province. *Soil Bull.* 32.

Zhu Xianmo. 1959. Preliminary system of Shaanxi soil classification. *J. Soil Sci.* 1: 23–27.

Zhu Xianmo. 1960. The effect of vegetation factor on soil-water erosion in the loess region. *Acta Pedologica Sinica* 8: 110–121.

Zhu Xianmo. 1964. Manural loessial soil (Loutu), pp. 12–70. Agricultural Publishing House, Beijing.

Zhu Xianmo, 1981. The main types of water erosion and the factors affecting water erosion. *Bull. Water Soil Conservation* 3: 1–9.

Zhu Xianmo. 1984a. Some problems on territory management in loess plateau. *Bull. Soil Water Cons.* 4: 1–6.

Zhu Xianmo. 1984b. Exploitation and protection of land resources in the Loess Plateau. *Scientia Geographica Sinica* 2: 97–104.

Zhu Xianmo. 1984c. Discussion on the strategic matter of soil and water erosion in loess plateau. *Bull. Soil Water Cons.* 1: 15–18.

Zhu Xianmo (Chief ed.). 1986a. Land resources in the Loess Plateau of China. Shaanxi Science and Technology Publishing House (a photography book without page numbers).

Zhu Xianmo. 1986b. Types of soil erosion and resistance of the soil to erosion in loess plateau of China. In: Soil Science Society of China (ed.) *Current Progress in Soil Research in People's Republic of China*, pp. 611–621. Jiangsu Science and Technology Publishing House.

Zhu Zhenda. 1987. Taklamakan desert area. *J. Desert Res.* 7: 16–28.

Zhu Zhenda, Wu Zheng, Liu Shu, and Di Xingmin. 1980. *An Introduction to China Desert* (Zhongguo shamo gailun), pp. 1–8, 20–29, 90–101. Science Press. Beijing.

Zhu Zhenda, Liu Shu, and Di Xingmin. 1989. Desertification and Rehabilitation in China (Zhongguo de shamohua jiqi zhili), pp. 1–27, 88–100, 120–126. Science Press, Beijing.

Zuo Shuzhen, and Chi Renli. 1987. A discussion on the soil tillage technique system on the county level in non-irrigated regions of North China—a study on the regional soil tillage technique system of Hailun county, Proceedings of the international symposium on dryland farming, vol. 2, pp. 385–388. Yangling, Shaanxi, China.

Index

MIX
Papier aus verantwortungsvollen Quellen
Paper from responsible sources
FSC® C105338

If you have any concerns about our products,
you can contact us on
ProductSafety@springernature.com

In case Publisher is established outside the EU,
the EU authorized representative is:
Springer Nature Customer Service Center GmbH
Europaplatz 3, 69115 Heidelberg, Germany

Printed by Libri Plureos GmbH
in Hamburg, Germany